Machine Learning Refined

Providing a unique approach to machine learning, this text contains fresh and intuitive, yet rigorous, descriptions of all fundamental concepts necessary to conduct research, build products, tinker, and play. By prioritizing geometric intuition, algorithmic thinking, and practical real-world applications in disciplines including computer vision, natural language processing, economics, neuroscience, recommender systems, physics, and biology, this text provides readers with both a lucid understanding of foundational material as well as the practical tools needed to solve real-world problems. With in-depth Python and MATLAB/OCTAVE-based computational exercises and a complete treatment of cutting edge numerical optimization techniques, this is an essential resource for students and an ideal reference for researchers and practitioners working in machine learning, computer science, electrical engineering, signal processing, and numerical optimization.

Key features:

- A presentation built on lucid geometric intuition
- A unique treatment of state-of-the-art numerical optimization techniques
- A fused introduction to logistic regression and support vector machines
- Inclusion of feature design and learning as major topics
- An unparalleled presentation of advanced topics through the lens of function approximation
- A refined description of deep neural networks and kernel methods

Jeremy Watt received his PhD in Computer Science and Electrical Engineering from Northwestern University. His research interests lie in machine learning and computer vision, as well as numerical optimization.

Reza Borhani received his PhD in Computer Science and Electrical Engineering from Northwestern University. His research interests lie in the design and analysis of algorithms for problems in machine learning and computer vision.

Aggelos K. Katsaggelos is a professor and holder of the Joseph Cummings chair in the Department of Electrical Engineering and Computer Science at Northwestern University, where he also heads the Image and Video Processing Laboratory.

Machine Learning Refined

Foundations, Algorithms, and Applications

JEREMY WATT, REZA BORHANI, AND
AGGELOS K. KATSAGGELOS
Northwestern University

CAMBRIDGE
UNIVERSITY PRESS

University Printing House, Cambridge CB2 8BS, United Kingdom

One Liberty Plaza, 20th Floor, New York, NY 10006, USA

477 Williamstown Road, Port Melbourne, VIC 3207, Australia

4843/24, 2nd Floor, Ansari Road, Daryaganj, Delhi - 110002, India

79 Anson Road, #06-04/06, Singapore 079906

Cambridge University Press is part of the University of Cambridge.

It furthers the University's mission by disseminating knowledge in the pursuit of education, learning and research at the highest international levels of excellence.

www.cambridge.org
Information on this title: www.cambridge.org/9781107123526

First published 2016

A catalogue record for this publication is available from the British Library

Library of Congress Cataloging in Publication data
Names: Watt, Jeremy, author. | Borhani, Reza. | Katsaggelos, Aggelos Konstantinos, 1956-
Title: Machine learning refined : foundations, algorithms, and applications / Jeremy Watt, Reza Borhani, Aggelos Katsaggelos.
Description: New York : Cambridge University Press, 2016.
Identifiers: LCCN 2015041122 | ISBN 9781107123526 (hardback)
Subjects: LCSH: Machine learning.
Classification: LCC Q325.5 .W38 2016 | DDC 006.3/1–dc23
LC record available at http://lccn.loc.gov/2015041122

ISBN 978-1-107-12352-6 Hardback

Additional resources for this publication at www.cambridge.org/watt

Contents

Preface

In the last decade the user base of machine learning has grown dramatically. From a relatively small circle in computer science, engineering, and mathematics departments the users of machine learning now include students and researchers from every corner of the academic universe, as well as members of industry, data scientists, entrepreneurs, and machine learning enthusiasts. The book before you is the result of a complete tearing down of the standard curriculum of machine learning into its most basic components, and a curated reassembly of those pieces (painstakingly polished and organized) that we feel will most benefit this broadening audience of learners. It contains fresh and intuitive yet rigorous descriptions of the most fundamental concepts necessary to conduct research, build products, tinker, and play.

Intended audience and book pedagogy

This book was written for readers interested in understanding the core concepts of machine learning from first principles to practical implementation. To make full use of the text one only needs a basic understanding of linear algebra and calculus (i.e., vector and matrix operations as well as the ability to compute the gradient and Hessian of a multivariate function), plus some prior exposure to fundamental concepts of computer programming (i.e., conditional and looping structures). It was written for first time learners of the subject, as well as for more knowledgeable readers who yearn for a more intuitive and serviceable treatment than what is currently available today.

To this end, throughout the text, in describing the fundamentals of each concept, we defer the use of probabilistic, statistical, and neurological views of the material in favor of a fresh and consistent geometric perspective. We believe that this not only permits a more intuitive understanding of many core concepts, but helps establish revealing connections between ideas often regarded as fundamentally distinct (e.g., the logistic regression and support vector machine classifiers, kernels, and feed-forward neural networks). We also place significant emphasis on the design and implementation of algorithms, and include many coding exercises for the reader to practice at the end of each chapter. This is because we strongly believe that the bulk of learning this subject takes place when learners "get their hands dirty" and code things up for themselves. In short, with this text we have aimed to create a learning experience for the reader where intuitive leaps precede intellectual ones and are tempered by their application.

What this book is about

The core concepts of our treatment of machine learning can be broadly summarized in four categories. *Predictive learning*, the first of these categories, comprises two kinds of tasks where we aim to either predict a continuous valued phenomenon (like the future location of a celestial body), or distinguish between distinct kinds of things (like different faces in an image). The second core concept, *feature design*, refers to a broad set of engineering and mathematical tools which are crucial to the successful performance of predictive learning models in practice. Throughout the text we will see that features are generated along a spectrum based on the level of our own understanding of a dataset. The third major concept, *function approximation*, is employed when we know too little about a dataset to produce proper features ourselves (and therefore must learn them strictly from the data itself). The final category, *numerical optimization*, powers the first three and is the engine that makes machine learning run in practice.

Overview of the book

This book is separated into three parts, with the latter parts building thematically on each preceding stage.

Part I: Fundamental tools and concepts

Here we detail the fundamentals of predictive modeling, numerical optimization, and feature design. After a general introduction in Chapter 1, Chapter 2 introduces the rudiments of numerical optimization, those critical tools used to properly tune predictive learning models. We then introduce predictive modeling in Chapters 3 and 4, where the regression and classification tasks are introduced, respectively. Along the way we also describe a number of examples where we have some level of knowledge about the underlying process generating the data we receive, which can be leveraged for the design of features.

Part 2: Tools for fully data-driven machine learning

In the absence of useful knowledge about our data we must broaden our perspective in order to design, or learn, features for regression and classification tasks. In Chapters 5 and 6 we review the classical tools of *function approximation*, and see how they are applied to deal with general regression and classification problems. We then end in Chapter 7 by describing several advanced topics related to the material in the preceding two chapters.

Part 3: Methods for large scale machine learning

In the final stage of the book we describe common procedures for scaling regression and classification algorithms to large datasets. We begin in Chapter 8 by introducing a number of advanced numerical optimization techniques. A continuation of the introduction in Chapter 2, these methods greatly enhance the power of predictive learning by means of more effective optimization algorithms. We then detail in Chapter 9 general techniques for properly lowering the dimension of input data, allowing us to deflate large datasets down to more manageable sizes.

Readers: how to use this book

As mentioned earlier, the only technical prerequisites for the effective use of this book are a basic understanding of linear algebra and vector calculus, as advanced concepts are introduced as necessary throughout the text, as well as some prior computer programming experience. Readers can find a brief tutorial on the Python and MATLAB/OCTAVE programming environments used for completing coding exercises, which introduces proper syntax for both languages as well as necessary libraries to download (for Python) as well as useful built-in functions (for MATLAB/OCTAVE), on the book website.

For self-study one may read all the chapters in order, as each builds on its direct predecessor. However, a solid understanding of the first six chapters is sufficient preparation for perusing individual topics of interest in the final three chapters of the text.

Instructors: how to use this book

The contents of this book have been used for a number of courses at Northwestern University, ranging from an introductory course for senior level undergraduate and beginning graduate students, to a specialized course on advanced numerical optimization for an audience largely consisting of PhD students. Therefore, with its treatment of foundations, applications, and algorithms this book is largely self-contained and can be used for a variety of machine learning courses. For example, it may be used as the basis for:

A single quarter or semester long senior undergraduate/beginning graduate level introduction to standard machine learning topics. This includes coverage of basic techniques from numerical optimization, regression/classification techniques and applications, elements of feature design and learning, and feed-forward neural networks. Chapters 1–6 provide the basis for such a course, with Chapters 7 and 9 (on kernel methods and dimension reduction/unsupervised learning techniques) being optimal add-ons.

A single quarter or semester long senior level undergraduate/graduate course on large scale optimization for machine learning. Chapters 2 and 6–8 provide the basis for a course on introductory and advanced optimization techniques for solving the applications and models introduced in the first two-thirds of the book.

1 Introduction

Machine learning is a rapidly growing field of study whose primary concern is the design and analysis of algorithms which enable computers to learn. While still a young discipline, with much more awaiting to be discovered than is currently known, today machine learning can be used to teach computers to perform a wide array of useful tasks. This includes tasks like the automatic detection of objects in images (a crucial component of driver-assisted and self-driving cars), speech recognition (which powers voice command technology), knowledge discovery in the medical sciences (used to improve our understanding of complex diseases), and predictive analytics (leveraged for sales and economic forecasting). In this chapter we give a high level introduction to the field of machine learning and the contents of this textbook. To get a big picture sense of how machine learning works we begin by discussing a simple toy machine learning problem: teaching a computer how to distinguish between pictures of cats from those with dogs. This will allow us to informally describe the procedures used to solve machine learning problems in general.

1.1 Teaching a computer to distinguish cats from dogs

To teach a child the difference between "cat" versus "dog", parents (almost!) never give their children some kind of formal scientific definition to distinguish the two; i.e., that a dog is a member of Canis Familiaris species from the broader class of Mammalia, and that a cat while being from the same class belongs to another species known as Felis Catus. No, instead the child is naturally presented with many images of what they are told are either "dogs" or "cats" until they fully grasp the two concepts. How do we know when a child can successfully distinguish between cats and dogs? Intuitively, when they encounter new (images of) cats and dogs, and can correctly identify each new example. Like human beings, computers can be taught how to perform this sort of task in a similar manner. This kind of task, where we aim to teach a computer to distinguish between different types of things, is referred to as a *classification* problem in machine learning.

1. Collecting data Like human beings, a computer must be trained to recognize the difference between these two types of animal by learning from a batch of examples, typically referred to as a *training set* of data. Figure 1.1 shows such a training set consisting

Fig. 1.1 A training set of six cats (left panel) and six dogs (right panel). This set is used to train a machine learning model that can distinguish between future images of cats and dogs. The images in this figure were taken from [31].

of a few images of different cats and dogs. Intuitively, the larger and more diverse the training set the better a computer (or human) can perform a learning task, since exposure to a wider breadth of examples gives the learner more experience.

2. Designing features Think for a moment about how you yourself tell the difference between images containing cats from those containing dogs. What do you look for in order to tell the two apart? You likely use color, size, the shape of the ears or nose, and/or some combination of these *features* in order to distinguish between the two. In other words, you do not just look at an image as simply a collection of many small square pixels. You pick out details, or features, from images like these in order to identify what it is you are looking at. This is true for computers as well. In order to successfully train a computer to perform this task (and any machine learning task more generally) we need to provide it with properly designed features or, ideally, have it find such features itself.

This is typically not a trivial task, as designing quality features can be very application dependent. For instance, a feature like "number of legs" would be unhelpful in discriminating between cats and dogs (since they both have four!), but quite helpful in telling cats and snakes apart. Moreover, extracting the features from a training dataset can also be challenging. For example, if some of our training images were blurry or taken from a perspective where we could not see the animal's head, the features we designed might not be properly extracted.

However, for the sake of simplicity with our toy problem here, suppose we can easily extract the following two features from each image in the training set:

1. *size of nose*, relative to the size of the head (ranging from small to big);
2. *shape of ears* (ranging from round to pointy).

Examining the training images shown in Fig. 1.1, we can see that cats all have *small* noses and *pointy* ears, while dogs all have *big* noses and *round* ears. Notice that with the current choice of features each image can now be represented by just two numbers:

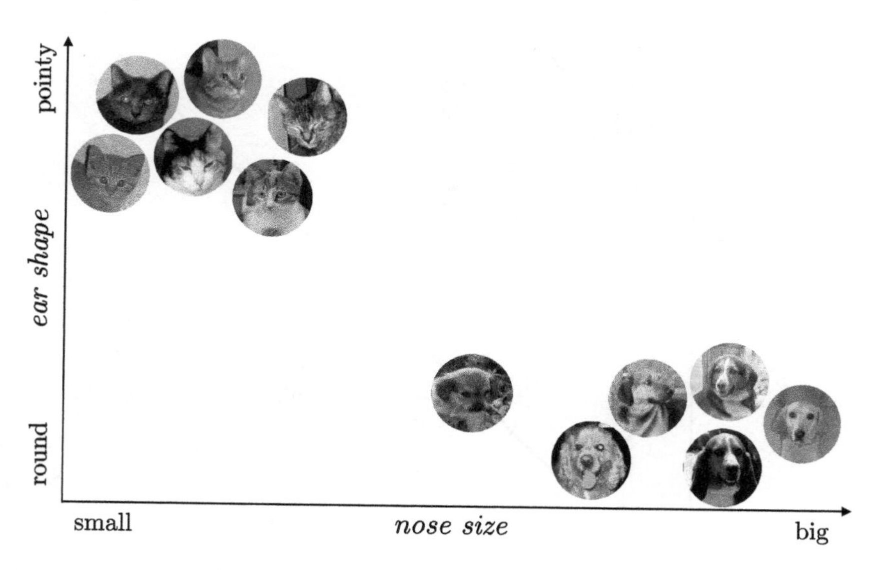

Fig. 1.2 Feature space representation of the training set where the horizontal and vertical axes represent the features "nose size" and "ear shape" respectively. The fact that the cats and dogs from our training set lie in distinct regions of the feature space reflects a good choice of features.

a number expressing the relative nose size, and another number capturing the pointy-ness or round-ness of ears. Therefore we now represent each image in our training set in a 2-dimensional *feature space* where the features "nose size" and "ear shape" are the horizontal and vertical coordinate axes respectively, as illustrated in Fig. 1.2. Because our designed features distinguish cats from dogs in our training set so well the feature representations of the cat images are all clumped together in one part of the space, while those of the dog images are clumped together in a different part of the space.

3. Training a model Now that we have a good feature representation of our training data the final act of teaching a computer how to distinguish between cats and dogs is a simple geometric problem: have the computer find a line or *linear model* that clearly separates the cats from the dogs in our carefully designed feature space.[1] Since a line (in a 2-dimensional space) has two parameters, a slope and an intercept, this means finding the right values for both. Because the parameters of this line must be determined based on the (feature representation) of the training data the process of determining proper parameters, which relies on a set of tools known as *numerical optimization,* is referred to as the training of a model.

Figure 1.3 shows a trained linear model (in black) which divides the feature space into cat and dog regions. Once this line has been determined, any future image whose feature representation lies above it (in the blue region) will be considered a cat by the computer, and likewise any representation that falls below the line (in the red region) will be considered a dog.

[1] While generally speaking we could instead find a curve or *nonlinear model* that separates the data, we will see that linear models are by far the most common choice in practice when features are designed properly.

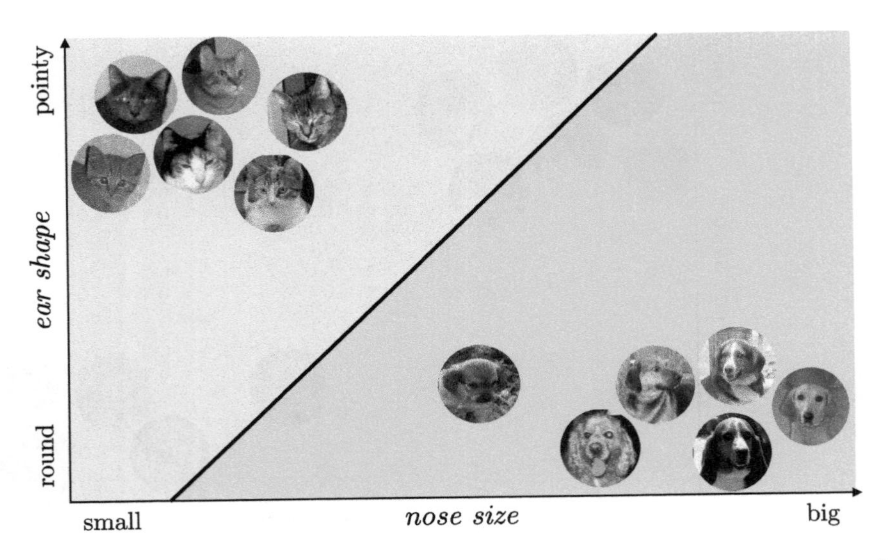

Fig. 1.3 A trained linear model (shown in black) perfectly separates the two classes of animal present in the training set. Any new image received in the future will be classified as a cat if its feature representation lies above this line (in the blue region), and a dog if the feature representation lies below this line (in the red region).

Fig. 1.4 A testing set of cat and dog images also taken from [31]. Note that one of the dogs, the Boston terrier on the top right, has both a short nose and pointy ears. Due to our chosen feature representation the computer will think this is a cat!

4. Testing the model To test the efficacy of our learner we now show the computer a batch of previously unseen images of cats and dogs (referred to generally as a *testing set* of data) and see how well it can identify the animal in each image. In Fig. 1.4 we show a sample testing set for the problem at hand, consisting of three new cat and dog images. To do this we take each new image, extract our designed features (nose size and ear shape), and simply check which side of our line the feature representation falls on. In this instance, as can be seen in Fig. 1.5 all of the new cats and all but one dog from the testing set have been identified correctly.

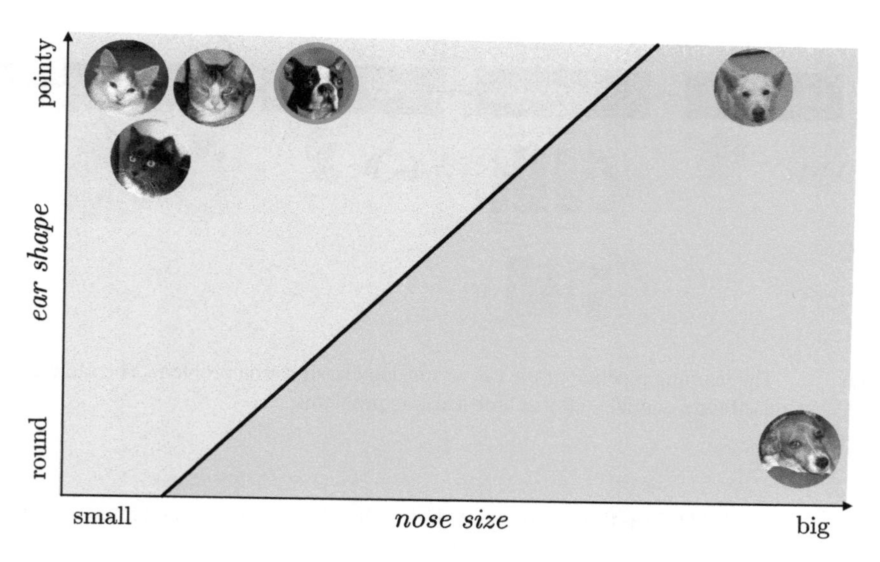

Fig. 1.5 Identification of (the feature representation of) our test images using our trained linear model. Notice that the Boston terrier is misclassified as a cat since it has pointy ears and a short nose, just like the cats in our training set.

The misidentification of the single dog (a Boston terrier) is due completely to our choice of features, which we designed based on the training set in Fig. 1.1. This dog has been misidentified simply because its features, a small nose and pointy ears, match those of the cats from our training set. So while it first appeared that a combination of nose size and ear shape could indeed distinguish cats from dogs, we now see that our training set was too small and not diverse enough for this choice of features to be completely effective.

To improve our learner we must begin again. First we should collect more data, forming a larger and more diverse training set. Then we will need to consider designing more discriminating features (perhaps eye color, tail shape, etc.) that further help distinguish cats from dogs. Finally we must train a new model using the designed features, and test it in the same manner to see if our new trained model is an improvement over the old one.

1.1.1 The pipeline of a typical machine learning problem

Let us now briefly review the previously described process, by which a trained model was created for the toy task of differentiating cats from dogs. The same process is used to perform essentially all machine learning tasks, and therefore it is worthwhile to pause for a moment and review the steps taken in solving typical machine learning problems. We enumerate these steps below to highlight their importance, which we refer to all together as the general pipeline for solving machine learning problems, and provide a picture that compactly summarizes the entire pipeline in Fig. 1.6.

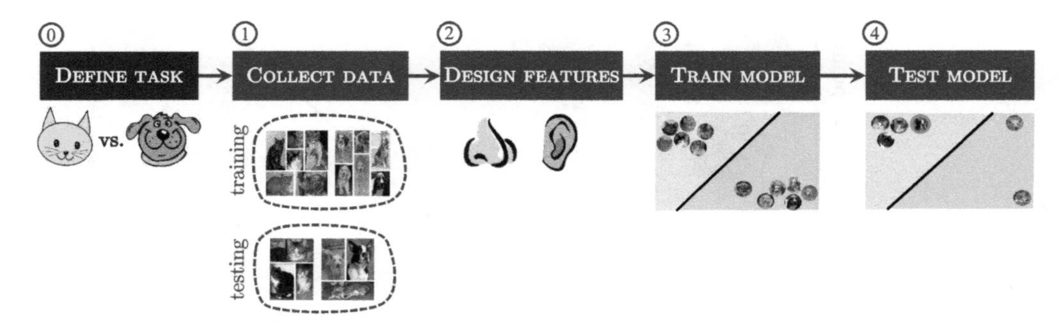

Fig. 1.6 The learning pipeline of the cat versus dog classification problem. The same general pipeline is used for essentially all machine learning problems.

(0) **Define the problem.** What is the task we want to teach a computer to do?

(1) **Collect data.** Gather data for training and testing sets. The larger and more diverse the data the better.

(2) **Design features.** What kind of features best describe the data?

(3) **Train the model.** Tune the parameters of an appropriate model on the training data using numerical optimization.

(4) **Test the model.** Evaluate the performance of the trained model on the testing data. If the results of this evaluation are poor, re-think the particular features used and gather more data if possible.

1.2 Predictive learning problems

Predictive learning problems constitute the majority of tasks machine learning can be used to solve today. Applicable to a wide array of situations and data types, in this section we introduce the two major predictive learning problems: *regression* and *classification*.

1.2.1 Regression

Suppose we wanted to predict the share price of a company that is about to go public (that is, when a company first starts offering its shares of stock to the public). Following the pipeline discussed in Section 1.1.1, we first gather a training set of data consisting of a number of corporations (preferably active in the same domain) with known share prices. Next, we need to design feature(s) that are thought to be relevant to the task at

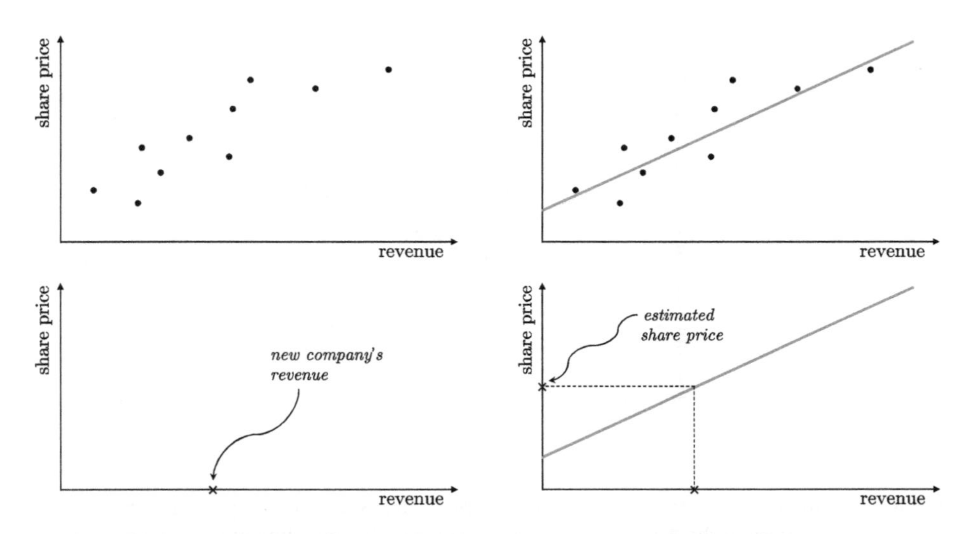

Fig. 1.7 (top left panel) A toy training dataset of ten corporations with their associated share price and revenue values. (top right panel) A linear model is fit to the data. This trend line models the overall trajectory of the points and can be used for prediction in the future as shown in the bottom left and bottom right panels.

hand. The company's revenue is one such potential feature, as we can expect that the higher the revenue the more expensive a share of stock should be.[2] Now in order to connect the share price to the revenue we train a linear model or *regression line* using our training data.

The top panels of Fig. 1.7 show a toy dataset comprising share price versus revenue information for ten companies, as well as a linear model fit to this data. Once the model is trained, the share price of a new company can be predicted based on its revenue, as depicted in the bottom panels of this figure. Finally, comparing the predicted price to the actual price for a testing set of data we can test the performance of our regression model and apply changes as needed (e.g., choosing a different feature). This sort of task, fitting a model to a set of training data so that predictions about a continuous-valued variable (e.g., share price) can be made, is referred to as *regression*. We now discuss some further examples of regression.

Example 1.1 The rise of student loan debt in the United States

Figure 1.8 shows the total student loan debt, that is money borrowed by students to pay for college tuition, room and board, etc., held by citizens of the United States from 2006 to 2014, measured quarterly. Over the eight year period reflected in this plot total student debt has tripled, totaling over one trillion dollars by the end of 2014. The regression line (in magenta) fit to this dataset represents the data quite well and, with its sharp positive slope, emphasizes the point that student debt is rising dangerously fast. Moreover, if

[2] Other potential features could include total assets, total equity, number of employees, years active, etc.

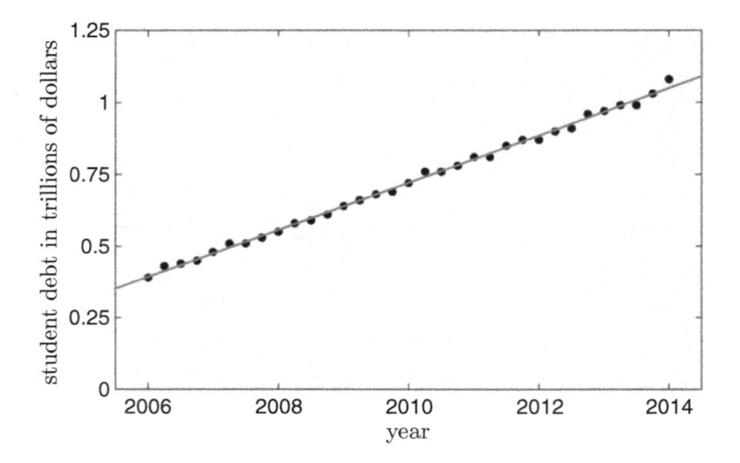

Fig. 1.8 Total student loan debt in the United States measured quarterly from 2006 to 2014. The rapid increase of the debt, measured by the slope of the trend line fit to the data, confirms the concerning claim that student debt is growing (dangerously) fast. The debt data shown in this figure was taken from [46].

this trend continues, we can use the regression line to predict that total student debt will reach a total of two trillion dollars by the year 2026.

Example 1.2 Revenue forecasting

In 1983, Academy Award winning screenwriter William Goldman coined the phrase "nobody knows anything" in his book *Adventures in the Screen Trade*, referring to his belief that at the time it was impossible to predict the success or failure of Hollywood movies. However, in the post-internet era of today, accurate estimation of box office revenue to be earned by upcoming movies is becoming possible. In particular, the quantity of internet searches for trailers, as well as the amount of discussion about a movie on social networks like Facebook and Twitter, have been shown to reliably predict a movie's opening weekend box office takings up to a month in advance (see e.g., [14, 62]). Sales forecasting for a range of products/services, including box office sales, is often performed using regression. Here the input feature can be for instance the volume of web searches for a movie trailer on a certain date, with the output being revenue made during the corresponding time period. Predicted revenue of a new movie can then be estimated using a regression model learned on such a dataset.

Example 1.3 Associating genes with quantitative traits

Genome-wide association (GWA) studies (Fig. 1.9) aim at understanding the connections between tens of thousands of genetic markers, taken from across the human genome of numerous subjects, with diseases like high blood pressure/cholesterol, heart

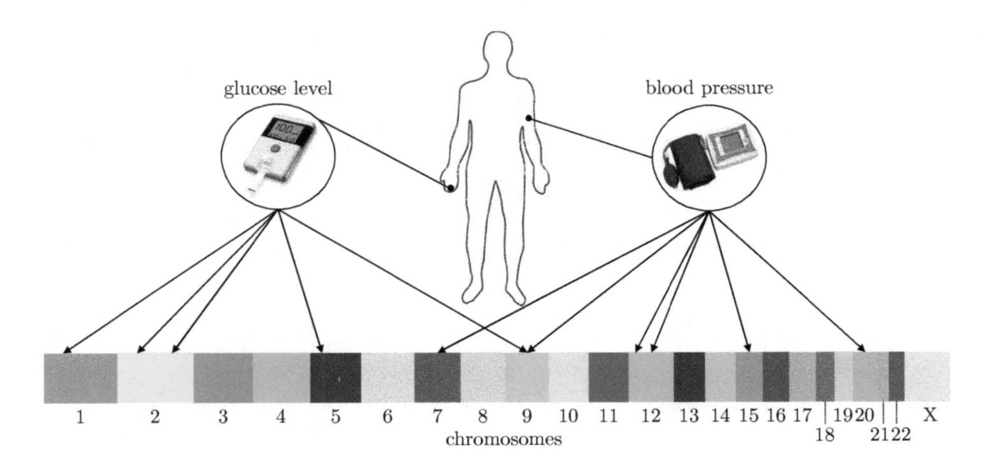

Fig. 1.9 Conceptual illustration of a GWA study employing regression, wherein a quantitative trait is to be associated with specific genomic locations.

disease, diabetes, various forms of cancer, and many others [26, 76, 80]. These studies are undertaken with the hope of one day producing gene-targeted therapies, like those used to treat diseases caused by a single gene (e.g., cystic fibrosis), that can help individuals with these multifactorial diseases. Regression as a commonly employed tool in GWA studies, is used to understand complex relationships between genetic markers (features) and quantitative traits like level of cholesterol or glucose (a continuous output variable).

1.2.2 Classification

The machine learning task of *classification* is similar in principle to that of regression. The key difference between the two is that instead of predicting a continuous-valued output (e.g., share price, blood pressure, etc.), with classification what we aim at predicting takes on discrete values or *classes*. Classification problems arise in a host of forms. For example *object recognition*, where different objects from a set of images are distinguished from one another (e.g., handwritten digits for the automatic sorting of mail or street signs for semi-autonomous and self-driving cars), is a very popular classification problem. The toy problem of distinguishing cats from dogs discussed in Section 1.1 was such a problem. Other common classification problems include speech recognition (recognizing different spoken words for voice recognition systems), determining the general sentiment of a social network like Twitter towards a particular product or service, as well as determining what kind of hand gesture someone is making from a finite set of possibilities (for use in e.g., controlling a computer without a mouse).

Geometrically speaking, a common way of viewing the task of classification is one of finding a separating line (or hyperplane in higher dimensions) that separates the two[3]

[3] Some classification problems (e.g., handwritten digit recognition) have naturally more than two classes for which we need a better model than a single line to separate the classes. We discuss in detail how multiclass classification is done later in Sections 4.4 and 6.3.

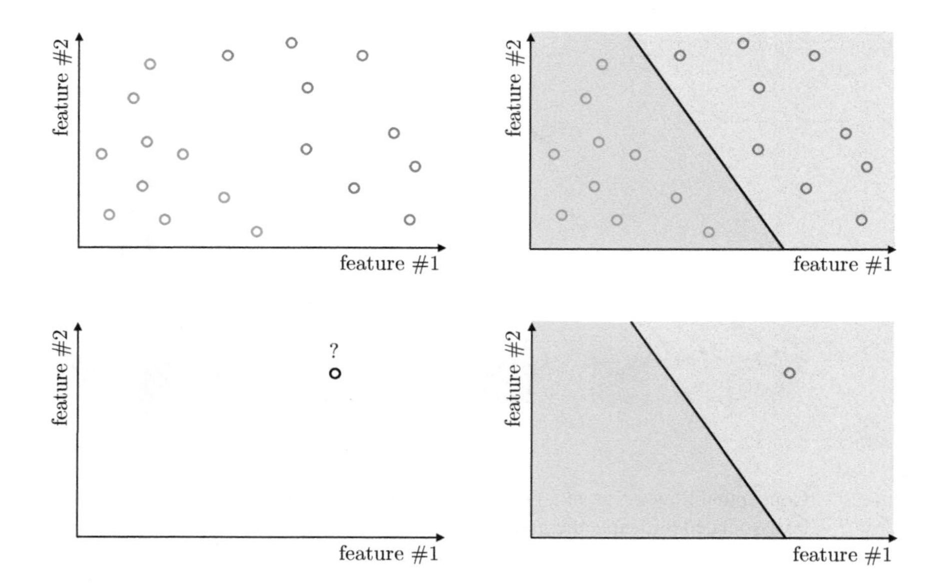

Fig. 1.10 (top left panel) A toy 2-dimensional training set consisting of two distinct classes, red and blue. (top right panel) A linear model is trained to separate the two classes. (bottom left panel) A test point whose class is unknown. (bottom right panel) The test point is classified as blue since it lies on the blue side of the trained linear classifier.

classes of data from a training set as best as possible. This is precisely the perspective on classification we took in describing the toy example in Section 1.1, where we used a line to separate (features extracted from) images of cats and dogs. New data from a testing set is then automatically classified by simply determining which side of the line/hyperplane the data lies on. Figure 1.10 illustrates the concept of a linear model or *classifier* used for performing classification on a 2-dimensional toy dataset.

Example 1.4 Object detection

Object detection, a common classification problem, is the task of automatically identifying a specific object in a set of images or videos. Popular object detection applications include the detection of faces in images for organizational purposes and camera focusing, pedestrians for autonomous driving vehicles,[4] and faulty components for automated quality control in electronics production. The same kind of machine learning framework, which we highlight here for the case of face detection, can be utilized for solving many such detection problems.

After training a linear classifier on a set of training data consisting of facial and non-facial images, faces are sought after in a new test image by sliding a (typically) square

[4] While the problem of detecting pedestrians is a particularly well-studied classification problem [29, 32, 53], a standard semi-autonomous or self-driving car will employ a number of detectors that scan the vehicle's surroundings for other important objects as well, like road markings, signs, and other cars on the road.

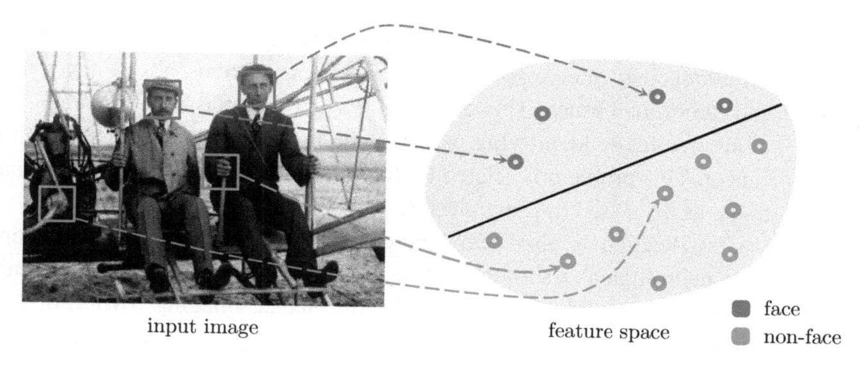

Fig. 1.11 To determine if any faces are present in a test image (in this instance an image of the Wright brothers, inventors of the airplane, sitting together in one of their first motorized flying machines in 1908) a small window is scanned across its entirety. The content inside the box at each instance is determined to be a face by checking which side of the learned classifier the feature representation of the content lies. In the figurative illustration shown here the area above and below the learned classifier (shown in black on the right) are the "face" and "non-face" sides of the classifier, respectively.

window over the entire image. At each location of the sliding window the image content inside is tested to see which side of the classifier it lies on (as illustrated in Fig. 1.11). If the (feature representation of the) content lies on the "face side" of the classifier the content is classified as a face.[5]

Example 1.5 Sentiment analysis

The rise of social media has significantly amplified the voice of consumers, providing them with an array of well-tended outlets on which to comment, discuss, and rate products and services. This has led many firms to seek out data intensive methods for gauging their customers' feelings towards recently released products, advertising campaigns, etc. Determining the aggregated feelings of a large base of customers, using text-based content like product reviews, tweets, and comments, is commonly referred to as *sentiment analysis*. Classification models are often used to perform sentiment analysis, learning to identify consumer data of either positive or negative feelings.

Example 1.6 Classification as a diagnostic tool in medicine

Cancer, in its many variations, remains among the most challenging diseases to diagnose and treat. Today it is believed that the culprit behind many types of cancers lies in accumulation of mutated genes, or in other words erroneous copies of an individual's

[5] In practice, to ensure that all faces at different distances from the camera are detected in a test image, typically windows of various sizes are used to scan as described here. If multiple detections are made around a single face they are then combined into a single highlighted window encasing the detected face.

DNA sequence. With the use of DNA microarray technology, geneticists are now able to simultaneously query expression levels of tens of thousands of genes from both healthy and tumorous tissues. This data can be used in a classification framework as a means of automatically identifying patients who have a genetic predisposition for contracting cancer. This problem is related to that of associating genes with quantitative biological traits, as discussed in Example 1.3.

Classification is also being increasingly used in the medical community to diagnose neurological disorders such as autism and attention deficit hyperactivity disorder (ADHD), using functional Magnetic Resonance Imaging (fMRI) of the human brain. These fMRI brain scans capture neural activity patterns localized in different regions of the brain over time as patients perform simple cognitive activities such as tracking a small visual object. The ultimate goal here is to train a diagnostic classification tool capable of distinguishing between patients who have a particular neurological disorder from those who do not, based solely on fMRI scans.

1.3 Feature design

As we have described in previous sections, *features* are those defining characteristics of a given dataset that allow for optimal learning. Indeed, well-designed features are absolutely crucial to the performance of both regression and classification schemes. However, broadly speaking the quality of features we can design is fundamentally dependent on our level of knowledge regarding the phenomenon we are studying. The more we understand (both intellectually and intuitively) the process generating the data we have at our fingertips, the better we can design features ourselves or, ideally, teach the computer to do this design work itself. At one extreme where we have near perfect understanding of the process generating our data, this knowledge having come from considerable intuitive, experimental, and mathematical reflection, the features we design allow near perfect performance. However, more often than not we know only a few facts, or perhaps none at all, about the data we are analyzing. The universe is an enormous and complicated place, and we have a solid understanding only of how a sliver of it works.

Below we give examples that highlight how our understanding of a phenomenon guides the design of features, from knowing quite a lot about a phenomenon to knowing just a few basic facts. A main thrust of the text will be to detail the current state of machine learning technology in dealing with this issue. One of the end goals of machine learning, which is still far from being solved adequately, is to develop effective tools for dealing with (finding patterns in) arbitrary kinds of data, a problem fundamentally having to do with finding good features.

Example 1.7 Galileo and uniform acceleration

In 1638 Galileo Galilei, infamous for his expulsion from the Catholic church for daring to claim that the earth orbited the sun and not the converse (as was the prevailing belief at

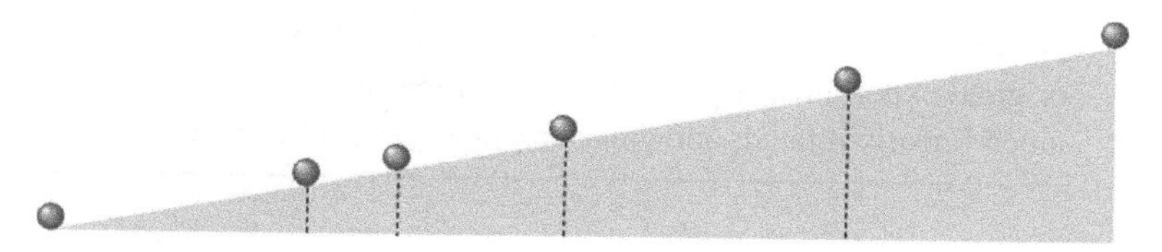

Fig. 1.12 Galileo's ramp experiment setup used for exploring the relationship between time and the distance an object falls due to gravity. To perform this experiment he repeatedly rolled a ball down a ramp and timed how long it took to get 1/4,1/2, 2/3, 3/4, and all the way down the ramp.

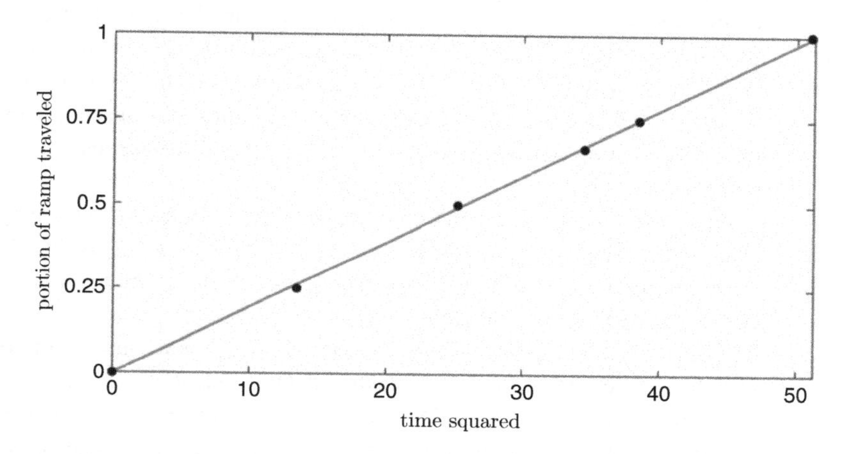

Fig. 1.13 Galileo's experimental data consisting of six points whose input is time and output is the fraction of the ramp traveled. Shown in the plot is the output with the feature time squared, along with a linear fit in magenta. In machine learning we call the variable "time squared" a feature of the original input variable "time."

the time) published his final book: *Dialogues Concerning Two New Sciences* [35]. In this book, written as a discourse among three men in the tradition of Aristotle, he described his experimental and philosophical evidence for the notion of uniformly accelerated physical motion. Specifically, Galileo (and others) had intuition that the acceleration of an object due to (the force we now know as) gravity is uniform in time, or in other words that the distance an object falls is directly proportional (i.e., linearly related) to the amount of time it has been traveling, squared. This relationship was empirically solidified using the following ingeniously simple experiment performed by Galileo.

Repeatedly rolling a metal ball down a grooved 51/2 meter long piece of wood set at an incline as shown in Fig. 1.12, Galileo timed how long the ball took to get 1/4,1/2, 2/3, 3/4, and all the way down the wood ramp.[6]

Data from a modern reenactment [75] of these experiments (averaged over 30 trials), results in the six data points shown in Fig. 1.13. However, here we show not the original input (time) and output (corresponding fraction of the ramp traveled) data, but the output

[6] A ramp was used, as opposed to simply dropping the ball vertically, because time keeping devices in the time of Galileo were not precise enough to make accurate measurements if the ball was simply dropped.

paired with the feature *time squared*, measured in milliliters of water[7] as in Galileo's original experiment. By using square of the time as a feature the dataset becomes very much linearly related, allowing for a near perfect linear regression fit.

Example 1.8 Feature design for visual object detection

A more modern example of feature design, where we have only partial understanding of the underlying process generating data, lies in the task of visual object detection (first introduced in Example 1.4). Unlike the case with Galileo and uniform acceleration described previously, here we do not know nearly as much about the underlying process of visual cognition in both an experimental and philosophical sense. However, even with only pieces of a complete understanding we can still design useful features for object detection.

One of the most crucial and commonly leveraged facts in designing features for visual classification tasks (as we will see later in Section 4.6.2) is that the distinguishing information in a natural image, that is an image a human being would normally be exposed to like a forest or outdoor scene, cityscapes, other people, animals, the insides of buildings, etc., is largely contained in the relatively small number of edges in an image [15, 16]. For example Fig. 1.14 shows a natural image along with an image consisting of its most prominent edges. The majority of the pixels in this image do not belong to any edges, yet with just the edges we can still tell what the image contains.

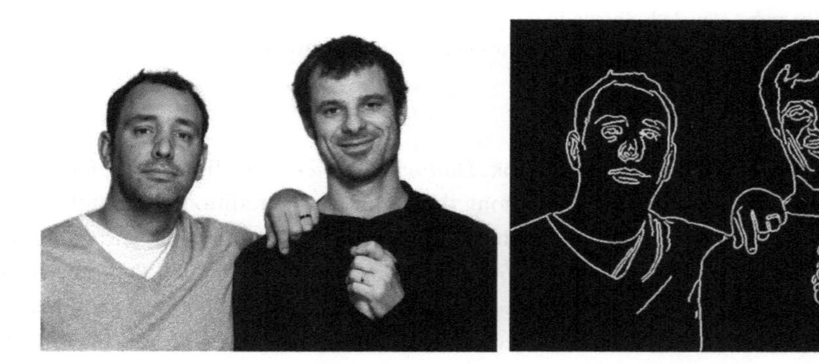

Fig. 1.14 (left panel) A natural image, in this instance of the two creators/writers of the television show South Park (this image is reproduced with permission of Jason Marck). (right panel) The edge detected version of this image, where the bright yellow pixels indicate large edge content, still describes the scene very well (in the sense that we can still tell there are two people in the image) using only a fraction of the information contained in the original image. Note that edges have been colored yellow for visualization purposes only.

[7] Chronological watches (personal timepieces that keep track of hours/minutes/seconds like we have today) did not exist in the time of Galileo. Instead time was measured by calculating the amount of water dripped from a spout into a small cup while each ball rolled down the ramp. This clever time keeping device was known as a "water clock."

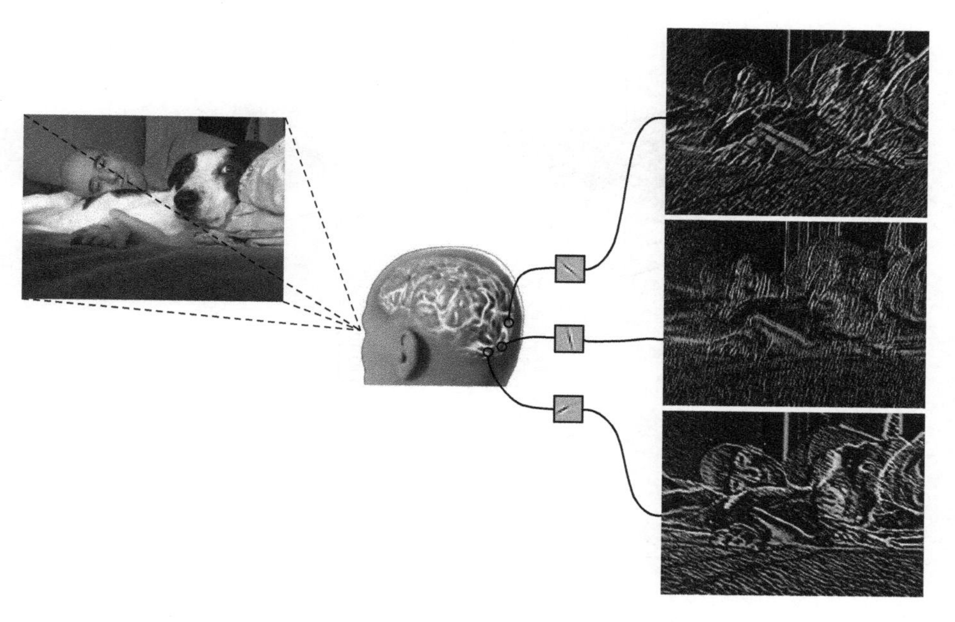

Fig. 1.15 Visual information is processed in an area of the brain where each neuron detects in the observed scene edges of a specific orientation and width. It is thought that what we (and other mammals) "see" is a processed interpolation of these edge detected images.

From visual studies performed largely on frogs, cats, and primates, where a subject is shown visual stimuli while electrical impulses are recorded in a small area in the subject's brain where visual information is processed, neuroscientists have determined that individual neurons involved roughly operate by identifying edges [41, 55]. Each neuron therefore acts as a small "edge detector," locating edges in an image of a specific orientation and thickness, as shown in Fig. 1.15. It is thought that by combining and processing these edge detected images, humans and other mammals "see."

1.4 Numerical optimization

As we will see throughout the remainder of the book, we can formalize the search for parameters of a learning model via well-defined mathematical functions. These functions, commonly referred to as *cost functions*, take in a specific set of model parameters and return a score indicating how well we would accomplish a given learning task using that choice of parameters. A high value indicates a choice of parameters that would give poor performance, while the opposite holds for a set of parameters providing a low value. For instance, recall the share price prediction example from Section 1.2.1, in which we aimed at learning a regression line to predict a company's share price based on its revenue. This line is fit properly by optimally tuning its two parameters: slope and intercept. Geometrically, this corresponds to finding the set of parameters providing the smallest value (called a *minimum*) of a 2-dimensional cost function, as shown in Fig. 1.16.

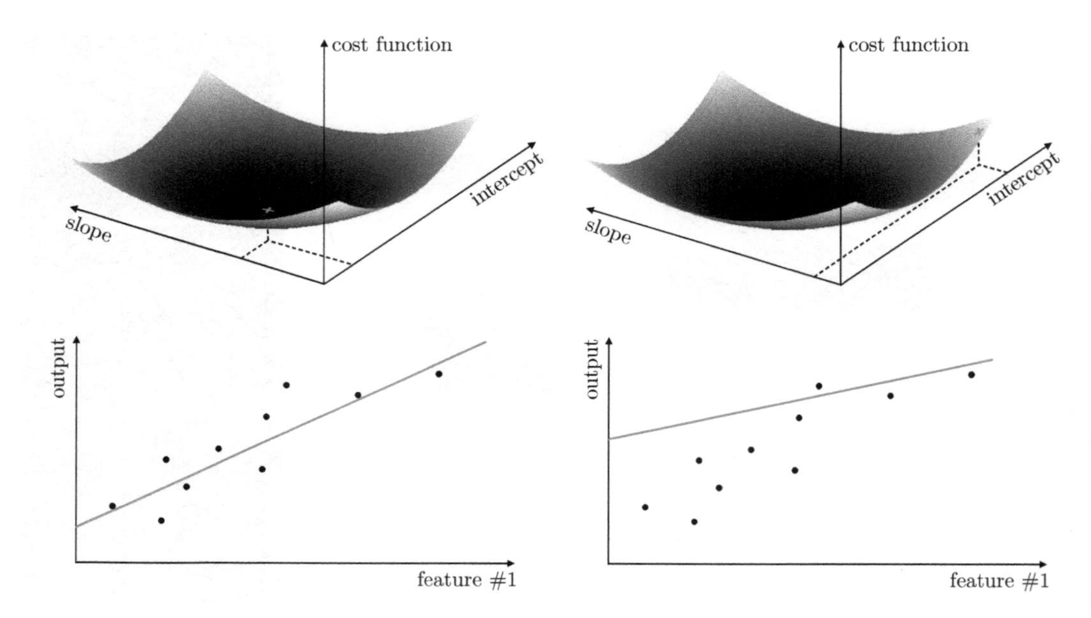

Fig. 1.16 (top panels) The 2-dimensional cost function associated with learning the slope and intercept parameters of a linear model for share price prediction based on revenue for a toy dataset first shown in Fig. 1.7. Also shown here are two different sets of parameter values, one (left) at the minimum of the cost function and the other (right) at a point with larger cost function value. (bottom panels) The linear model corresponding to each set of parameters in the top panel. The set of parameters resulting in the best performance are found at the minimum of the cost surface.

This concept plays a similarly fundamental role with classification as well. Recall the toy problem of distinguishing between images of cats and dogs, described in Section 1.1. There we discussed how a linear classifier is trained on the feature representations of a training set by finding ideal values for its two parameters, which are again slope and intercept. The ideal setting for these parameters again corresponds with the minimum of a cost function, as illustrated in Fig. 1.17.

Because a low value corresponds to a high performing model in the case of both regression and classification, we will always look to *minimize* cost functions in order to find the ideal parameters of their associated learning models. As the study of computational methods for minimizing formal mathematical functions, the tools of *numerical optimization* therefore play a fundamental role throughout the text.

1.5 Summary

In this chapter we have given a broad overview of machine learning, with an emphasis on critical concepts we will see repeatedly throughout the book. We began in Section 1.1 where we described a prototypical machine learning problem, as well as the steps typically taken to solve such a problem (summarized in Fig. 1.6). In Section 1.2 we then introduced the two fundamental problems of machine learning: *regression*

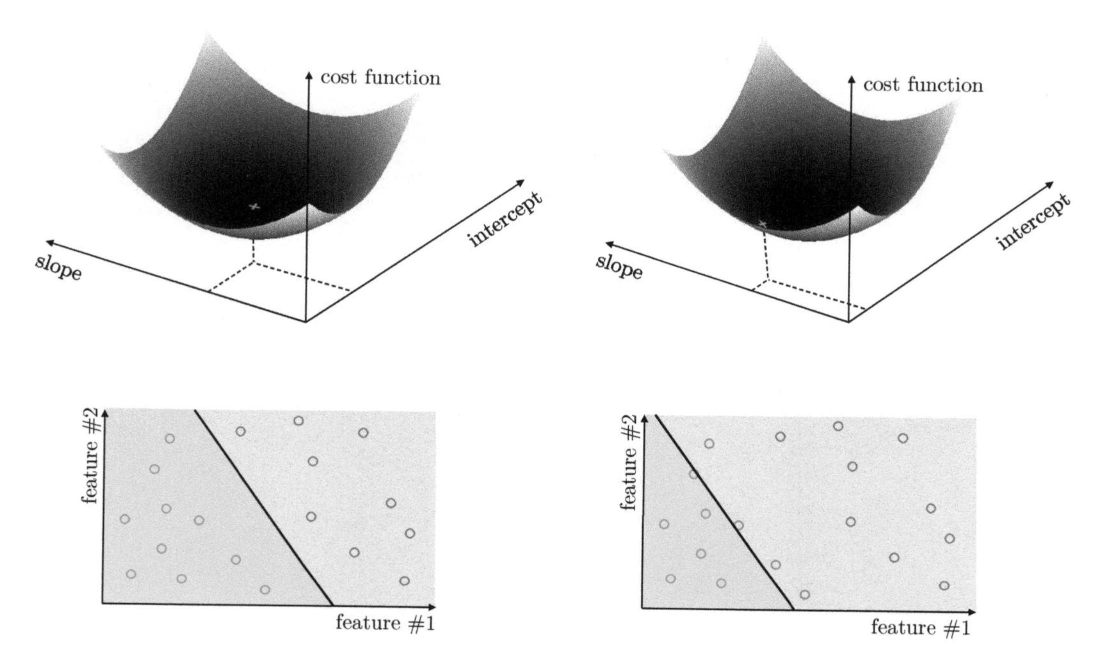

Fig. 1.17 (top panels) The 2-dimensional cost function associated with learning the slope and intercept parameters of a linear model separating two classes of data. Also shown here are two different sets of parameter values, one (left) corresponding to the minimum of the cost function and the other (right) corresponding to a point with larger cost function value. (bottom panels) The linear classifiers corresponding to each set of parameters in the top panels. The optimal set of parameters, i.e., those giving the minimum value of the associated cost function, allow for the best performance.

and *classification*, detailing a number of applications of both. Next, Section 1.3 introduced the notion of *features*, or those uniquely descriptive elements of data that are crucial to effective learning. Finally in Section 1.4 we motivated the need for *numerical optimization* due to the pursuit of ideal parameters of a learning model having direct correspondence to the geometric problem of finding the smallest value of an associated cost function (summarized pictorially in Fig. 1.16 and 1.17).

Part I

Fundamental tools and concepts

Overview of Part I

In the three chapters that follow we describe in significant detail the basic concepts of machine learning introduced in Chapter 1, beginning with an introduction to several fundamental tools of numerical optimization in Chapter 2. This includes a thorough description of calculus-defined optimality, as well as the widely used gradient descent and Newton's method algorithms. Discussing these essential tools first will enable us to immediately and effectively deal with all of the formal learning problems we will see throughout the entirety of the text. Chapters 3 and 4 then introduce linear regression and classification respectively, the two predictive learning problems which form the bedrock of modern machine learning. We motivate both problems naturally, letting illustrations and geometric intuition guide our formal derivations, and in each case describe (through a variety of examples) how knowledge is used to forge effective features.

2 Fundamentals of numerical optimization

In this chapter we review fundamental concepts from the field of numerical optimization that will be used throughout the text in order to determine optimal parameters for learning models (as described in Section 1.4) via the minimization of a differentiable function. Specifically, we describe two basic but widely used algorithms, known as *gradient descent* and *Newton's method*, beginning with a review of several important ideas from calculus that provide the mathematical foundation for both methods.

2.1 Calculus-defined optimality

In this section we briefly review how calculus is used to describe the local geometry of a function, as well as its minima or lowest points. As we will see later in the chapter, powerful numerical algorithms can be built using these simple concepts.

2.1.1 Taylor series approximations

To glean some basic insight regarding the geometry of a many times differentiable function $g(w)$ near a point v we may form a *linear* approximation to the function near this point. This is just a tangent line passing through the point $(v, g(v))$, as illustrated in Fig. 2.1, which contains the first derivative information $g'(v)$.

Such a linear approximation (also known as a first order Taylor series) is written as

$$h(w) = g(v) + g'(v)(w - v). \tag{2.1}$$

Note that indeed this function is **a)** linear in w, **b)** tangent to $g(w)$ at v since $h(v) = g(v)$ and because it contains the first derivative information of g at v i.e., $h'(v) = g'(v)$. This linear approximation holds particularly well near v because the derivative contains slope information.

To understand even more about g near v we may form a quadratic approximation (also illustrated in Fig. 2.1) that contains both first and second derivative information $g'(v)$ and $g''(v)$. This quadratic, referred to as the second order Taylor series approximation, is written as

$$h(w) = g(v) + g'(v)(w - v) + \frac{1}{2}g''(v)(w - v)^2. \tag{2.2}$$

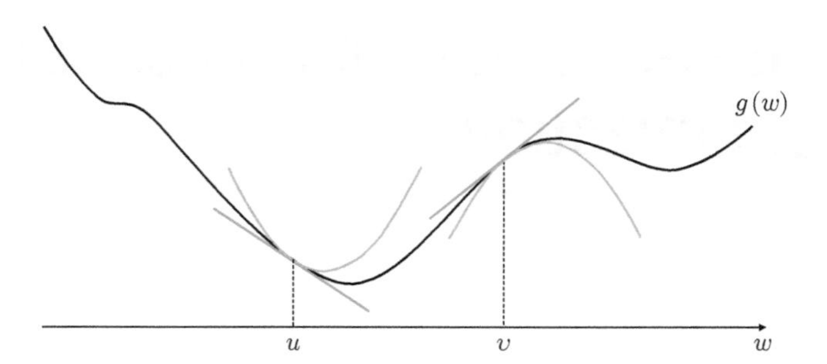

Fig. 2.1 Linear (in green) and quadratic (in blue) approximations to a differentiable function $g(w)$ at two points: $w = u$ and $w = v$. Often these linear and quadratic approximations are equivalently referred to as first and second order Taylor series approximations, respectively.

This quadratic contains the same tangency and first order information of the linear approximation (i.e., $h(v) = g(v)$ and $h'(v) = g'(v)$) with additional second order derivative information as well at g near v since $h''(v) = g''(v)$. The second order Taylor series approximation more closely resembles the underlying function around v because the second derivative contains so-called curvature information.

We may likewise define linear and quadratic approximations for a many times differentiable function $g(\mathbf{w})$ of vector valued input $\mathbf{w} = \begin{bmatrix} w_1 & w_2 & \cdots & w_N \end{bmatrix}^T$. In general we may formally write the linear approximation as

$$h(\mathbf{w}) = g(\mathbf{v}) + \nabla g(\mathbf{v})^T (\mathbf{w} - \mathbf{v}), \tag{2.3}$$

where $\nabla g(\mathbf{v}) = \begin{bmatrix} \frac{\partial}{\partial w_1} g(\mathbf{v}) & \frac{\partial}{\partial w_2} g(\mathbf{v}) & \cdots & \frac{\partial}{\partial w_N} g(\mathbf{v}) \end{bmatrix}^T$ is the $N \times 1$ gradient of partial derivatives (which reduces to $g'(v) = \frac{\partial}{\partial w} g(v)$ in the case $N = 1$). We may also generally write the quadratic approximation as

$$h(\mathbf{w}) = g(\mathbf{v}) + \nabla g(\mathbf{v})^T (\mathbf{w} - \mathbf{v}) + \frac{1}{2} (\mathbf{w} - \mathbf{v})^T \nabla^2 g(\mathbf{v}) (\mathbf{w} - \mathbf{v}), \tag{2.4}$$

where $\nabla^2 g(\mathbf{v})$ is the $N \times N$ symmetric Hessian matrix of second derivatives (which is just the second derivative $g''(v) = \frac{\partial^2}{\partial w^2} g(v)$ when $N = 1$) defined as

$$\nabla^2 g(\mathbf{v}) = \begin{bmatrix} \frac{\partial^2}{\partial w_1 \partial w_1} g(\mathbf{v}) & \frac{\partial^2}{\partial w_1 \partial w_2} g(\mathbf{v}) & \cdots & \frac{\partial^2}{\partial w_1 \partial w_N} g(\mathbf{v}) \\ \frac{\partial^2}{\partial w_2 \partial w_1} g(\mathbf{v}) & \frac{\partial^2}{\partial w_2 \partial w_2} g(\mathbf{v}) & \cdots & \frac{\partial^2}{\partial w_2 \partial w_N} g(\mathbf{v}) \\ \vdots & \vdots & \ddots & \vdots \\ \frac{\partial^2}{\partial w_N \partial w_1} g(\mathbf{v}) & \frac{\partial^2}{\partial w_N \partial w_2} g(\mathbf{v}) & \cdots & \frac{\partial^2}{\partial w_N \partial w_N} g(\mathbf{v}) \end{bmatrix}. \tag{2.5}$$

2.1.2 The first order condition for optimality

Minimum values of a function g are naturally located at "valley floors" where the line or hyperplane tangent to the function has zero slope. Because the derivative/gradient contains this slope information, calculus thereby provides a convenient way of finding

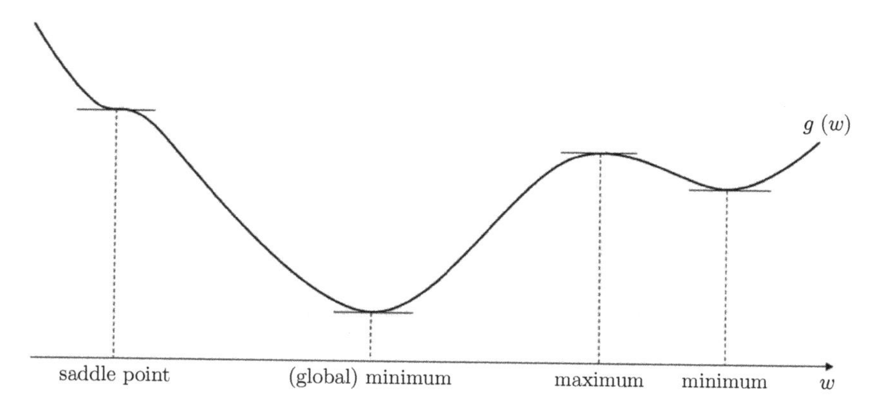

Fig. 2.2 Stationary points of the general function g include minima, maxima, and saddle points. At all such points the gradient is zero.

minimum values of g. In $N = 1$ dimension any point v where $g'(v) = 0$ is a potential minimum. Analogously with general N-dimensional input any point $\mathbf{v}$ where $\nabla g(\mathbf{v}) = \mathbf{0}_{N\times 1}$ is a potential minimum as well. Note that the condition $\nabla g(\mathbf{v}) = \mathbf{0}_{N\times 1}$ can be equivalently written as a system of N equations:

$$\begin{aligned} \frac{\partial}{\partial w_1} g &= 0, \\ \frac{\partial}{\partial w_2} g &= 0, \\ &\vdots \\ \frac{\partial}{\partial w_N} g &= 0. \end{aligned} \tag{2.6}$$

However, for a general function g minima are not the only points that satisfy this condition. As illustrated in Fig. 2.2, a function's maxima as well as saddle points (i.e., points at which the curvature of the function changes from negative to positive or vice-versa) are also points at which the function has a vanishing gradient. Together minima, maxima, and saddle points are referred to as *stationary points* of a function.

In short, while calculus provides us with a useful method for determining minima of a general function g, this method unfortunately determines other undesirable points (maxima and saddle points) as well.[1] Regardless, as we will see later in this chapter the condition $\nabla g(\mathbf{w}) = \mathbf{0}_{N\times 1}$ is a hugely important tool for determining minima, generally referred to as the *first order condition for optimality*, or in short the first order condition.

> A stationary point $\mathbf{v}$ of a function g (including minima, maxima, and saddle points) satisfies the first order condition $\nabla g(\mathbf{v}) = \mathbf{0}_{N\times 1}$.

[1] Although there is a second order condition for optimality that can be used to distinguish between various types of stationary points, it is not often used in practice since it is much easier to construct optimization schemes based solely on the first order condition stated here.

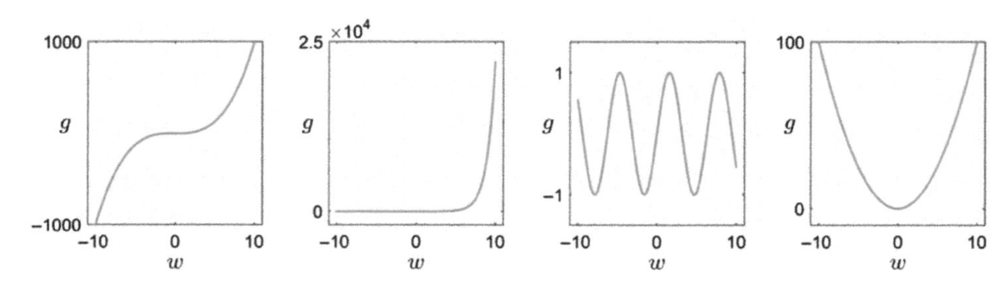

Fig. 2.3 From left to right, plots of functions $g(w) = w^3$, $g(w) = e^w$, $g(w) = \sin(w)$, and $g(w) = w^2$.

Example 2.1 Stationary points of simple functions

In this example we use the first order condition for optimality to compute stationary points of functions $g(w) = w^3$, $g(w) = e^w$, $g(w) = \sin(w)$, $g(w) = w^2$, and $g(\mathbf{w}) = \frac{1}{2}\mathbf{w}^T\mathbf{Q}\mathbf{w} + \mathbf{r}^T\mathbf{w} + d$.

- $g(w) = w^3$, plotted in the first panel of Fig. 2.3, the first order condition gives $g'(w) = 3w^2 = 0$ with a saddle point at $w = 0$.
- $g(w) = e^w$, plotted in the second panel of Fig. 2.3, the first order condition gives $g'(w) = e^w = 0$ which is only satisfied as w goes to $-\infty$, giving a minimum.
- $g(w) = \sin(w)$, plotted in the third panel of Fig. 2.3, the first order condition gives stationary points wherever $g'(w) = \cos(w) = 0$ which occurs at odd integer multiples of $\pi/2$, i.e., maxima at $w = \frac{(4n+1)\pi}{2}$ and minima at $w = \frac{(4n+3)\pi}{2}$ where n is any integer.
- $g(w) = w^2$, plotted in the fourth panel of Fig. 2.3, the first order condition gives $g'(w) = 2w = 0$ with a minimum at $w = 0$.
- $g(\mathbf{w}) = \frac{1}{2}\mathbf{w}^T\mathbf{Q}\mathbf{w} + \mathbf{r}^T\mathbf{w} + d$ where $\mathbf{Q}$ is an $N \times N$ symmetric matrix (i.e., $\mathbf{Q} = \mathbf{Q}^T$), $\mathbf{r}$ is an $N \times 1$ vector, and d is a scalar. Then $\nabla g(\mathbf{w}) = \mathbf{Q}\mathbf{w} + \mathbf{r}$ and thus stationary points exist for all solutions to the linear system of equations $\mathbf{Q}\mathbf{w} = -\mathbf{r}$.

2.1.3 The convenience of convexity

As discussed in Section 1.4, solving a machine learning problem eventually reduces to finding the minimum of an associated cost function. Of all (potentially many) minima of a cost function, we are especially interested in the one that provides the lowest possible value of the function, known as the *global minimum*. For a special family of functions, referred to as convex functions, the first order condition is particularly useful because *all stationary points of a convex function are global minima.* In other words, convex functions are free of maxima and saddle points as well as non-global minima.

To determine if a function g is *convex* (facing upward, as the function shown in Fig. 2.1 is at the point u) or *concave* (facing downward, as the function shown in Fig. 2.1

is at the point v) at a point v we check its curvature or second derivative information there:

$$\begin{array}{lll} g''(v) \geq 0 & \Longleftrightarrow & g \text{ is convex at } v \\ g''(v) \leq 0 & \Longleftrightarrow & g \text{ is concave at } v. \end{array} \tag{2.7}$$

Here, if a statement on one side of the symbol $\Longleftrightarrow$ (which reads "if and only if") is true then the statement on the other side is true as well (likewise if one is false then the other is false as well). Similarly, for general N an analogous statement can be made regarding the eigenvalues of $\nabla^2 g(\mathbf{v})$, i.e., g is convex (or concave) at $\mathbf{v}$ if and only if the Hessian matrix evaluated at this point has all non-negative (or non-positive) *eigenvalues*, in which case the Hessian is called positive semi-definite (or negative semi-definite).

Based on this rule, $g(w)$ is convex everywhere, a convex function, if its second derivative $g''(w)$ is always non-negative. Likewise $g(\mathbf{w})$ is convex if $\nabla^2 g(\mathbf{w})$ always has non-negative eigenvalues. This is generally referred to as the *second order definition of convexity*.[2]

A twice differentiable function is convex if and only if $g''(w) \geq 0$ for all w (or $\nabla^2 g(\mathbf{w})$ has non-negative eigenvalues for all $\mathbf{w}$)

Example 2.2 Convexity of simple functions with scalar input

In this example we use the second order definition of convexity to verify whether each of the functions shown in Fig. 2.3 is convex or not:

- $g(w) = w^3$ has second derivative $g''(w) = 6w$ which is not always non-negative, hence g is not convex.
- $g(w) = e^w$ has second derivative $g''(w) = e^w$ which is positive for any choice of w, and so g is convex.
- $g(w) = \sin(w)$ has second derivative $g''(w) = -\sin(w)$. Since this is not always non-negative g is non-convex.
- $g(w) = w^2$ has second derivative $g''(w) = 2$, and so g is convex.

Example 2.3 Convexity of a quadratic function with vector input

In N-dimensions a quadratic function in $\mathbf{w}$ takes the form

$$g(\mathbf{w}) = \frac{1}{2}\mathbf{w}^T\mathbf{Q}\mathbf{w} + \mathbf{r}^T\mathbf{w} + d, \tag{2.8}$$

[2] While there are a number of ways to formally check that a function is convex, we will see that the second order approach is especially convenient. The interested reader can see Appendix D for additional information regarding convex functions.

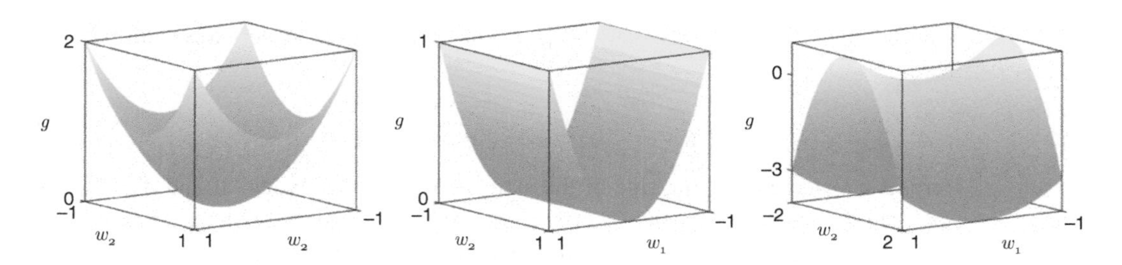

Fig. 2.4 Three quadratic functions of the form $g(\mathbf{w}) = \frac{1}{2}\mathbf{w}^T\mathbf{Q}\mathbf{w} + \mathbf{r}^T\mathbf{w} + d$ generated by different instances of matrix $\mathbf{Q}$ in Example 2.3. In all three cases $\mathbf{r} = \mathbf{0}_{2\times1}$ and $d = 0$. As can be visually verified, only the first two functions are convex. The last "saddle-looking" function on the right has a saddle point at zero!

with its Hessian given by

$$\nabla^2 g(\mathbf{w}) = \frac{1}{2}\left(\mathbf{Q} + \mathbf{Q}^T\right). \tag{2.9}$$

Note that if $\mathbf{Q}$ is symmetric, then $\nabla^2 g(\mathbf{w}) = \mathbf{Q}$. Also note that $\mathbf{r}$ and d have no influence on the convexity of g. We now verify convexity for three simple instances of $\mathbf{Q}$ where $N = 2$. We discuss a convenient way to determine the convexity of more general vector input functions in the exercises.

- When $\mathbf{Q} = \begin{bmatrix} 2 & 0 \\ 0 & 2 \end{bmatrix}$ the Hessian $\nabla^2 g(\mathbf{w}) = \mathbf{Q}$ has two eigenvalues equaling 2, so the corresponding quadratic, shown in the left panel of Fig. 2.4, is convex.
- When $\mathbf{Q} = \begin{bmatrix} 2 & 0 \\ 0 & 0 \end{bmatrix}$ again the Hessian $\nabla^2 g(\mathbf{w}) = \mathbf{Q}$ has two eigenvalues (2 and 0), so the corresponding quadratic, shown in the middle panel of Fig. 2.4, is convex.
- When $\mathbf{Q} = \begin{bmatrix} 2 & 0 \\ 0 & -2 \end{bmatrix}$ again the Hessian is $\nabla^2 g(\mathbf{w}) = \mathbf{Q}$ and has eigenvalues 2 and -2, so the corresponding quadratic, shown in the right panel of Fig. 2.4, is non-convex.

2.2 Numerical methods for optimization

In this section we introduce two basic but widely used numerical techniques, known as *gradient descent* and *Newton's method*, for finding minima of a function $g(\mathbf{w})$. The formal manner of describing the minimization of a function g is commonly written as

$$\underset{\mathbf{w}}{\text{minimize}} \;\; g(\mathbf{w}), \tag{2.10}$$

which is simply shorthand for saying "minimize g over all input values $\mathbf{w}$." The solution to (2.10), referred to as the *optimal* $\mathbf{w}$, is typically denoted as $\mathbf{w}^\star$. While both gradient descent and Newton's method operate sequentially by finding points at which g gets smaller and smaller, both methods are only guaranteed to find stationary points of g, i.e., those points satisfying the first order condition (discussed in Section 2.1.2). Thus one can also consider these techniques for numerically solving the system of N equations $\nabla g(\mathbf{w}) = \mathbf{0}_{N\times 1}$.

2.2.1 The big picture

All numerical optimization schemes for minimization of a general function g work as follows:

(1) Start the minimization process from some *initial point* $\mathbf{w}^0$.

(2) Take *iterative* steps denoted by $\mathbf{w}^1, \mathbf{w}^2, \ldots$, going "downhill" towards a stationary point of g.

(3) Repeat step (2) until the sequence of points converges to a stationary point of g.

This idea is illustrated in Fig. 2.5 for the minimization of a non-convex function. Note that since this function has three stationary points, the one we reach by traveling downhill depends entirely on where we begin the optimization process. Ideally we would like to find the *global minimum*, or the lowest of the function's minima, which for a general non-convex function requires that we run the procedure several times with different initializations (or starting points).

As we will see in later chapters, many important machine learning cost functions are convex and hence have only global minima, as in Fig. 2.6, in which case any initialization will recover a global minimum.

The numerical methods discussed here halt at a stationary point $\mathbf{w}$, that is a point where $\nabla g(\mathbf{w}) = \mathbf{0}_{N\times 1}$, which as we have previously seen may or may not constitute a minimum of g if g is non-convex. However, this issue does not at all preclude the use of non-convex cost functions in machine learning (or other scientific disciplines), it is simply worth being aware of.

2.2.2 Stopping condition

One of several *stopping conditions* may be selected to halt numerical algorithms that seek stationary points of a given function g. The two most commonly used stopping criteria are:

(1) When a pre-specified number of iterations are complete.

(2) When the gradient is small enough, i.e., $\left\|\nabla g\left(\mathbf{w}^k\right)\right\|_2 < \epsilon$ for some small $\epsilon > 0$.

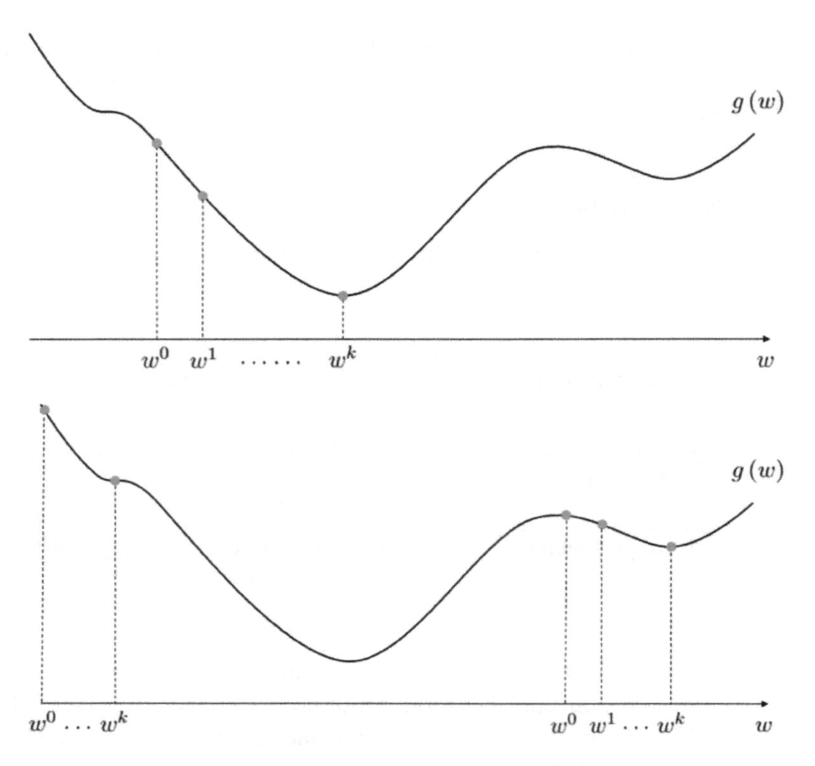

Fig. 2.5 The stationary point of a non-convex function found via numerical optimization is dependent on the choice of initial point w^0. In the top panel our initialization leads us to find the global minimum, while in the bottom panel the two different initializations lead to a saddle point on the left, and a non-global minimum on the right.

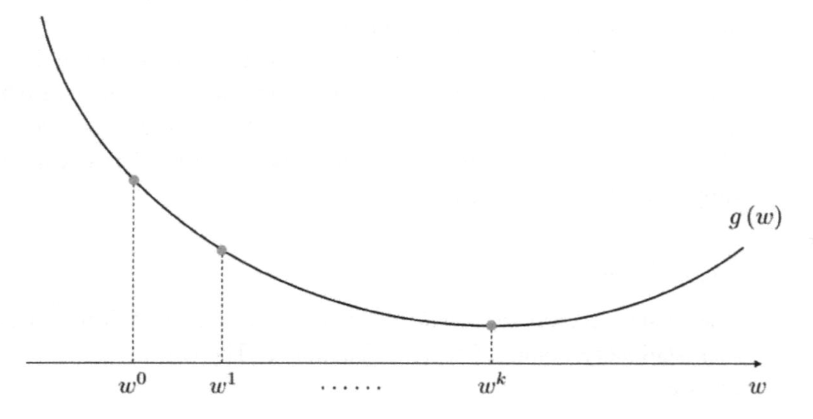

Fig. 2.6 A global minimum of a convex function is found via numerical optimization regardless of our choice of the initial point w^0.

Perhaps the most naive stopping condition for a numerical algorithm is to halt the procedure after a pre-defined number of iterations. Note that this extremely simple condition does not provide any convergence guarantee, and hence is typically used in practice in conjunction with other stopping criteria as a necessary cap on the number of iterations when the convergence is achieved slowly. The second condition directly translates our desire to finding a stationary point at which the gradient is by definition zero. One could also stop the procedure when continuing it does not considerably decrease the objective function (or the stationary point itself) from one iteration to the next.

2.2.3 Gradient descent

The defining characteristic differentiating various numerical optimization methods is the way iterative steps are taken for reducing the value of g. The two classic methods, gradient descent and Newton's method, use local models for the function at each step in order to find smaller and smaller values of the function. As illustrated in Fig. 2.7, the basic idea with gradient descent is to build a linear model of the function g, determine the "downward" direction on this hyperplane, travel a short distance along this direction, hop back on to the function g, and repeat until convergence. Starting at an initial point $\mathbf{w}^0$ and by carefully choosing how far we travel at each step, the gradient descent procedure produces a sequence of points $\mathbf{w}^1$, $\mathbf{w}^2$, $\mathbf{w}^3 \ldots$, that shrinks the value of g at each step and eventually reaches a stationary point of g.

Formally, beginning at an initial point $\mathbf{w}^0$ the linear model of g at this point is given precisely by the first order Taylor series approximation in (2.3) centered at $\mathbf{w}^0$:

$$h(\mathbf{w}) = g\left(\mathbf{w}^0\right) + \nabla g\left(\mathbf{w}^0\right)^T\left(\mathbf{w} - \mathbf{w}^0\right). \tag{2.11}$$

We now take our first step by traveling in the direction in which the tangent hyperplane most sharply angles downward (referred to as the *steepest descent direction*). Using a

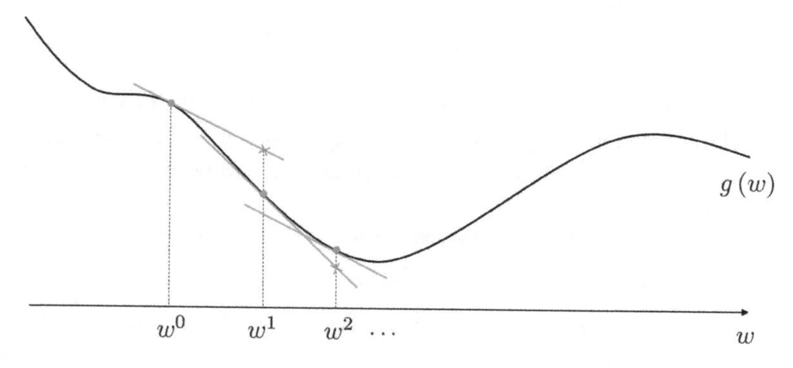

Fig. 2.7 With the gradient descent method, we travel in the downward direction of a linear approximation, hop back onto the function, and repeat in order to find a stationary point of g.

simple calculus-based argument[3] it can be shown that this steepest descent direction is given precisely as $-\nabla g\left(\mathbf{w}^0\right)$. Thus we descend in the direction of the negative gradient (hence the name of the algorithm, *gradient descent*) taking our first step to a point $\mathbf{w}^1$ where

$$\mathbf{w}^1 = \mathbf{w}^0 - \alpha_1 \nabla g\left(\mathbf{w}^0\right). \tag{2.12}$$

Here α_1 is a positive constant, called a *step length* (sometimes referred to as a *learning rate*), that controls how far we descend in the negative gradient direction, from our initial point $\mathbf{w}^0$. We then repeat this procedure constructing the first order Taylor series approximation at $\mathbf{w}^1$, travel in its steepest descent direction, which is again given by the negative gradient $-\nabla g\left(\mathbf{w}^1\right)$, taking our second step to the point $\mathbf{w}^2$ where

$$\mathbf{w}^2 = \mathbf{w}^1 - \alpha_2 \nabla g\left(\mathbf{w}^1\right). \tag{2.13}$$

Once again α_2 is a small positive step length, perhaps different from α_1, that controls the length of our travel along the negative gradient from $\mathbf{w}^1$. This entire procedure is repeated with the kth step being given analogously as

$$\mathbf{w}^k = \mathbf{w}^{k-1} - \alpha_k \nabla g\left(\mathbf{w}^{k-1}\right), \tag{2.14}$$

where α_k is the step length associated with the kth gradient descent step. Note that this procedure only stops when at some iteration $\nabla g\left(\mathbf{w}^{k-1}\right) \approx \mathbf{0}_{N\times 1}$, that is when we have approximately satisfied the first order condition and essentially reached a stationary point $\mathbf{w}^k$ of g. For easy reference we give the gradient descent procedure (with a fixed given step length) in Algorithm 2.1.

Algorithm 2.1 Gradient descent (with fixed step length)

Input: differentiable function g, fixed step length α, and initial point $\mathbf{w}^0$
$k = 1$
Repeat until stopping condition is met:
  $\mathbf{w}^k = \mathbf{w}^{k-1} - \alpha \nabla g\left(\mathbf{w}^{k-1}\right)$
  $k \leftarrow k + 1$

How do we choose a proper value for the step length at each iteration? As illustrated in the top panel of Fig. 2.8, it cannot be set too large, because then the algorithm travels too far at each step and will bounce around and perhaps never converge. On the other

[3] Note that for a unit length direction $\mathbf{d}$, h in (2.11) can be written as $h\left(\mathbf{d}\right) = g\left(\mathbf{w}^0\right) - \nabla g\left(\mathbf{w}^0\right)^T \mathbf{w}^0 + \nabla g\left(\mathbf{w}^0\right)^T \mathbf{d}$, where the first two terms on the right hand side are constant with respect to $\mathbf{d}$. Thus the unit length direction $\mathbf{d}$ that minimizes the inner product $\nabla g\left(\mathbf{w}^0\right)^T \mathbf{d}$ will result in the sharpest descent in h. From the inner product rule (see Appendix A) this is smallest when $\mathbf{d} = -\frac{\nabla g\left(\mathbf{w}^0\right)}{\left\|\nabla g\left(\mathbf{w}^0\right)\right\|_2}$, and so the steepest descent direction is indeed $-\nabla g\left(\mathbf{w}^0\right)$.

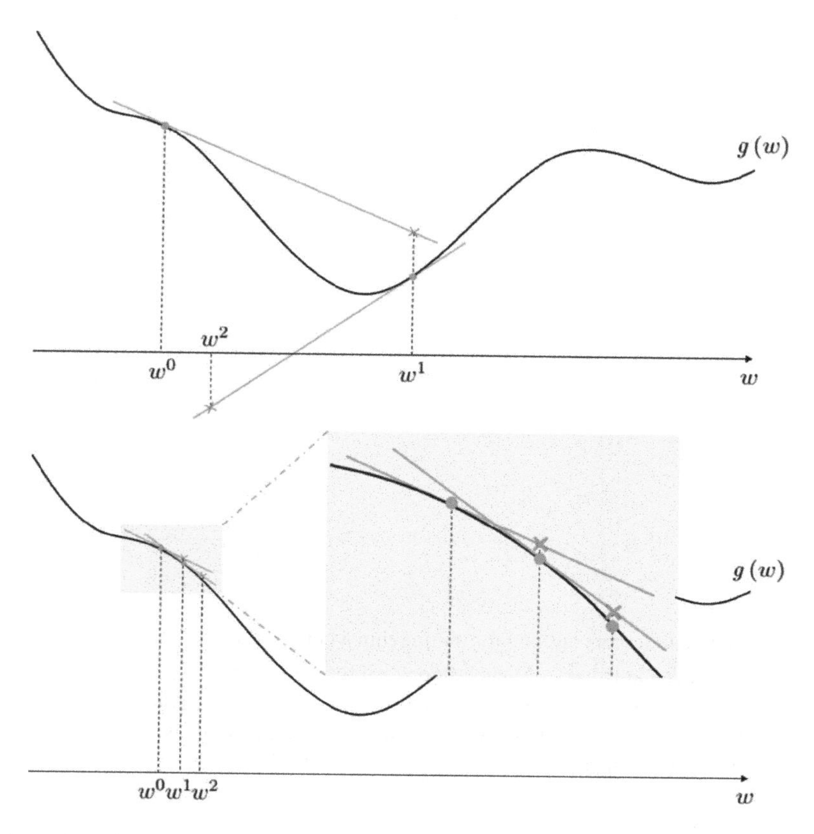

Fig. 2.8 The effect of step length on the convergence of gradient descent. (top) Too large a step length and we may overshoot the minimum and possibly never converge. (bottom) A small step length makes the gradient descent converge to the minimum very slowly.

hand, if α_k is made too small at each step as illustrated in the bottom panel of Fig. 2.8, the algorithm crawls downward far too slowly and may never reach the stationary point in our lifetime!

For machine learning problems it is common practice to choose a value for the step length by trial and error. In other words, just try a range of values, for each performing a complete run of the gradient descent procedure with the step length at every iteration fixed at this value. One can then choose the particular step length value that provides (the fastest) convergence (you can practice this in Exercise 2.14). There are also formal methods that do not rely on trial and error for determining appropriate step lengths which guarantee convergence. These methods are detailed in Sections 8.1 and 8.2.

The step length for gradient descent can be chosen by trial and error and fixed for all iterations, or determined using the formal schemes described in Sections 8.1 and 8.2.

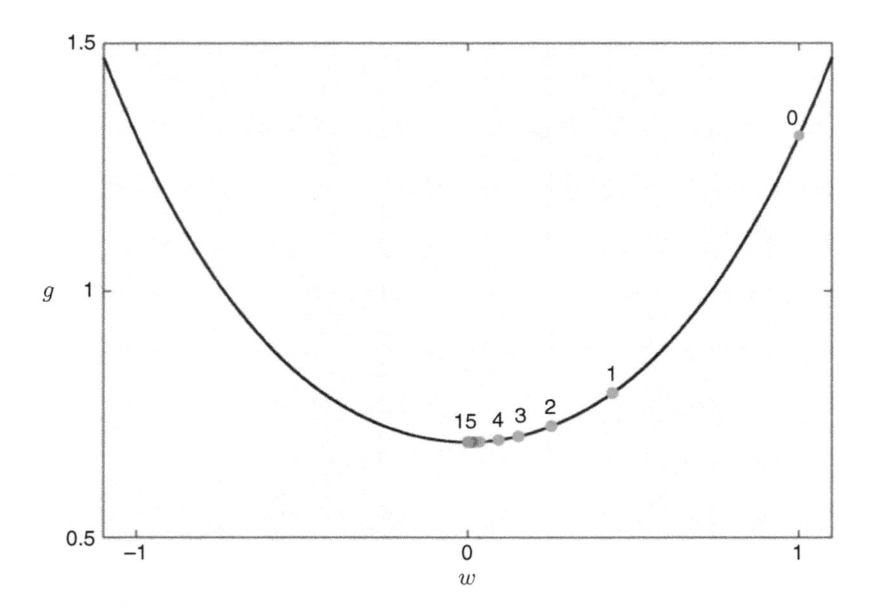

Fig. 2.9 Gradient descent applied for minimizing a convex function with scalar input (see Example 2.4) initialized at $w^0 = 1$. At each step w^k of the procedure, the point $\left(w^k, g\left(w^k\right)\right)$ is colored green and numbered on the function itself, with the final point colored red. Note that only the first five points as well as the final point are numbered in this figure, since the rest blur together as the minimum is approached.

Example 2.4 Gradient descent for a cost function with scalar input

Suppose that $g(w) = \log\left(1 + e^{w^2}\right)$, whose first derivative is given as $g'(w) = \frac{2e^{w^2}w}{1+e^{w^2}}$. As illustrated in Fig. 2.9, g is convex, and with the initial point $w^0 = 1$ and a step length fixed at $\alpha = 10^{-1}$ (determined by trial and error) gradient descent requires a fair number of steps to converge to the global minimum of g. Specifically, it takes 15 steps for gradient descent to reach a point at which the absolute value of the derivative falls below $\epsilon = 10^{-3}$.

Example 2.5 Gradient descent for a cost function with vector input

Take $g(\mathbf{w}) = -\cos\left(2\pi\mathbf{w}^T\mathbf{w}\right) + 2\mathbf{w}^T\mathbf{w}$, where $\mathbf{w}$ is a 2-dimensional vector, and the gradient of g is given as $\nabla g(\mathbf{w}) = 4\pi\sin\left(2\pi\mathbf{w}^T\mathbf{w}\right)\mathbf{w} + 4\mathbf{w}$. In Fig. 2.10 we show the objective value of gradient descent steps $g\left(\mathbf{w}^k\right)$ with two starting points where a fixed step length of $\alpha = 10^{-3}$ (determined by trial and error) was used for all iterations of each run. In this instance one of the starting points ($\mathbf{w}^0 = \begin{bmatrix} -0.7 & 0 \end{bmatrix}^T$) allows gradient descent to reach the global minimum of the function, while the other ($\mathbf{w}^0 = \begin{bmatrix} 0.85 & 0.85 \end{bmatrix}^T$) ends up at a local minimum of the surface.

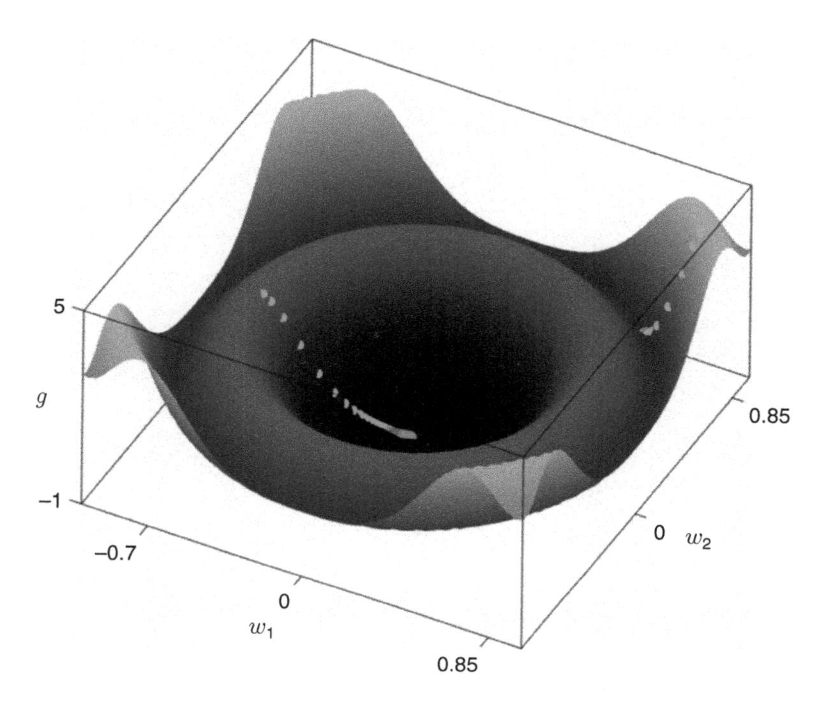

Fig. 2.10 Gradient descent applied for minimizing a non-convex function with 2-dimensional input (see Example 2.5). Of the two initializations used here only one leads to the global minimum. For each run the point $\left(w_1^k, w_2^k, g\left(w_1^k, w_2^k\right)\right)$ is colored green at each step of the procedure, with the final point colored red.

2.2.4 Newton's method

Like gradient descent, Newton's method works by using approximations to a function at each step in order to lower its value. However, with Newton's method a quadratic approximation, again generated via the Taylor series approximation, is used. As illustrated in the top panel of Fig. 2.11, starting at an initial point $\mathbf{w}^0$ Newton's method produces a sequence of points $\mathbf{w}^1$, $\mathbf{w}^2, \ldots$, that minimizes g by repeatedly creating a quadratic approximation to the function, traveling to a stationary point of this quadratic, and hopping back onto the function. Because Newton's method uses quadratic as opposed to linear approximations at each step, with a quadratic more closely mimicking the associated function, it is often much more effective than gradient descent (in the sense that it requires far fewer steps for convergence [24, 50]). However this reliance on quadratic information makes Newton's method more difficult to use with non-convex functions,[4] since at concave portions of such a function the algorithm can climb to a maximum, as illustrated in the bottom panel of Fig. 2.11.

[4] A number of procedures exist that adjust Newton's method at concave portions of a function in order to make it more effective for use with non-convex functions. See Exercise 3.14 as well as [50, 59] for further information.

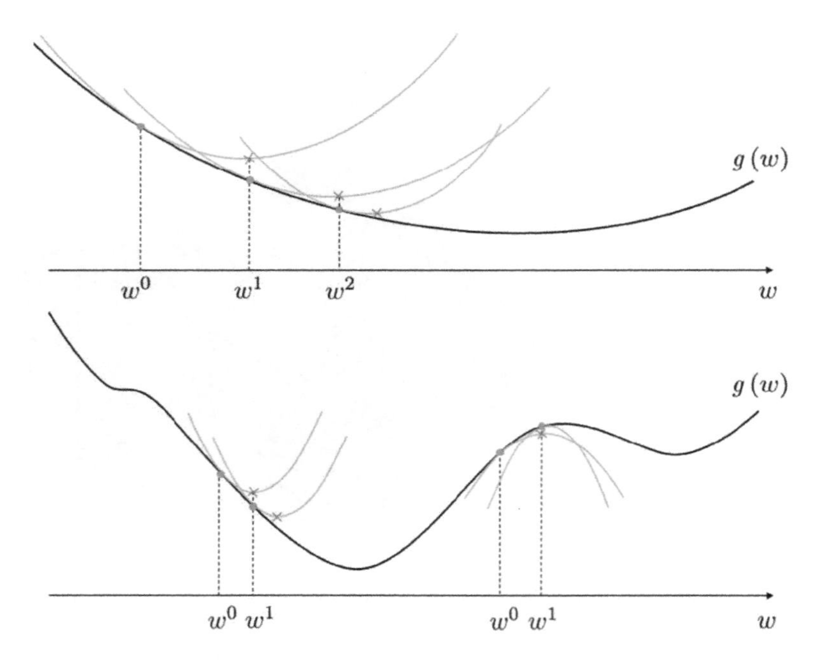

Fig. 2.11 Newton's method illustrated. To find a minimum of g Newton's method hops down the stationary points of quadratic approximations generated by g's second order Taylor series. (top panel) For convex functions these quadratic approximations are themselves always convex (upward facing) and so their stationary points are minima, and the sequence leads to a minimum of the original function. (bottom panel) For non-convex functions, quadratic approximations can be concave or convex depending on where they are constructed, leading the algorithm to possibly converge to a maximum.

Formally, beginning at a point $\mathbf{w}^0$ the quadratic model of g at this point is given precisely by the second order Taylor series approximation in (2.4) centered at $\mathbf{w}^0$,

$$h(\mathbf{w}) = g\left(\mathbf{w}^0\right) + \nabla g\left(\mathbf{w}^0\right)^T \left(\mathbf{w} - \mathbf{w}^0\right) + \frac{1}{2}\left(\mathbf{w} - \mathbf{w}^0\right)^T \nabla^2 g\left(\mathbf{w}^0\right)\left(\mathbf{w} - \mathbf{w}^0\right). \quad (2.15)$$

We now wish to travel to a stationary point of this quadratic which is a minimum in the case where g is convex.[5] To do this we can use the first order condition (see Section 2.1.2) by setting the gradient of h to zero and solving for $\mathbf{w}$. This gives the $N \times N$ system of linear equations[6]

$$\nabla^2 g\left(\mathbf{w}^0\right)\mathbf{w} = \nabla^2 g\left(\mathbf{w}^0\right)\mathbf{w}^0 - \nabla g\left(\mathbf{w}^0\right). \quad (2.16)$$

[5] A common way of adjusting Newton's method for use with non-convex functions is to add a so-called "regularizer," described in Section 3.3.2, to the original function. See e.g., [50] for further details.

[6] Setting the gradient of h to zero we have $\nabla h(\mathbf{w}) = \nabla g\left(\mathbf{w}^0\right) + \nabla^2 g\left(\mathbf{w}^0\right)\left(\mathbf{w} - \mathbf{w}^0\right) = \mathbf{0}_{N\times 1}$. Solving for $\mathbf{w}$ then gives the linear system of equations $\nabla^2 g\left(\mathbf{w}^0\right)\mathbf{w} = \nabla^2 g\left(\mathbf{w}^0\right)\mathbf{w}^0 - \nabla g\left(\mathbf{w}^0\right)$, which can be written more familiarly as $\mathbf{A}\mathbf{w} = \mathbf{b}$ where $\mathbf{A}_{N\times N} = \nabla^2 g\left(\mathbf{w}^0\right)$ and $\mathbf{b}_{N\times 1} = \nabla^2 g\left(\mathbf{w}^0\right)\mathbf{w}^0 - \nabla g\left(\mathbf{w}^0\right)$ are a fixed matrix and vector, respectively.

A solution to this system of equations gives the first point $\mathbf{w}^1$ traveled to by Newton's method. To take the next step we repeat this procedure, forming a quadratic Taylor series approximation of g (this time centered at $\mathbf{w}^1$) and determine a stationary point of this quadratic by checking the first order condition. This leads to the same kind of linear system of equations,

$$\nabla^2 g\left(\mathbf{w}^1\right)\mathbf{w} = \nabla^2 g\left(\mathbf{w}^1\right)\mathbf{w}^1 - \nabla g\left(\mathbf{w}^1\right), \tag{2.17}$$

a solution of which provides the second step to the point $\mathbf{w}^2$. This entire procedure is repeated until convergence, with the kth Newton step $\mathbf{w}^k$ defined as a stationary point of the quadratic approximation centered at $\mathbf{w}^{k-1}$,

$$\begin{aligned} h(\mathbf{w}) = g\left(\mathbf{w}^{k-1}\right) + \nabla g\left(\mathbf{w}^{k-1}\right)^T\left(\mathbf{w} - \mathbf{w}^{k-1}\right) \\ + \frac{1}{2}\left(\mathbf{w} - \mathbf{w}^{k-1}\right)^T \nabla^2 g\left(\mathbf{w}^{k-1}\right)\left(\mathbf{w} - \mathbf{w}^{k-1}\right), \end{aligned} \tag{2.18}$$

which, again applying the first order condition for optimality and solving for $\mathbf{w}$, gives a linear system of equations,

$$\nabla^2 g\left(\mathbf{w}^{k-1}\right)\mathbf{w} = \nabla^2 g\left(\mathbf{w}^{k-1}\right)\mathbf{w}^{k-1} - \nabla g\left(\mathbf{w}^{k-1}\right). \tag{2.19}$$

As with gradient descent, note that these steps halt when at some kth step we have $\nabla g\left(\mathbf{w}^{k-1}\right) \approx \mathbf{0}_{N\times 1}$, i.e., when we have approximately satisfied the first order condition, essentially recovering a stationary point $\mathbf{w}^k$ of g. For convenience we give the Newton's method scheme in Algorithm 2.2.

Algorithm 2.2 Newton's method

Input: twice differentiable function g, and initial point $\mathbf{w}^0$
$k = 1$
Repeat until stopping condition is met:
  Solve the system $\nabla^2 g\left(\mathbf{w}^{k-1}\right)\mathbf{w}^k = \nabla^2 g\left(\mathbf{w}^{k-1}\right)\mathbf{w}^{k-1} - \nabla g\left(\mathbf{w}^{k-1}\right)$ for $\mathbf{w}^k$.
  $k \leftarrow k + 1$

Note that in cases where the matrix $\nabla^2 g\left(\mathbf{w}^{k-1}\right)$ is invertible we may write the solution to the system in (2.19) algebraically as

$$\mathbf{w}^k = \mathbf{w}^{k-1} - \left[\nabla^2 g\left(\mathbf{w}^{k-1}\right)\right]^{-1}\nabla g\left(\mathbf{w}^{k-1}\right), \tag{2.20}$$

which makes the Newton step look like a gradient step in (2.14), replacing the step length with the inverted Hessian matrix. When the Hessian is not invertible (and there are infinitely many solutions to the system in (2.19)) we may employ the so-called pseudo-inverse (see Appendix C) of the Hessian, denoted as $\left[\nabla^2 g\left(\mathbf{w}^{k-1}\right)\right]^\dagger$, and write the update analogously as

$$\mathbf{w}^k = \mathbf{w}^{k-1} - \left[\nabla^2 g\left(\mathbf{w}^{k-1}\right)\right]^\dagger\nabla g\left(\mathbf{w}^{k-1}\right). \tag{2.21}$$

While not the most computationally efficient[7] method of solving the Newton system, the solution above is always the *smallest* one possible and so (practically speaking) it can be a useful choice when the alternative is an unfamiliar numerical linear algebra solver (e.g., one which may return very large solutions to the system in (2.19), producing numerical instability in subsequent Newton steps).

As previously mentioned, because it uses more precise second order information, Newton's method converges in a much smaller number of steps than gradient descent. This comes at the cost of having to store a Hessian matrix and solve a corresponding linear system at each step, which with modern computational resources is not typically problematic for functions with up to several thousand input variables.[8] For such (especially convex) functions the standard Newton's method described is highly effective.

Example 2.6 Newton's method for a cost function with scalar input

Let us consider again the function $g(w) = \log\left(1 + e^{w^2}\right)$ from Example 2.4, whose second derivative is given as $g''(w) = \frac{2e^{w^2}\left(2w^2+e^{w^2}+1\right)}{\left(1+e^{w^2}\right)^2}$. Note that $g''(w) > 0$ for all w, and thus the kth Newton step in (2.20) for a scalar w reduces to

$$w^k = w^{k-1} - \frac{g'\left(w^{k-1}\right)}{g''\left(w^{k-1}\right)}. \tag{2.22}$$

As illustrated in Fig. 2.12, beginning at point $w^0 = 1$ we need only three Newton steps to reach the minimum of g (where the absolute value of the derivative falls below $\epsilon = 10^{-3}$). This is significantly fewer steps than the gradient descent procedure shown in Fig. 2.9.

Example 2.7 Newton's method for a cost function with vector input

Let g be the quadratic function $g(\mathbf{w}) = \frac{1}{2}\mathbf{w}^T\mathbf{Q}\mathbf{w}+\mathbf{r}^T\mathbf{w}+d$ where $\mathbf{Q} = \begin{bmatrix} 1 & 0.75 \\ 0.75 & 1 \end{bmatrix}$, $\mathbf{r} = \begin{bmatrix} 1 & 1 \end{bmatrix}^T$, and $d = 0$. As shown in Fig. 2.13, with the initial point

[7] It always more computationally efficient to find $\mathbf{w}^k$ by directly solving the linear system in (2.19) using numerical linear algebra software, rather than by calculating the inverse or pseudo-inverse of the Hessian and forming the explicit update in (2.20) or (2.21). See e.g., [24] for further details.

[8] For larger dimensional input, storing the Hessian matrix itself can become problematic, let alone solving the associated linear system at each step. For example with a 10 000-dimensional input $\mathbf{w}$ the corresponding Hessian matrix will be of size $10\,000 \times 10\,000$, with 10^8 values to store for the Hessian matrix alone. Several approaches exist which aim at addressing both the storage and computation issues associated with these large linear systems, e.g., subsampling methods (termed quasi-Newton methods) [50, 59], methods for exploiting special structure in the second order linear system if it is present [24], and conjugate gradient methods (particularly useful for sparse Hessians) [73].

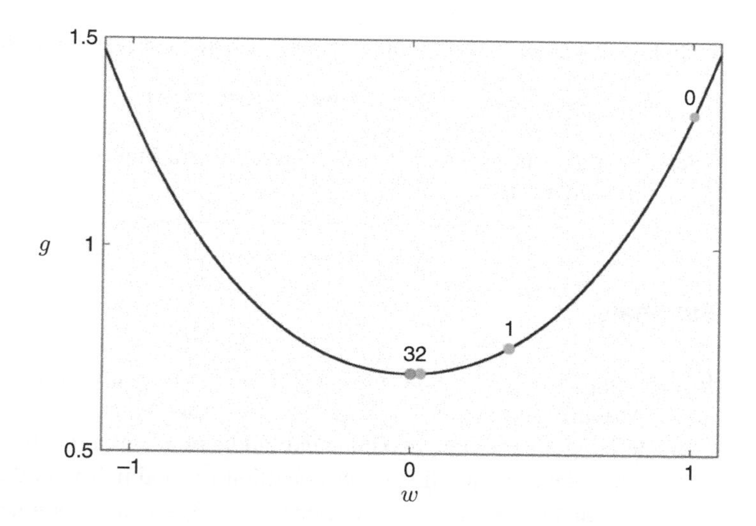

Fig. 2.12 Newton's method applied for minimizing a convex function with scalar input (see Example 2.6) initialized at $w^0 = 1$.

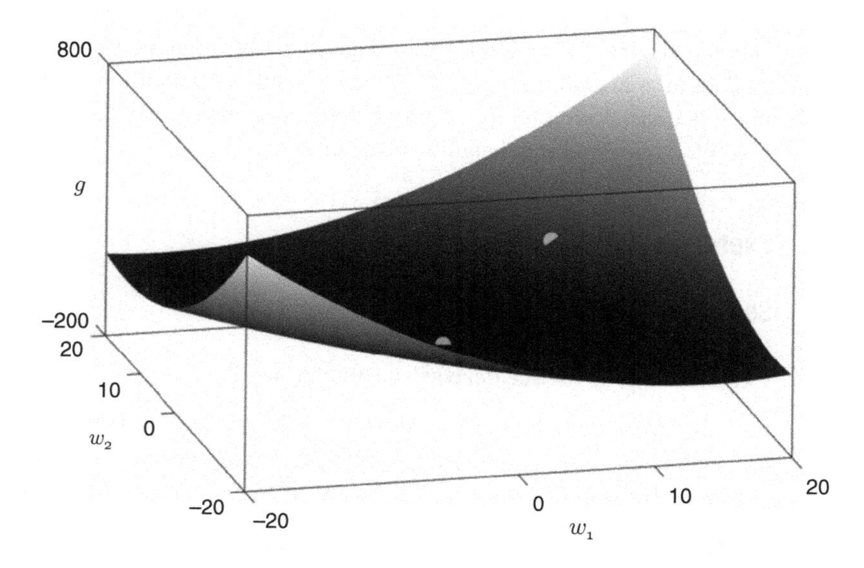

Fig. 2.13 Newton's method applied for minimizing a convex quadratic function with 2-dimensional input (see Example 2.7). Initialized at any point, only one Newton step is required to reach the minimum.

$\mathbf{w}^0 = \begin{bmatrix} 10 & 10 \end{bmatrix}^T$ only one Newton step is required to reach the global minimum of g. This is in fact the case regardless of the initial point chosen since g is quadratic, and thus at any point its quadratic Taylor series approximation is *itself*. This can also be seen by plugging $\nabla g\left(\mathbf{w}^0\right) = \mathbf{Q}\mathbf{w}^0 + \mathbf{r}$ and $\nabla^2 g\left(\mathbf{w}^0\right) = \mathbf{Q}$ into (2.19), giving

$$\mathbf{Q}\mathbf{w}^1 = \mathbf{Q}\mathbf{w}^0 - \left(\mathbf{Q}\mathbf{w}^0 + \mathbf{r}\right) = -\mathbf{r}. \tag{2.23}$$

Note that using (2.19) to compute $\mathbf{w}^2$ results in the exact same linear system since

$$\mathbf{Q}\mathbf{w}^2 = \mathbf{Q}\mathbf{w}^1 - \left(\mathbf{Q}\mathbf{w}^1 + \mathbf{r}\right) = -\mathbf{r}. \tag{2.24}$$

Therefore only one Newton step is required to find the minimum of g at $-\mathbf{Q}^{-1}\mathbf{r} = \begin{bmatrix} -0.57 & -0.57 \end{bmatrix}^T$.

2.3 Summary

In this chapter we have seen how to formalize the search for the minima of a general function g. In Section 2.1.2 we saw how calculus provides a useful condition for characterizing the minima, maxima, and saddle points of a function (together known as stationary points) via the first order condition for optimality. In the convenient case of a convex function, as discussed in Section 2.1.3, all such stationary points are global minima of the function. As described in Section 2.2, numerical algorithms aim at minimizing a function, but are only guaranteed to converge to stationary points. Two commonly used numerical methods, gradient descent and Newton's method, employ first and second order Taylor series expansions of a function respectively, in order to produce a converging sequence. Newton's method, which is easier to apply to convex functions, converges in far fewer steps than gradient descent and, unlike gradient descent, requires no step length to be determined.

2.4 Exercises

Section 2.1 exercises

Exercises 2.1 Practice derivative calculations I

Compute the first and second derivatives of the following functions (remember to use the product/chain rules where necessary). *Hint: see appendix for more information on how to compute first and second derivatives if these concepts are unfamiliar.*

a) $g(w) = \frac{1}{2}qw^2 + rw + d$ where q, r, and d are constants;

b) $g(w) = -\cos\left(2\pi w^2\right) + w^2$;

c) $g(w) = \sum\limits_{p=1}^{P} \log\left(1 + e^{-a_p w}\right)$ where $a_1 \ldots a_P$ are constants.

Exercises 2.2 Practice derivative calculations II

Compute the gradient and Hessian matrix of the following (remember to use the product/chain rules where necessary). Note that here $\mathbf{w}$ is an $N \times 1$ dimensional vector in all three cases. *Hint: see appendix for more information on how to compute gradients and Hessians if these concepts are unfamiliar.*

a) $g(\mathbf{w}) = \frac{1}{2}\mathbf{w}^T\mathbf{Q}\mathbf{w} + \mathbf{r}^T\mathbf{w} + d$, here $\mathbf{Q}$ is an $N \times N$ symmetric matrix (i.e., $\mathbf{Q} = \mathbf{Q}^T$), $\mathbf{r}$ is an $N \times 1$ vector, and d is a scalar;

b) $g(\mathbf{w}) = -\cos\left(2\pi\mathbf{w}^T\mathbf{w}\right) + \mathbf{w}^T\mathbf{w}$;

c) $g(\mathbf{w}) = \sum_{p=1}^{P} \log\left(1 + e^{-\mathbf{a}_p^T\mathbf{w}}\right)$ where $\mathbf{a}_1 \ldots \mathbf{a}_p$ are $N \times 1$ vectors.

Exercises 2.3 Outer products and outer product matrices

Let $\mathbf{x}$ and $\mathbf{y}$ be $N \times 1$ and $M \times 1$ vectors respectively. The *outer product* of $\mathbf{x}$ and $\mathbf{y}$, written as $\mathbf{x}\mathbf{y}^T$, is the $N \times M$ matrix defined as

$$\mathbf{x}\,\mathbf{y}^T = \begin{bmatrix} x_1y_1 & x_1y_2 & \cdots & x_1y_M \\ x_2y_1 & x_2y_2 & \cdots & x_2y_M \\ \vdots & \vdots & \ddots & \vdots \\ x_Ny_1 & x_Ny_2 & \cdots & x_Ny_M \end{bmatrix}. \tag{2.25}$$

Suppose that $\mathbf{X}$ is an $N \times P$ and $\mathbf{Y}$ is an $M \times P$ matrix and $\mathbf{x}_p$ and $\mathbf{y}_p$ are the pth columns of $\mathbf{X}$ and $\mathbf{Y}$ respectively, verify that $\mathbf{X}\mathbf{Y}^T = \sum_{p=1}^{P} \mathbf{x}_p\mathbf{y}_p^T$ where $\mathbf{x}_p\mathbf{y}_p^T$ is the outer product of $\mathbf{x}_p$ and $\mathbf{y}_p$.

Exercises 2.4 Taylor series calculations

Write out the first and second order Taylor series approximations for the following functions:

a) $g(w) = \log\left(1 + e^{w^2}\right)$ near a point v;

b) $g(\mathbf{w}) = \frac{1}{2}\mathbf{w}^T\mathbf{Q}\mathbf{w} + \mathbf{r}^T\mathbf{w} + d$ where $\mathbf{Q}$ is an $N \times N$ symmetric matrix, $\mathbf{r}$ is an $N \times 1$ vector, and d is a constant. In particular show that the second order Taylor series approximation $h(\mathbf{w})$ centered at any point $\mathbf{v}$ is precisely the function $g(\mathbf{w})$ itself. Why is this?

Exercises 2.5 First order Taylor series geometry

Verify that the normal vector to the tangent hyperplane generated by the first order Taylor series approximation centered at a point $\mathbf{v}$ shown in Equation (2.3) takes the form $\mathbf{n} = \begin{bmatrix} 1 \\ -\nabla g(\mathbf{v}) \end{bmatrix}$.

Exercises 2.6 First order condition calculations

Use the first order condition to find all stationary points of each function below.

a) $g(w) = w\log(w) + (1 - w)\log(1 - w)$ where w lies between 0 and 1.

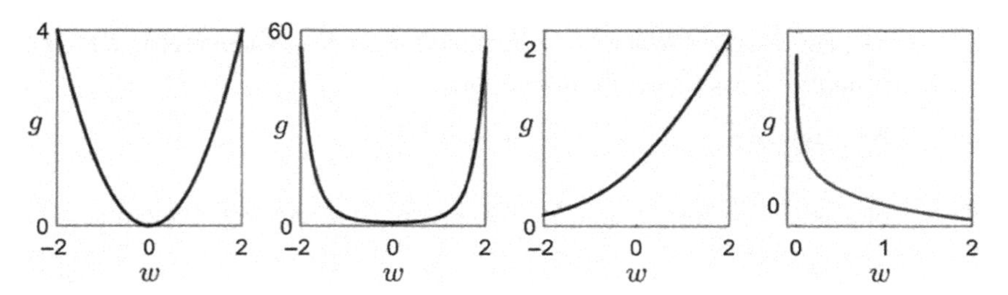

Fig. 2.14 From left to right, plots of four convex functions $g(w) = w^2$, e^{w^2}, $\log(1 + e^w)$, and $-\log(w)$.

b) $g(\mathbf{w}) = \frac{1}{2}\mathbf{w}^T\mathbf{Q}\mathbf{w} + \mathbf{r}^T\mathbf{w} + d$ where $\mathbf{w}$ is 2-dimensional, $\mathbf{Q} = \begin{bmatrix} 1 & 0.75 \\ 0.75 & 1 \end{bmatrix}$ and $\mathbf{r} = \begin{bmatrix} 1 & 1 \end{bmatrix}^T$ and $d = 5$.

Exercises 2.7 Second order convexity calculations

In Fig. 2.14 we show several one-dimensional examples of convex functions. Confirm using the second order definition of convexity that each is indeed convex.

Exercises 2.8 A non-convex function whose only stationary point is a global minimum

a) Use the first order condition to determine the stationary point of $g(w) = w\tanh(w)$ where $\tanh(w)$ is the hyperbolic tangent function. To do this you might find it helpful to graph the first derivative $\frac{\partial}{\partial w} g(w)$ and see where it crosses the w axis. Plot the function to verify that the stationary point you find is the global minimum of the function.

b) Use the second order definition of convexity to show that g is non-convex. *Hint: you can plot the second derivative* $\frac{\partial^2}{\partial w^2} g(w)$.

Exercises 2.9 How to determine whether or not the eigenvalues of a symmetric matrix Q are all nonnegative

In this exercise we investigate an alternative approach to checking that the eigenvalues of a square symmetric matrix $\mathbf{Q}$ (e.g., like a Hessian matrix) are all nonnegative, which does not involve explicitly computing the eigenvalues themselves, and is significantly easier to employ in practice.

a) Let $\mathbf{Q}$ be an $N \times N$ symmetric matrix. Show that if $\mathbf{Q}$ has all nonnegative eigenvalues then the quantity $\mathbf{z}^T\mathbf{Q}\mathbf{z} \geq 0$ for all $\mathbf{z}$. *Hint: use the eigen-decomposition of* $\mathbf{Q} = \mathbf{E}\mathbf{D}\mathbf{E}^T = \sum_{n=1}^{N} \mathbf{e}_n\mathbf{e}_n^T d_n$ *where the* $N \times N$ *orthogonal matrix* $\mathbf{E}$ *contains eigenvectors* $\mathbf{e}_n$ *of* $\mathbf{Q}$ *as its columns, and* $\mathbf{D} = \text{diag}(d_1 \ldots d_N)$ *is a diagonal matrix containing the eigenvalues of* $\mathbf{Q}$ *(see Appendix C for more on the eigenvalue decomposition).*

b) Show the converse. That if an $N \times N$ square symmetric matrix $\mathbf{Q}$ satisfies $\mathbf{z}^T\mathbf{Q}\mathbf{z} \geq 0$ for all $\mathbf{z}$ then it must have all nonnegative eigenvalues.

c) Use this method to verify that the second order definition of convexity holds for the quadratic function $g(\mathbf{w}) = \frac{1}{2}\mathbf{w}^T\mathbf{Q}\mathbf{w} + \mathbf{r}^T\mathbf{w} + d$, where $\mathbf{Q} = \begin{bmatrix} 1 & 1 \\ 1 & 1 \end{bmatrix}$, $\mathbf{r} = \begin{bmatrix} 1 & 1 \end{bmatrix}^T$, and $d = 1$.

d) Show that the eigenvalues of $\mathbf{Q} + \lambda\mathbf{I}_{N\times N}$ can all be made to be positive by setting λ large enough. What is the smallest value of λ that will make this happen?

Exercises 2.10 Outer product matrices have all nonnegative eigenvalues

a) Use the method described in Exercise 2.9 to verify that for any N length vector $\mathbf{x}$ the $N \times N$ outer product matrix $\mathbf{x}\mathbf{x}^T$ has all nonnegative eigenvalues.

b) Similarly show for any set of P vectors $\mathbf{x}_1 \ldots \mathbf{x}_P$ of length N that the sum of outer product matrices $\sum_{p=1}^{P} \delta_p \mathbf{x}_p \mathbf{x}_p^T$ has all nonnegative eigenvalues if each $\delta_p \geq 0$.

c) Show that the matrix $\sum_{p=1}^{P} \delta_p \mathbf{x}_p \mathbf{x}_p^T + \lambda\mathbf{I}_{N\times N}$ where each $\delta_p \geq 0$ and $\lambda > 0$ has all positive eigenvalues.

Exercises 2.11 An easier way to check the second order definition of convexity

Recall that the second order definition of convexity for a vector input function $g(\mathbf{w})$ requires that we verify whether or not the eigenvalues of $\nabla^2 g(\mathbf{w})$ are nonnegative for each input $\mathbf{w}$. However, to explicitly compute the eigenvalues of the Hessian in order to check this is a cumbersome or even impossible task for all but the nicest of functions. Here we use the result of Exercise 2.9 to express the second order definition of convexity in a way that is often much easier to employ in practice.

a) Using the result of Exercise 2.9 show that the second order definition of convexity for vector input functions $g(\mathbf{w})$, which has been previously stated as holding if the eigenvalues of the Hessian $\nabla^2 g(\mathbf{w})$ are nonnegative at every $\mathbf{w}$, equivalently holds if the quantity $\mathbf{z}^T\left(\nabla^2 g(\mathbf{w})\right)\mathbf{z} \geq 0$ holds at each $\mathbf{w}$ for all $\mathbf{z}$.

b) Use this manner of expressing the second order definition of convexity to verify that the general quadratic function $g(\mathbf{w}) = \frac{1}{2}\mathbf{w}^T\mathbf{Q}\mathbf{w} + \mathbf{r}^T\mathbf{w} + d$, where $\mathbf{Q}$ is symmetric and known to have all nonnegative eigenvalues and $\mathbf{r}$ and d are arbitrary, always defines a convex function.

c) Verify that $g(\mathbf{w}) = -\cos\left(2\pi\mathbf{w}^T\mathbf{w}\right) + \mathbf{w}^T\mathbf{w}$ is non-convex by showing that it does *not* satisfy the second order definition of convexity.

Section 2.2 exercises

Exercises 2.12 Play with gradient descent code

Play with the gradient descent demo file ***convex_grad_surrogate*** which illustrates the consequences of using a fixed step length to minimize the simple convex function shown in the left panel of Fig. 2.4. Try changing the initial point and fixed step-length to see how the gradient descent path changes. For example, find a steplength that causes the algorithm to diverge (meaning that the steps go off to infinity). Also look inside the gradient descent subfunction and see how it mirrors Algorithm 2.1.

You can also play with ***non-convex_grad_surrogate*** which shows the demo, this time for a curvy non-convex function.

Exercises 2.13 Code up gradient descent

In this exercise you will reproduce Fig. 2.10 by using gradient descent in order to minimize the function $g(\mathbf{w}) = -\cos\left(2\pi\mathbf{w}^T\mathbf{w}\right) + \mathbf{w}^T\mathbf{w}$. Use the wrapper ***two_d_grad_wrapper_hw*** to perform gradient descent, filling in the form of the gradient in the subfunction

[in,out] = gradient_descent(alpha,w0).

This subfunction performs gradient descent and is complete with exception of the gradient. Here *alpha* is a fixed step length and w0 is the initial point (both provided in the wrapper), the *in* variable contains each gradient step taken i.e., *in* $= \left\{\mathbf{w}^k = \mathbf{w}^{k-1} - \alpha\nabla g\left(\mathbf{w}^{k-1}\right)\right\}_{k=0}^{K}$ and corresponding *out* is a vector of the objective function evaluated at these steps, i.e., *out* $= \left\{g\left(\mathbf{w}^k\right)\right\}_{k=0}^{K}$ where K is the total number of steps taken. These are collected so they may be printed on the cost function surface for viewing.

Exercises 2.14 Tune fixed step length for gradient descent

When minimizing a function with high dimensional input $g(\mathbf{w})$ using any numerical method, it is helpful to store each iteration and the corresponding objective value at each iteration $g\left(\mathbf{w}^k\right)$ to make sure your algorithm is converging properly. In this example you will use gradient descent to minimize a simple function, and plot the objective values at each step, comparing the effect of different step sizes on convergence rate of the algorithm.

Suppose $g(\mathbf{w}) = \mathbf{w}^T\mathbf{w}$ where $\mathbf{w}$ is an $N = 10$ dimensional input vector. This is just a generalization of the simple parabola in one dimension. g is convex with a single global minimum at $\mathbf{w} = \mathbf{0}_{N\times 1}$. Code up gradient descent with a maximum iteration stopping criterion of 100 iterations (with no other stopping conditions). Using the initial point $\mathbf{w}^0 = 10 \cdot \mathbf{1}_{N\times 1}$ run gradient descent with step lengths $\alpha_1 = 0.001$, $\alpha_2 = 0.1$ and $\alpha_3 = 1.001$ and record the objective function value $g\left(\mathbf{w}^k\right)$ at each iteration of each run. Plot these on a single graph like the one shown in Fig. 2.15.

Make sure to use the gradient descent as described in Algorithm 2.1 with only the maximum iteration stopping condition.

Exercises 2.15 Geometry of gradient descent step

The distance between the $(k-1)$ th and kth gradient step can easily be calculated as $\left\|\mathbf{w}^k - \mathbf{w}^{k-1}\right\|_2 = \left\|\left(\mathbf{w}^{k-1} - \alpha\nabla g\left(\mathbf{w}^{k-1}\right)\right) - \mathbf{w}^{k-1}\right\|_2 = \alpha\left\|\nabla g\left(\mathbf{w}^{k-1}\right)\right\|_2$. In this

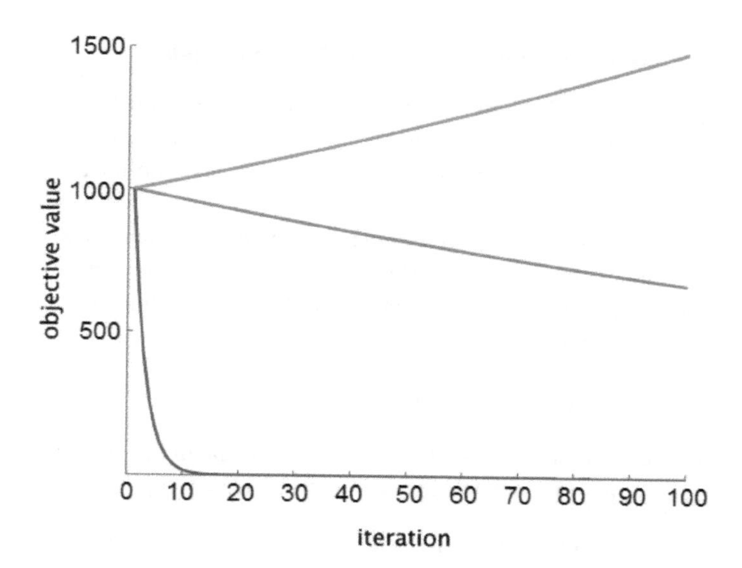

Fig. 2.15 Three runs of gradient descent with different step lengths (see text for further details). The smallest step length (in magenta) causes the algorithm to converge very slowly, while the largest (in green) causes it to diverge as the objective value here is increasing. The middle step length value (in blue) causes the algorithm to converge very rapidly to the unique solution of the problem.

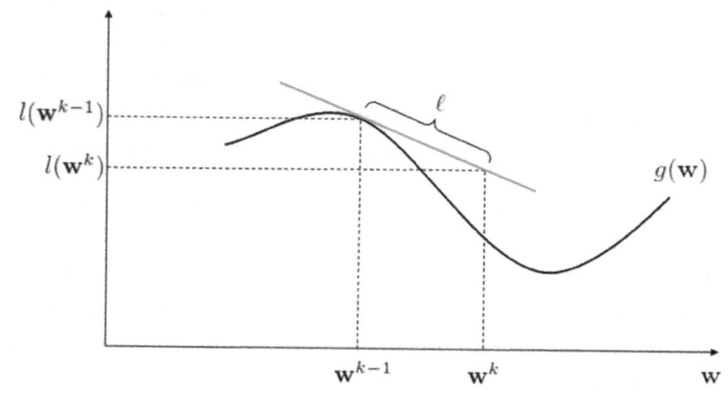

Fig. 2.16 The geometry of a gradient descent step.

exercise you will compute the corresponding length traveled along the $(k-1)$ th linear surrogate $l(\mathbf{w}) = g\left(\mathbf{w}^{k-1}\right) + \nabla g\left(\mathbf{w}^{k-1}\right)^T\left(\mathbf{w} - \mathbf{w}^{k-1}\right)$.

In Fig. 2.16 we show a detailed description of the precise geometry involved in taking the kth gradient descent step with a fixed step length α. Use the details of this picture to show that the corresponding length traveled along the linear surrogate, i.e., $\ell = \left\|\begin{bmatrix} \mathbf{w}^k \\ l\left(\mathbf{w}^k\right) \end{bmatrix} - \begin{bmatrix} \mathbf{w}^{k-1} \\ l\left(\mathbf{w}^{k-1}\right) \end{bmatrix}\right\|_2$, is given precisely as

$$\ell = \alpha\sqrt{1 + \left\|\nabla g\left(\mathbf{w}^{k-1}\right)\right\|_2^2}\left\|\nabla g\left(\mathbf{w}^{k-1}\right)\right\|_2. \tag{2.26}$$

Hint: use the Pythagorean theorem.

Exercises 2.16 Play with Newton's method code

Play with the Newton's method demo file ***convex_newt_demo*** which illustrates Newton's method applied to minimizing the simple one-dimensional function discussed in Example 2.6. Look inside the Newton's method subfunction and see how it mirrors Algorithm 2.2.

Here also play with ***non-convex_newt_demo*** which shows the same kind of demo for a non-convex function. Here note that, as illustrated in Fig. 2.11, if you begin on a concave portion of the function Newton's method will indeed climb to a local maximum of the function!

Exercises 2.17 Code up Newton's method

a) Use the first order condition to determine the unique stationary point of the function $g(\mathbf{w}) = \log\left(1 + e^{\mathbf{w}^T\mathbf{w}}\right)$ where $N = 2$ i.e., $\mathbf{w} = \begin{bmatrix} w_1 & w_2 \end{bmatrix}^T$.

b) Make a surface plot of the function $g(\mathbf{w})$ or use the second order definition of convexity to verify that $g(\mathbf{w})$ is convex, implying that the stationary point found in part **a)** is a global minimum. *Hint: to check the second order definition use Exercise 2.10.*

c) Perform Newton's method to find the minimum of the function $g(\mathbf{w})$ determined in part **a)**. Initialize your algorithm at $\mathbf{w}^0 = \mathbf{1}_{N\times1}$ and make a plot of the function value $g\left(\mathbf{w}^k\right)$ for ten iterations of Newton's method, as was done in Exercise 2.14 with gradient descent, in order to verify that your algorithm works properly and is converging.

Make sure to follow the Newton's method algorithm as described in Algorithm 2.2 with only the maximum iteration stopping condition, and use the pseudo-inverse solution to each Newton system in your implementation as given in (2.21).

d) Now run your Newton's method code from part **c)** again, this time initializing at the point $\mathbf{w}^0 = 4 \cdot \mathbf{1}_{N\times1}$. While this initialization is further away from the unique minimum of $g(\mathbf{w})$ than the one used in part **c)**, your Newton's method algorithm should converge *faster* starting at this point. At first glance this result seems very counter-intuitive, as we (rightly) expect that an initial point closer to a minimum will provoke more rapid convergence of Newton's method!

Can you explain why this result actually makes sense for the particular function $g(\mathbf{w})$ we are minimizing here? Or, in other words, why the minimum of the second order Taylor series approximation of $g(\mathbf{w})$ centered at $\mathbf{w}^0 = 4 \cdot \mathbf{1}_{N\times1}$ is essentially the minimum of $g(\mathbf{w})$ itself? *Hint: use the fact for large values of t that log $\left(1 + e^t\right) \approx t$, and that the second order Taylor series approximation of a quadratic function (like the one given in part **b)** of Exercise 2.4) is just the quadratic function itself.*

3 Regression

In this chapter we formally describe the regression problem, or the fitting of a representative line or curve (in higher dimensions a hyperplane or general surface) to a set of input/output data points as first broadly detailed in Section 1.2.1. Regression in general may be performed for a variety of reasons: to produce a so-called trend line (or curve) that can be used to help visually summarize, drive home a particular point about the data under study, or to learn a model so that precise predictions can be made regarding output values in the future. Here we also discuss more formally the notion of feature design for regression, in particular focusing on rare low dimensional instances (like the one outlined in Example 1.7) when very specific feature transformations of the data can be proposed. We finally end by discussing regression problems that have non-convex cost functions associated with them and a commonly used approach, called ℓ_2 regularization, for ameliorating some of the problems associated with the minimization of such functions.

3.1 The basics of linear regression

With linear regression we aim to fit a line (or hyperplane in higher dimensions) to a scattering of data. In this section we describe the fundamental concepts underlying this procedure.

3.1.1 Notation and modeling

Data for regression problems comes in the form of a training set of P input/output observation pairs:

$$\{(\mathbf{x}_1, y_1), (\mathbf{x}_2, y_2), \ldots, (\mathbf{x}_P, y_P)\}, \tag{3.1}$$

or $\left\{(\mathbf{x}_p, y_p)\right\}_{p=1}^{P}$ for short, where $\mathbf{x}_p$ and y_p denote the pth input and output respectively. In many instances of regression, like the one discussed in Example 1.1, the input to regression problems is scalar-valued (the output will always be considered scalar-valued here) and hence the linear regression problem is geometrically speaking one of fitting a line to the associated scatter of data points in 2-dimensional space. In general, however, each input $\mathbf{x}_p$ may be a column vector of length N

$$\mathbf{x}_p = \begin{bmatrix} x_{1,p} \\ x_{2,p} \\ \vdots \\ x_{N,p} \end{bmatrix}, \tag{3.2}$$

in which case the linear regression problem is analogously one of fitting a hyperplane to a scatter of points in $N + 1$ dimensional space.

In the case of scalar input, fitting a line to the data (see Fig. 3.1) requires we determine a slope w and bias (or "y-intercept") b so that the approximate linear relationship holds between the input/output data,

$$b + x_p w \approx y_p, \quad p = 1, \ldots, P. \tag{3.3}$$

Note that we have used the *approximately equal* sign in (3.3) because we cannot be sure that all data lies completely on a single line. More generally, when the input dimension is $N \geq 1$, then we have a bias and N associated weights,

$$\mathbf{w} = \begin{bmatrix} w_1 \\ w_2 \\ \vdots \\ w_N \end{bmatrix}, \tag{3.4}$$

to tune properly in order to fit a hyperplane (see Fig. 3.1). Likewise the linear relationship in (3.3) is then more generally given as

$$b + \mathbf{x}_p^T \mathbf{w} \approx y_p, \quad p = 1, \ldots, P. \tag{3.5}$$

The elements of an input vector $\mathbf{x}_p$ are referred to as *input features* to a regression problem. For instance the student debt data described in Example 1.1 has only one feature: *year*. Conversely in the GDP growth rate data described in Example 3.1 the first element of the input feature vector might contain the feature *unemployment rate* (that is,

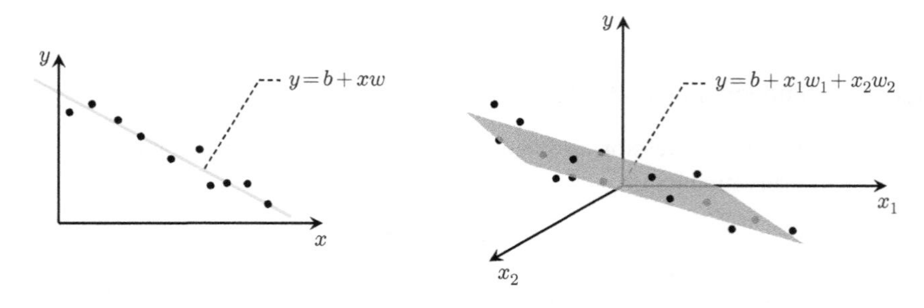

Fig. 3.1 (left panel) A dataset in two dimensions along with a well-fitting line. A line in two dimensions is defined as $b + xw = y$, where b is referred to as the bias and w the weight, and a point (x_p, y_p) lies close to it if $b + x_p w \approx y_p$. (right panel) A simulated 3-dimensional dataset along with a well-fitting hyperplane. A hyperplane is defined as $b + \mathbf{x}^T \mathbf{w} = y$, where again b is called the bias and $\mathbf{w}$ the weight vector, and a point $(\mathbf{x}_p, y_p)$ lies close to it if $b + \mathbf{x}_p^T \mathbf{w} \approx y_p$.

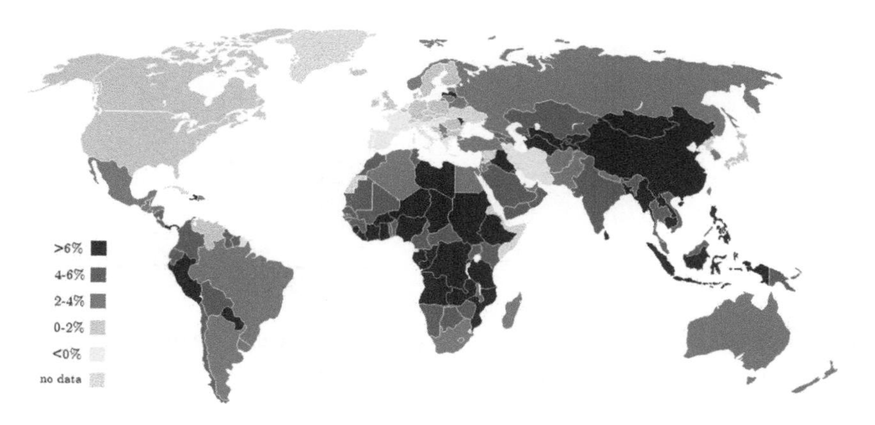

Fig. 3.2 A map of the world where countries are color-coded by their GDP growth rates (the darker the color the higher the growth rate) as reported by the International Monetary Fund (IMF) in 2013.

the unemployment data from each country under study), the second might contain the feature *education level*, and so on.

Example 3.1 Predicting Gross Domestic Product growth rates

As an example of a regression problem with vector-valued input consider the problem of predicting the growth rate of a country's Gross Domestic Product (GDP), which is the value of all goods and services produced within a country during a single year. Economists are often interested in understanding factors (e.g., unemployment rate, education level, population count, land area, income level, investment rate, life expectancy, etc.,) which determine a country's GDP growth rate in order to inform better financial policy making. To understand how these various *features* of a country relate to its GDP growth rate economists often perform linear regression [33, 72].

In Fig. 3.2 we show a heat map of the world where countries are color-coded based on their GDP growth rate in 2013, reported by the International Monetary Fund (IMF) (data used in this figure was taken from [12]).

3.1.2 The Least Squares cost function for linear regression

To find the parameters of the hyperplane which best fits a regression dataset, it is common practice to first form the *Least Squares cost function*. For a given set of parameters $(b, \mathbf{w})$ this cost function computes the total squared error between the associated hyperplane and the data (as illustrated pictorially in Fig. 3.3), giving a good measure of how well the particular linear model fits the dataset. Naturally then the best fitting hyperplane is the one whose parameters minimize this error.

Because we aim to have the system of equations in (3.5) hold as well as possible, to form the desired cost we simply square the difference (or error) between the linear model $b + \mathbf{x}_p^T \mathbf{w}$ and the corresponding output y_p over the entire dataset. This gives the Least Squares cost function

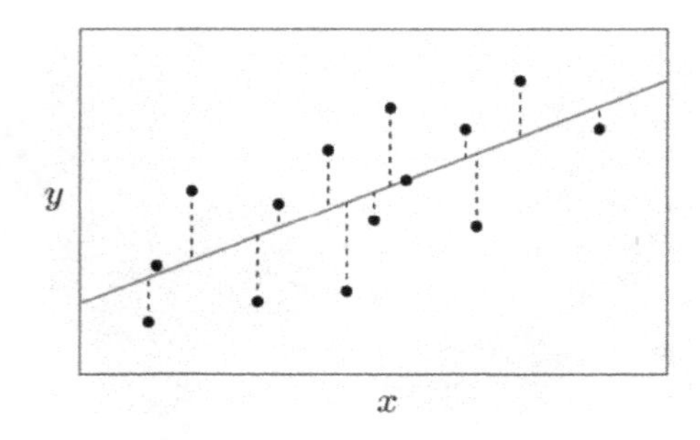

Fig. 3.3 A simulated 2-dimensional training dataset along with a line (in magenta) fit to the data using the Least Squares framework, which aims at recovering the line that minimizes the total squared length of the dashed error bars.

$$g(b, \mathbf{w}) = \sum_{p=1}^{P} \left(b + \mathbf{x}_p^T \mathbf{w} - y_p\right)^2. \tag{3.6}$$

We of course want to find a parameter pair $(b, \mathbf{w})$ that provides a small value for $g(b, \mathbf{w})$ since the larger this value the larger the squared error between the corresponding linear model and the data, and hence the poorer we represent the given data. Therefore we aim to *minimize* g over the bias and weight vector in order to recover the best pair $(b, \mathbf{w})$, which is written formally (see Section 2.2) as

$$\underset{b, \mathbf{w}}{\text{minimize}} \sum_{p=1}^{P} \left(b + \mathbf{x}_p^T \mathbf{w} - y_p\right)^2. \tag{3.7}$$

By checking the second order definition of convexity (see Exercise 3.3) we can easily see that the Least Squares cost function in (3.6) is convex. Figure 3.4 illustrates the Least Squares cost associated with the student loan data in Example 1.1, whose "upward bending" shape confirms its convexity visually in the instance of that particular dataset.

3.1.3 Minimization of the Least Squares cost function

Now that we have a minimization problem to solve we can employ the tools described in Chapter 2. To perform calculations it will first be convenient to use the following more compact notation:

$$\tilde{\mathbf{x}}_p = \begin{bmatrix} 1 \\ \mathbf{x}_p \end{bmatrix} \quad \tilde{\mathbf{w}} = \begin{bmatrix} b \\ \mathbf{w} \end{bmatrix}. \tag{3.8}$$

With this notation we can rewrite the cost function shown in (3.6) in terms of the single vector $\tilde{\mathbf{w}}$ of parameters as

$$g(\tilde{\mathbf{w}}) = \sum_{p=1}^{P} \left(\tilde{\mathbf{x}}_p^T \tilde{\mathbf{w}} - y_p\right)^2. \tag{3.9}$$

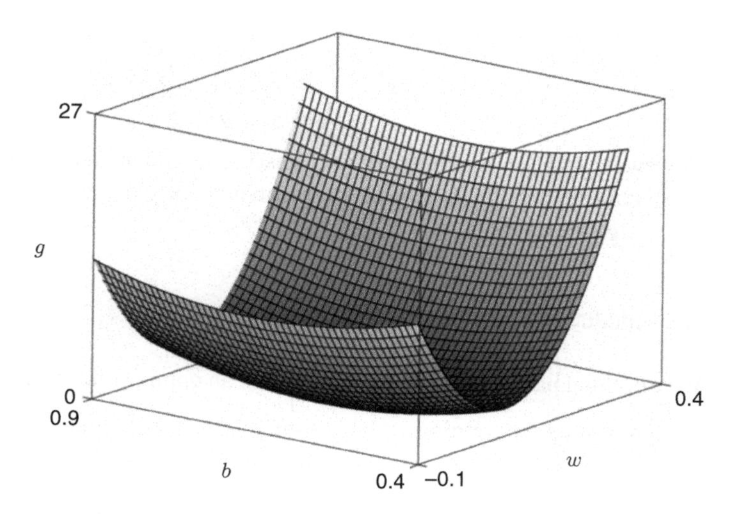

Fig. 3.4 The surface generated by the Least Squares cost function using the student loan debt data shown in Fig. 1.8, is clearly convex. However, regardless of the dataset, the Least Squares cost for linear regression is always convex.

To compute the gradient of this cost we simply apply the chain rule from calculus, which gives

$$\nabla g\left(\tilde{\mathbf{w}}\right) = 2\sum_{p=1}^{P}\tilde{\mathbf{x}}_p\left(\tilde{\mathbf{x}}_p^T\tilde{\mathbf{w}} - y_p\right) = 2\left(\sum_{p=1}^{P}\tilde{\mathbf{x}}_p\tilde{\mathbf{x}}_p^T\right)\tilde{\mathbf{w}} - 2\sum_{p=1}^{P}\tilde{\mathbf{x}}_p y_p. \tag{3.10}$$

Using this we can perform gradient descent to minimize the cost. However, in this (rare) instance we can actually solve the first order system directly in order to recover a global minimum. Setting the gradient above to zero and solving for $\tilde{\mathbf{w}}$ gives the system of linear equations

$$\left(\sum_{p=1}^{P}\tilde{\mathbf{x}}_p\tilde{\mathbf{x}}_p^T\right)\tilde{\mathbf{w}} = \sum_{p=1}^{P}\tilde{\mathbf{x}}_p y_p. \tag{3.11}$$

In particular one algebraic solution to this system,[1] if the matrix $\sum_{p=1}^{P}\tilde{\mathbf{x}}_p\tilde{\mathbf{x}}_p^T$ is invertible,[2] may be written as

[1] By setting the input vectors $\tilde{\mathbf{x}}_p$ columnwise to form the matrix $\tilde{\mathbf{X}}$ and by stacking the output y_p into the column vector $\mathbf{y}$ we may write the linear system in Equation (3.11) equivalently as $\tilde{\mathbf{X}}\tilde{\mathbf{X}}^T\tilde{\mathbf{w}} = \tilde{\mathbf{X}}\mathbf{y}$.

[2] In instances where the linear system in (3.11) has more than one solution, or in other words when $\sum_{p=1}^{P}\tilde{\mathbf{x}}_p\tilde{\mathbf{x}}_p^T$ is not invertible, one can choose the solution with the smallest length (or ℓ_2 norm), sometimes written as $\tilde{\mathbf{w}}^\star = \left(\sum_{p=1}^{P}\tilde{\mathbf{x}}_p\tilde{\mathbf{x}}_p^T\right)^\dagger \sum_{p=1}^{P}\tilde{\mathbf{x}}_p y_p$, where $(\cdot)^\dagger$ denotes the pseudo-inverse of its input matrix. See Appendix C for further details.

$$\tilde{\mathbf{w}}^{\star}=\left(\sum_{p=1}^{P}\tilde{\mathbf{x}}_p\tilde{\mathbf{x}}_p^T\right)^{-1}\sum_{p=1}^{P}\tilde{\mathbf{x}}_p y_p. \tag{3.12}$$

However, while an algebraically expressed solution is appealing it is typically more computationally efficient in practice to solve the original linear system using numerical linear algebra software.

3.1.4 The efficacy of a learned model

With optimal parameters $\tilde{\mathbf{w}}^{\star}=\begin{bmatrix} b^{\star} \\ \mathbf{w}^{\star}\end{bmatrix}$ we can compute the efficacy of the linear model in representing the training set by computing the mean squared error (or MSE),

$$\text{MSE}=\frac{1}{P}\sum_{p=1}^{P}\left(b^{\star}+\mathbf{x}_p^T\mathbf{w}^{\star}-y_p\right)^2. \tag{3.13}$$

When possible it is also a good idea to compute the MSE of a learned regression model on a set of new testing data, i.e., data that was not used to learn the model itself, to provide some assurance that the learned model will perform well on future data points. This is explored further in Chapter 5 in the context of *cross-validation*.

3.1.5 Predicting the value of new input data

With optimal parameters $(b^{\star},\mathbf{w}^{\star})$, found by minimizing the Least Squares cost, we can predict the output y_{new} of a new input feature $\mathbf{x}_{\text{new}}$ by simply plugging the new input into the tuned linear model and estimating the associated output as

$$y_{\text{new}}=b^{\star}+\mathbf{x}_{\text{new}}^T\mathbf{w}^{\star}. \tag{3.14}$$

This is illustrated pictorially on a toy dataset for the case when $N=1$ in Fig. 3.5.

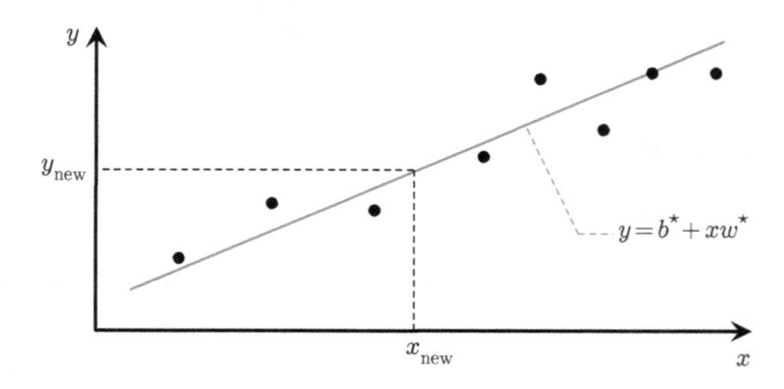

Fig. 3.5 Once a line/hyperplane has been fit to a dataset via minimizing the Least Squares cost function it may be used to predict the output value of future input. Here a line has been fit to a two-dimensional dataset in this manner, giving optimal parameters $b^{\star}$ and $w^{\star}$, and the output value of a new point x_{new} is created using the learned linear model as $y_{\text{new}}=b^{\star}+x_{\text{new}}w^{\star}$.

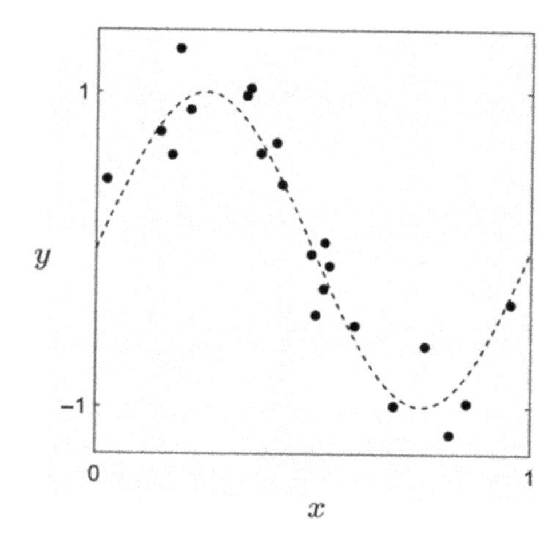

Fig. 3.6 A simulated regression dataset where the relationship between the input feature x and the output y is not linear. However, because we can visualize this dataset we can see that there is clearly a structured nonlinear relationship between its input and output. Our knowledge in this instance, based on our ability to visualize the data, allows us to design a new feature for the data and formulate a corresponding function (shown here in dashed black) that appears to be generating the data.

3.2 Knowledge-driven feature design for regression

In many regression problems the relationship between input feature(s) and output values is nonlinear, as in Fig. 3.6, which illustrates a simulated dataset where the scalar feature x and the output y are related in a nonlinear fashion. In such instances a linear model would clearly fail at representing how the input and output are related. In this brief section we present two simple examples through which we discuss how to fit a nonlinear model to the data when we have significant understanding or *knowledge* about the data itself. This knowledge may originate from our prior understanding or intuition about the phenomenon under study or simply our ability to visualize low dimensional data. As we now see, based on this knowledge we can propose an appropriate nonlinear feature transformation which allows us to employ the linear regression framework as described in the previous section.

Example 3.2 Sinusoidal pattern

In the left panel of Fig. 3.7 we show a simulated regression dataset (first shown in Fig. 3.6) consisting of $P = 21$ data points. Visually analyzing this data it appears to trace out (with some noise) one period of a sine wave over the interval $[0, 1]$. Therefore we can reasonably propose that some weighted version of the sinusoidal function $f(x) = \sin(2\pi x)$, i.e., $y = b + f(x)\,w$ where b and w are respectively a bias and weight to learn, will properly describe this data. In machine learning the function $f(x)$, in this instance a sinusoid, is referred to as a *feature transformation* of the original input. One could of

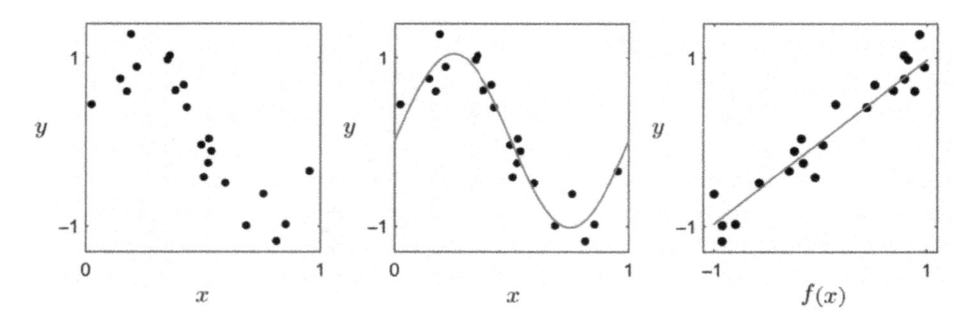

Fig. 3.7 (left panel) A simulated regression dataset. (middle panel) A weighted form of a simple sinusoid $y = b + f(x)\,w$ (in magenta), where $f(x) = \sin(2\pi x)$ and where b and w are tuned properly, describes the data quite well. (right panel) Fitting a sinusoid in the original feature space is equivalent to fitting a line in the transformed feature space where the input feature has undergone feature transformation $x \longrightarrow f(x) = \sin(2\pi x)$.

course propose a more curvy feature, but the sinusoid seems to explain the data fairly well while remaining relatively simple.

By fitting a simple weighted sinusoid to the data, we would like to find the parameter pair (b, w) so that

$$b + f\left(x_p\right) w = b + \sin\left(2\pi x_p\right) w \approx y_p, \quad p = 1, \ldots, P. \tag{3.15}$$

Note that while this is nonlinear in the input x, it is still linear in both b and w. In other words, the relationship between the output y and the new feature $f(x) = \sin(2\pi x)$ is linear. Plotting $\left\{\left(f\left(x_p\right), y_p\right)\right\}_{p=1}^{P}$ in the *transformed feature space* (i.e., the space whose input is given by the new feature $f(x)$ and whose output is still y) shown in the right panel of Fig. 3.7, we can see that the new feature and given output are now indeed linearly related.

After creating the new features for the data by transforming the input as $f_p = f\left(x_p\right) = \sin\left(2\pi x_p\right)$, we may solve for the parameter pair by minimizing the Least Squares cost function formed by summing the squared error between the model containing each transformed input $b + f_p w$ and the corresponding output value y_p (so that (3.15) holds as well as possible) as

$$\underset{b,\,w}{\text{minimize}} \sum_{p=1}^{P} \left(b + f_p w - y_p\right)^2. \tag{3.16}$$

The cost function here is still convex, and can be minimized by a numerical scheme like gradient descent or by directly solving its first order system to recover a global minimum. By using the compact notation

$$\tilde{\mathbf{f}}_p = \begin{bmatrix} 1 \\ f_p \end{bmatrix}, \quad \tilde{\mathbf{w}} = \begin{bmatrix} b \\ w \end{bmatrix}, \tag{3.17}$$

we can rewrite the cost function in terms of the single vector $\tilde{\mathbf{w}}$ of parameters as

$$g\left(\tilde{\mathbf{w}}\right) = \sum_{p=1}^{P} \left(\tilde{\mathbf{f}}_p^T \tilde{\mathbf{w}} - y_p\right)^2. \tag{3.18}$$

Mirroring the discussion in Section 3.1.3, setting the gradient of the above to zero then gives the linear system of equations to solve

$$\left(\sum_{p=1}^{P} \tilde{\mathbf{f}}_p \tilde{\mathbf{f}}_p^T\right) \tilde{\mathbf{w}} = \sum_{p=1}^{P} \tilde{\mathbf{f}}_p y_p. \tag{3.19}$$

In the middle and right panels of Fig. 3.7 we show the resulting fit to the data $y = b^{\star} + f(x)\, w^{\star}$ in magenta, where $b^{\star}$ and $w^{\star}$ are recovered by solving this system. We refer to this fit as the *estimated data generating function* since it is our estimation of the underlying continuous function generating this dataset (shown in dashed black in Fig. 3.6). Note that this fit is a sinusoid in the original feature space (middle panel), and a line in the transformed feature space (right panel).

Example 3.3 Galileo and uniform acceleration

Recall Galileo's acceleration experiment, first described in Example 1.7. In the left panel of Fig. 3.8 we show data consisting of $P = 6$ data points from a modern reenactment of this experiment [75], where the input x denotes the time and the output y denotes the corresponding portion of the ramp traversed. Several centuries ago Galileo saw data very similar looking to the data shown here. To describe the data he saw, Galileo proposed the relation $y = f(x)\, w$, where $f(x) = x^2$ is a simple quadratic feature and w is some weight to be tuned to the data. Note that in this specific example there is no need to add

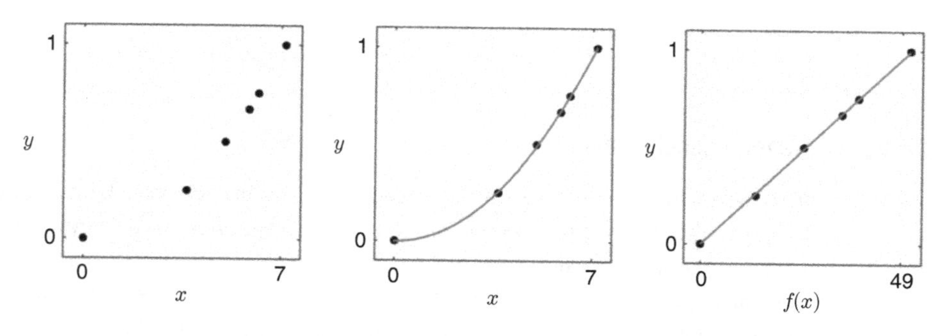

Fig. 3.8 Data from a modern reenactment of Galileo's ramp experiment. (left panel) The raw data seems to reflect a quadratic relationship between the input and output variables. (middle panel) A weighted form of a simple quadratic feature $y = f(x)\, w$ (in magenta) where $f(x) = x^2$ and where w is tuned properly, describes the data quite well. (right panel) Fitting a quadratic to the data in the original feature space is equivalent to fitting a line to the data in a transformed feature space wherein the input feature has undergone feature transformation $x \longrightarrow f(x) = x^2$.

a bias parameter b to the quadratic model since at time zero the ball has not moved at all, and hence the output must be precisely zero.

Looking at how the data is distributed in the left panel of Fig. 3.8, we too can intuit that such a quadratic feature of the input appears to be a reasonable guess at what lies beneath the data shown.

By fitting a simple weighted quadratic to the data, we would like to find parameter w such that

$$f\left(x_p\right) w = x_p^2 w \approx y_p, \quad p = 1, \ldots, P. \tag{3.20}$$

Although the relationship between the input feature x and output y is nonlinear, this model is still linear in the weight w. Thus we may tune w by minimizing the Least Squares cost function after forming the new features. That is, transforming each input as $f_p = f\left(x_p\right) = x_p^2$ we can find the optimal w by solving

$$\underset{w}{\text{minimize}} \sum_{p=1}^{P} \left(f_p w - y_p\right)^2. \tag{3.21}$$

The cost function in (3.21) is again convex and we may solve for the optimal w by simply checking the first order condition. Setting the derivative of the cost function equal to zero and solving for w, after a small amount of algebraic rearrangement, gives

$$w^{\star} = \frac{\sum_{p=1}^{P} f_p y_p}{\sum_{p=1}^{P} f_p^2}. \tag{3.22}$$

We show in the middle panel of Fig. 3.8 the weighted quadratic fit $y = f(x) w^{\star}$ (our estimated data generating function) to the data (in magenta) in the original feature space. In the right panel of this figure we show the same fit, this time in the transformed feature space where the fit is linear.

3.2.1 General conclusions

The two examples discussed above are very special. In each case, by using our ability to visualize the data we have been able to design an excellent new feature $f(x)$ explicitly using common algebraic functions. As we have seen in these two examples, a properly designed feature (or set of features more generally) for linear regression is one that provides a good *nonlinear* fit in the original space while, simultaneously, a good *linear* fit in the transformed feature space. In other words, a properly designed set of features for linear regression produces a good linear fit to the feature-transformed data.[3]

[3] Technically speaking there is one subtle yet important caveat to the use of the word "good" in this statement, this being that we do not want to "overfit" the data (an issue we discuss at length in Chapter 5). However, for now this issue will not concern us.

A properly designed feature (or set of features) for linear regression provides a good *nonlinear* fit in the original feature space and, simultaneously, a good *linear* fit in the transformed feature space.

However, just because we can visualize a low dimensional regression dataset does not mean we can easily design a proper feature "by eye" as we have done in Examples 3.2 and 3.3. For instance, in Fig. 3.9 we show a simulated dataset built by randomly taking $P = 30$ inputs x_p on the interval $[0, 1]$, evaluating each through a rather wild function[4] $y(x)$ (shown in dashed black in the figure), and then adding a small amount of noise to each output. Here even though we can clearly see a structured nonlinear relationship in the data, it is not immediately obvious how to formulate a proper feature $f(x)$ to recover the original data generating function $y(x)$. No common algebraic function (e.g., a quadratic, a sine wave, an exponential, etc.,) seems to be a reasonable candidate and hence our knowledge, in this case the fact that we can visualize the data itself, is not enough to form a proper feature (or set of features) for this data.

For vector-valued input we can say something very similar. While we can imagine forming a proper set of features for a dataset with vector-valued input, the fact that we cannot visualize the data prohibits us from "seeing" the right sort of feature(s) to use.

In fact rarely in practice can we use our knowledge of a dataset to construct perfect features. Often we may only be able to make a rough guess at a proper feature transformation given our intuition about the data at hand, or can make no educated guess at all. Thankfully, we can *learn* feature transformations automatically from the data itself that can ameliorate this problem. This process will be described in Chapter 5.

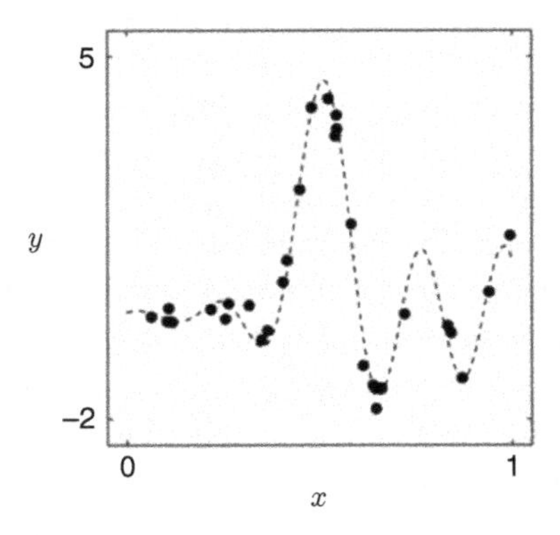

Fig. 3.9 A simulated dataset generated as noisy samples of a data generating function $y(x)$. We show $y(x)$ here in dashed black. Unlike the previous two cases in Fig. 3.7 and 3.8, it is not so clear what sort of function would serve as a proper feature $f(x)$ here.

[4] Here $y(x) = e^{3x}\frac{\sin\left(3\pi^2(x-0.5)\right)}{3\pi^2(x-0.5)}$.

3.3 Nonlinear regression and ℓ_2 regularization

In the previous section we discussed examples of regression where the nonlinear relationship in a given dataset could be determined by intuiting a nonlinear feature transformation, and where (once the data is transformed) the associated cost functions remained convex and linear in their parameters. Because the input/output relationship associated to each of these examples was linear in its parameters (see (3.15) and (3.20)), each is still referred to as a *linear* regression problem. In this section we explore the consequences of employing nonlinear models for regression where the corresponding cost function is non-convex and the input/output relationship nonlinear in its parameters (referred to as instances of *nonlinear* regression). We also introduce a common tool for partially ameliorating the practical inconveniences of non-convex cost functions, referred to as the ℓ_2 regularizer. Specifically we describe how the ℓ_2 regularizer helps numerical optimization techniques avoid poor stationary points of non-convex cost functions. Because of this utility, regularization is often used with non-convex cost functions, as we will see later with e.g., neural networks in Chapters 5–7.[5]

While the themes of this section are broadly applicable, for the purpose of clarity we frame our discussion of nonlinear regression and ℓ_2 regularization around a single classic example referred to as *logistic regression* (which we will also see arise in the context of classification in the next chapter). Further examples of nonlinear regression are explored in the chapter exercises.

3.3.1 Logistic regression

At the heart of the classic logistic regression problem is the so-called *logistic sigmoid function*, illustrated in the left panel of Fig. 3.10, and defined mathematically as

$$\sigma(t) = \frac{1}{1 + e^{-t}}, \tag{3.23}$$

where t can take on any real value. Invented in the early 19th century by mathematician Pierre François Verhulst [79], this function was designed in his pursuit of modeling how a population (of microbes, animal species, etc.,) grows over time, taking into account the realistic assumption that regardless of the kind of organism under study, the system in which it lives has only a finite amount of resources.[6] Thus, as a result, there should be a strict cap on the total population in any biological system. According to Verhulst's model, the initial stages of growth should follow an exponential trend until a saturation level where, due to lack of required resources (e.g., space, food, etc.), the growth stabilizes and levels off.[7]

[5] Additionally, ℓ_2 regularization also arises in the context of the (convex) support vector machine classifier, as we will see in Section 4.3. Another popular use of ℓ_2 regularization is discussed later in Section 7.3 in the context of 'cross-validation'.

[6] Beyond its classical use in modeling population growth, we will see in the next chapter that logistic regression can also be used for the task of classification.

[7] Like any good mathematician, Verhulst first phrased this ideal population growth model in terms of a differential equation. Denoting the desired function f and the maximum population level of the system as

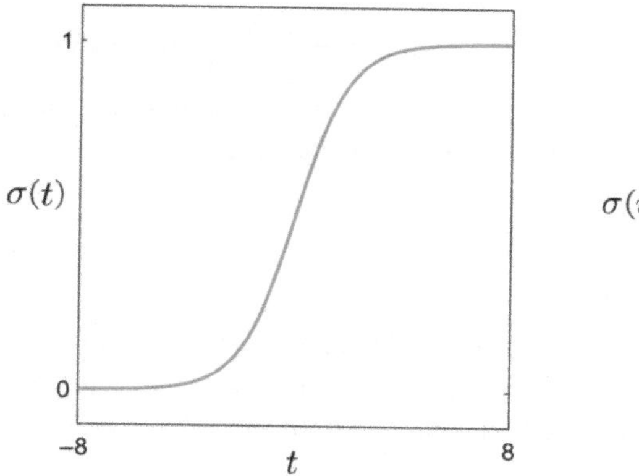

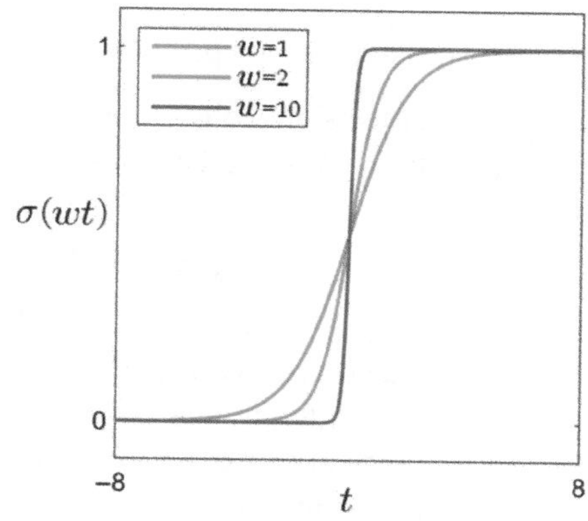

Fig. 3.10 (left panel) Plot of the logistic sigmoid function defined in (3.23). Note that the output of this function is always between 0 and 1. (right panel) By increasing the weight w of the sigmoid function $\sigma(wt)$ from $w = 1$ (red) to $w = 2$ (green) and finally to $w = 10$ (blue), the sigmoid becomes an increasingly good approximator of a "step function," that is a function that only takes on the values 0 and 1 with a sharp transition between the two.

If a dataset of P points $\left\{(x_p, y_p)\right\}_{p=1}^{P}$ is roughly distributed like a sigmoid function, then this data satisfies

$$\sigma\left(b + x_p w\right) \approx y_p, \quad p = 1, \ldots, P, \tag{3.24}$$

where b and w are parameters which must be properly tuned. The weight w, as illustrated in the right panel of Fig. 3.10, controls how quickly the system saturates, and the bias term b shifts the curve left and right along the horizontal axis. Likewise when the input is N-dimensional the system of equations given in (3.24) may be written analogously as

$$\sigma\left(b + \mathbf{x}_p^T \mathbf{w}\right) \approx y_p, \quad p = 1, \ldots, P, \tag{3.25}$$

where as usual $\mathbf{x}_p = \begin{bmatrix} x_{1,p} & x_{2,p} & \ldots & x_{N,p} \end{bmatrix}^T$ and $\mathbf{w} = \begin{bmatrix} w_1 & w_2 & \ldots & w_N \end{bmatrix}^T$. Note that unlike the analogous set of equations with linear regression given in Equation (3.5), each of these equations is nonlinear[8] in b and $\mathbf{w}$. These nonlinearities lead to a non-convex Least Squares cost function which is formed by summing the squared differences of Equation (3.25) over all p,

$$g\left(b, \mathbf{w}\right) = \sum_{p=1}^{P} \left(\sigma\left(b + \mathbf{x}_p^T \mathbf{w}\right) - y_p\right)^2. \tag{3.26}$$

1, he supposed that the population growth rate $\frac{df}{dt}$ should, at any time t, be proportional to both the current population level f as well as the remaining capacity left in the system $1 - f$. Together this gives the differential equation $\frac{df}{dt} = f\left(1 - f\right)$. One can check by substitution that the logistic sigmoid function satisfies this relationship with initial condition $f\left(0\right) = 1/2$.

[8] In certain circumstances this system may be transformed into one that is linear in its parameters. See Exercise 3.10 for further details.

Using the compact notation $\tilde{\mathbf{x}}_p = \begin{bmatrix} 1 \\ \mathbf{x}_p \end{bmatrix}$ and $\tilde{\mathbf{w}} = \begin{bmatrix} b \\ \mathbf{w} \end{bmatrix}$, the fact that the derivative of the sigmoid is given as $\sigma'(t) = \sigma(t)(1 - \sigma(t))$ (see footnote 7), and the chain rule from calculus, the gradient of this cost can be calculated as

$$\nabla g(\tilde{\mathbf{w}}) = 2\sum_{p=1}^{P} \left(\sigma\left(\tilde{\mathbf{x}}_p^T \tilde{\mathbf{w}}\right) - y_p\right) \sigma\left(\tilde{\mathbf{x}}_p^T \tilde{\mathbf{w}}\right)\left(1 - \sigma\left(\tilde{\mathbf{x}}_p^T \tilde{\mathbf{w}}\right)\right) \tilde{\mathbf{x}}_p. \tag{3.27}$$

Due to the many nonlinearities involved in the above system of equations, solving the first order system directly is a fruitless venture, instead a numerical technique (i.e., gradient descent or Newton's method) must be used to find a useful minimum of the associated cost function.

Example 3.4 Bacterial growth

In the left panel of Fig. 3.11 we show a real dataset consisting of $P = 9$ data points corresponding to the normalized cell concentration[9] of a particular bacteria, *Lactobacillus delbrueckii*, in spatially constrained laboratory conditions over the period of 24 hours. Also shown in this panel are two sigmoidal fits (shown in magenta and green) found via minimizing the Least Squares cost in (3.26) using gradient descent. In the middle panel we show the surface of the cost function which is clearly non-convex, having stationary points in the large flat region colored orange as well as a global minimum in the long narrow valley highlighted in dark blue. Two paths taken by initializing gradient descent at different values are shown in magenta and green, respectively, on the surface itself. While the initialization of the magenta path in the yellow-green area of the surface

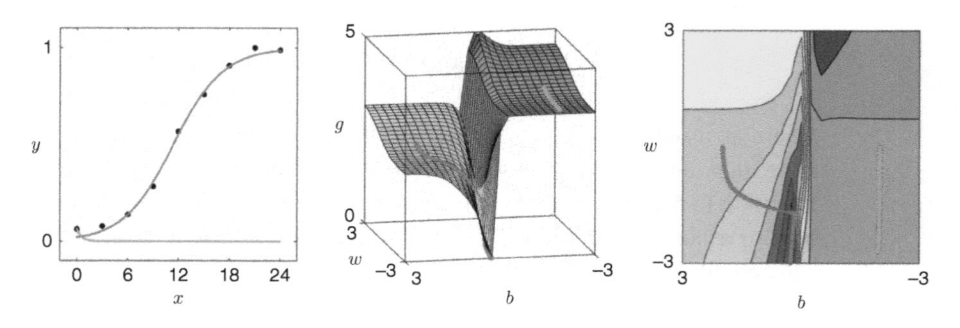

Fig. 3.11 (left panel) A dataset along with two sigmoidal fits (shown in magenta and green), each found via minimizing the Least Squares cost in (3.26) using gradient descent with a different initialization. A surface (middle) and contour (right) plot of this cost function, along with the paths taken by the two runs of gradient descent. Each path has been colored to match the resulting sigmoidal fit produced in the left panel (see text for further details). Data in this figure is taken from [48].

[9] Cell concentration is measured as the mass of organism per unit volume. Here we have normalized the cell concentration values so that they lie strictly in the interval (0, 1).

leads to the global minimum, which corresponds with the good sigmoidal fit in magenta shown in the left panel, the initialization of the green path in the large flat orange region leads to a poor solution, with corresponding poor fit shown in green in the left panel. In the right panel we show the contour plot of the same surface (along with the two gradient descent paths) that more clearly shows the long narrow valley containing the desired global minimum of the surface.

3.3.2 Non-convex cost functions and ℓ_2 regularization

The problematic flat areas posed by non-convex cost functions like the one shown in Fig. 3.11 can be ameliorated by the addition of a *regularizer*. A regularizer is a simple convex function that is often added to such a cost function, slightly convexifying it and thereby helping numerical optimization techniques avoid poor solutions in its flat areas. One of the most common regularizers used in practice is the squared ℓ_2 norm of the weights $\|\mathbf{w}\|_2^2 = \sum_{n=1}^{N} w_n^2$, referred to as the ℓ_2 regularizer. To regularize a cost function $g(b, \mathbf{w})$ with this regularizer we simply add it to g giving the regularized cost function

$$g(b, \mathbf{w}) + \lambda \|\mathbf{w}\|_2^2 . \tag{3.28}$$

Here $\lambda \geq 0$ is a parameter (set by the user in practice) that controls the strength of each term, the original cost function and the regularizer, in the final sum. For example, if $\lambda = 0$ we have our original cost. On the other hand, if λ is set very large then the regularizer drowns out the cost and we have $g(b, \mathbf{w}) + \lambda \|\mathbf{w}\|_2^2 \approx \lambda \|\mathbf{w}\|_2^2$. Typically in practice λ is set fairly small (e.g., $\lambda = 0.1$ or smaller).

In Fig. 3.12 we show two simple examples of non-convex cost functions which exemplify how the ℓ_2 regularizer can help numerical techniques avoid many (but not all) poor solutions in practice.

The first non-convex cost function,[10] shown in the top left panel of Fig. 3.12, is defined over a symmetric interval about the origin and has three large flat areas containing undesirable stationary points. This kind of non-convex function is highly problematic because if an algorithm like gradient descent or Newton's method is initialized at any point lying in these flat regions it will immediately halt. In the top right panel we show an ℓ_2 regularized version of the same cost. Note how regularization slightly convexifies the entire cost function, and in particular how it forces the flat regions to curve upwards. Now if e.g., gradient descent is initialized in either of the two flat regions on the left or right sides it will in fact travel downwards and reach a minimum. Note that both minima have slightly changed position from the original cost, but as long as λ is set relatively small this small change does not typically make a difference in practice. Note in this instance, however, that regularization has not helped with the

[10] Here the cost is defined as $g(w) = \max^2\left(0, e^{-w}\sin\left(4\pi\left(w - 0.1\right)\right)\right)$, and λ has been set fairly high at $\lambda = 1$ for illustrative purposes only.

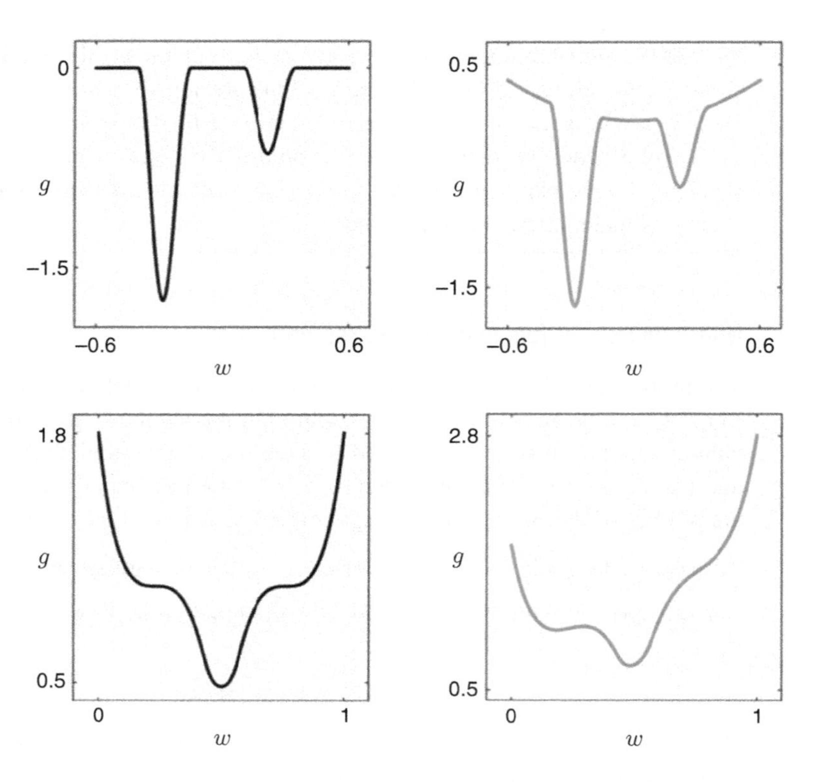

Fig. 3.12 Two examples of non-convex functions with flat regions (top left panel) and saddle points (bottom left panel) where numerical optimization methods can halt undesirably. Using the ℓ_2 regularizer we can slightly convexify each (right panels), which can help avoid some of these undesirable solutions. See text for further details.

problem of gradient descent halting at a poor solution if initialized in the middle flat region of the original cost. That is, by regularizing we have actually created a local minimum near the middle of the original flat region in the regularized cost function, and so if gradient descent is initialized in this region it will halt at this undesirable solution.

The second non-convex cost function,[11] shown in the bottom left panel of Fig. 3.12, is defined over the unit interval and has two saddle points at which the derivative is zero and so at which e.g., gradient descent, will halt undesirably if initialized at a point corresponding to any region on the far left or right. In the lower right panel we show the ℓ_2 regularized cost which no longer has an issue with the saddle point on the right, as the region surrounding it has been curved upwards. However the saddle point on the left is still problematic, as regularizing the original cost has created a local minimum near the point that will cause gradient descent to continue to halt at an undesirable solution.

[11] Here the cost is defined as $g(w) = \max^2\left(0, (3w - 2.3)^3 + 1\right) + \max^2\left(0, (-3w + 0.7)^3 + 1\right)$, and λ has been set fairly high at $\lambda = 1$ for illustrative purposes only.

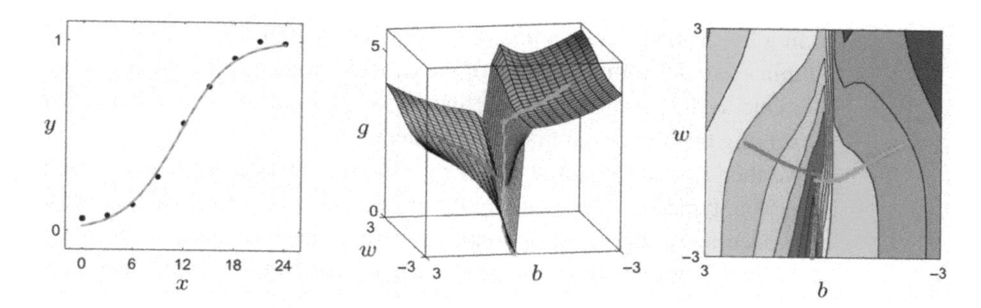

Fig. 3.13 A regularized version of Fig. 3.11. (left panel) Plot of the bacterial growth dataset along with two overlapping sigmoidal fits (shown in magenta and green) found via minimizing the ℓ_2 regularized Least Squares cost for logistic regression in (3.29) using gradient descent. (middle and right panels) The surface and contour plot of the regularized cost function along with the paths (in magenta and green) of gradient descent with same two initializations as shown in Fig. 3.11. While the surface is still non-convex, the large flat region that originally led the initialization of the green path to a poor solution with the unregularized cost has been curved upwards by the regularizer, allowing the green run of gradient descent to reach the global minimum of the problem. Data in this figure is taken from [48].

Example 3.5 **ℓ_2 regularized Least Squares for logistic regression**

We saw in Fig. 3.11 that the initialization of gradient descent in the flat orange region resulted in a poor fit to the bacterial growth data. A second version of all three panels from this figure is duplicated in Fig. 3.13, only here we add the ℓ_2 regularizer with $\lambda = 0.1$ to the original Least Squares logistic cost in (3.26). Formally, this ℓ_2 regularized Least Squares cost function is written as

$$g\left(b, \mathbf{w}\right) = \sum_{p=1}^{P} \left(\sigma \left(b + \mathbf{x}_p^T \mathbf{w} \right) - y_p \right)^2 + \lambda \left\| \mathbf{w} \right\|_2^2. \tag{3.29}$$

Once again in order to minimize this cost we can employ gradient descent (see Exercise 3.13). Comparing the regularized surface in Fig. 3.13 to the original shown in Fig. 3.11 we can see that regularizing the original cost curves the flat regions of the surface upwards, helping gradient descent avoid poor solutions when initialized in these areas. Now both initializations first shown in Fig. 3.11 lead gradient descent to the global minimum of the surface.

3.4 Summary

Linear regression is a fundamental predictive learning problem which aims at determining the relationship between continuous-valued input and output data via the fitting of an appropriate model that is linear in its parameters. In this chapter we first saw how to fit a linear model (i.e., a line or hyperplane in higher dimensions) to a given dataset, culminating in the minimization of the Least Squares cost function at the end of Section 3.1. Due to the parameters being linearly related, this cost function may be minimized by solving the associated first order system.

We then saw in Section 3.2 how in some rare instances our understanding of a phenomenon, typically due to our ability to visualize a low dimensional dataset, can permit us to accurately suggest an appropriate feature transformation to describe our data. This provides a proper nonlinear fit to the original data while simultaneously fitting linearly to the data in an associated transformed feature space.

Finally, using the classical example of logistic regression, we saw in Section 3.3 how a nonlinear regression model typically involves the need to minimize an associated non-convex cost function. We then saw how ℓ_2 regularization is used as a way of "convexifying" a non-convex cost function to help gradient descent avoid some undesirable stationary points of such a function.

3.5 Exercises

Section 3.1 exercises

Exercises 3.1 Fitting a regression line to the student debt data

Fit a linear model to the U.S. student loan debt dataset shown in Fig. 1.8, called *student_debt_data.csv*, by solving the associated linear regression Least Squares problem. If this linear trend continues what will the total student debt be in 2050?

Exercises 3.2 Kleiber's law and linear regression

After collecting and plotting a considerable amount of data comparing the body mass versus metabolic rate (a measure of at rest energy expenditure) of a variety of animals, early 20th century biologist Max Kleiber noted an interesting relationship between the two values. Denoting by x_p and y_p the body mass (in kg) and metabolic rate (in kJ/day) of a given animal respectively, treating the body mass as the input feature Kleiber noted (by visual inspection) that the natural logs of these two values were linearly related. That is,

$$w_0 + \log\left(x_p\right) w_1 \approx \log\left(y_p\right). \tag{3.30}$$

In Fig. 3.14 we show a large collection of transformed data points $\left\{\left(\log\left(x_p\right), \log\left(y_p\right)\right)\right\}_{p=1}^{P}$, each representing an animal ranging from a small black-chinned hummingbird in the bottom left corner to a large walrus in the top right corner.

a) Fit a linear model to the data shown in Fig. 3.14 (called *kleibers_law_data.csv*). Make sure to take the log of both arguments!

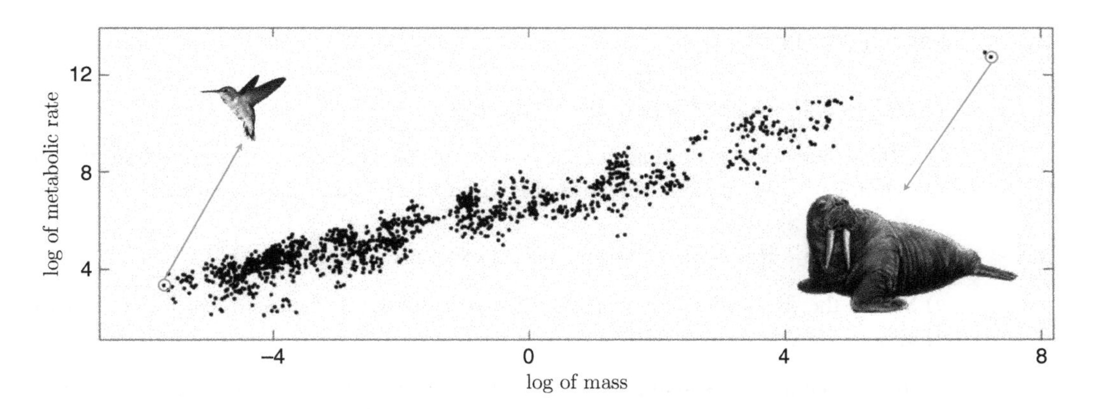

Fig. 3.14 A large set of body mass/metabolic rate data points, transformed by taking the log of each value, for various animals over a wide range of different masses.

b) Use the optimal parameters you found in part (a) along with the properties of the log function to write the nonlinear relationship between the body mass x and the metabolic rate y.

c) Use your fitted line to determine how many calories an animal weighing 10 kg requires (note each calorie is equivalent to 4.18 J).

Exercises 3.3 The Least Squares cost for linear regression is convex

Show that the Least Squares cost function for linear regression written compactly as in Section 3.1.3,

$$g\left(\tilde{\mathbf{w}}\right) = \sum_{p=1}^{P} \left(\tilde{\mathbf{x}}_p^T \tilde{\mathbf{w}} - y_p\right)^2, \tag{3.31}$$

is a convex quadratic function by completing the following steps.

a) Show that $g\left(\tilde{\mathbf{w}}\right)$ can be written as a quadratic function of the form

$$g\left(\tilde{\mathbf{w}}\right) = \frac{1}{2}\tilde{\mathbf{w}}^T \mathbf{Q} \tilde{\mathbf{w}} + \mathbf{r}^T \tilde{\mathbf{w}} + d \tag{3.32}$$

by determining proper $\mathbf{Q}$, $\mathbf{r}$, and d.

b) Show that $\mathbf{Q}$ has all nonnegative eigenvalues (*hint: see Exercise 2.10*).

c) Verify that $\nabla^2 g\left(\tilde{\mathbf{w}}\right) = \mathbf{Q}$ and so that g satisfies the second order definition of convexity, and is therefore convex.

d) Show that applying a single Newton step (see Section 2.2.4) to minimize the Least Squares cost function leads to precisely the first order system of linear equations discussed in Section 3.1.3, i.e., to the system $\left(\sum_{p=1}^{P} \tilde{\mathbf{x}}_p \tilde{\mathbf{x}}_p^T\right) \tilde{\mathbf{w}} = \sum_{p=1}^{P} \tilde{\mathbf{x}}_p y_p$. (This is because the Least Squares cost is a quadratic function, as in Example 2.7.)

Section 3.2 exercises

Exercises 3.4 Reproduce Galileo's example

Use the value for the optimal weight shown in Equation (3.22) to reproduce the fits shown in Fig. 3.8. The data shown in this figure is located in the file *Galileo_data.csv*.

Exercises 3.5 Reproduce the sinusoidal example

a) Set up the first order system associated with the Least Squares cost function being minimized in Equation (3.16).

b) Reproduce the sinusoidal and associated linear fit shown in the middle and right panels of Fig. 3.7 by solving for the proper weights via the first order system you determined in part **a)**. The dataset shown in this figure is called *sinusoid_example_data.csv*.

Exercises 3.6 Galileo's extended ramp experiment

In this exercise we modify Galileo's ramp experiment, discussed in Example 3.3, to explore the relationship between the angle x of the ramp and the distance y that the ball travels during a certain fixed amount of time. In Fig. 3.15 we plot six simulated measurements corresponding to six different angle values x (measured in degrees).

a) Propose a suitable nonlinear feature transformation for this dataset such that the relationship between the new feature you form and the distance traveled is linear in its weights. *Hint: there is no need for a bias parameter b here.*

b) Formulate and minimize a Least Squares cost function using your new feature for a proper weight w, the data (located in the file *another_ramp_experiment.csv*). Plot the resulting fit in the data space.

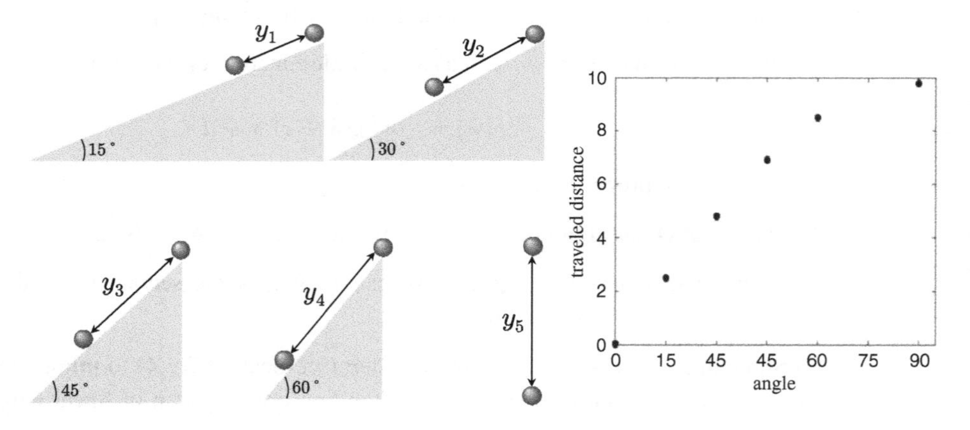

Fig. 3.15 An extended set of data from Galileo's ramp experiment, first described in Example 3.3. See text for details.

Exercises 3.7 Moore's law and the power of future computers

Gordon Moore, co-founder of Intel corporation, predicted in a 1965 paper[12] that the number of transistors on an integrated circuit would double approximately every two years. This conjecture, referred to nowadays as Moore's law, has proven to be sufficiently accurate over the past five decades. Since the processing power of computers is directly related to the number of transistors in their CPUs, Moore's law provides a trend model to predict the computing power of future microprocessors. Figure 3.16 plots the transistor counts of several microprocessors versus the year they were released, starting from Intel 4004 in 1971 with only 2300 transistors, to Intel's Xeon E7 introduced in 2014 with more than 4.3 billion transistors.

a) Propose an exponential-based transformation of the Moore's law dataset shown in Fig. 3.16 so that the transformed input/output data is related linearly. *Hint: to produce a linear relationship you will end up having to transform the output, not the input.*

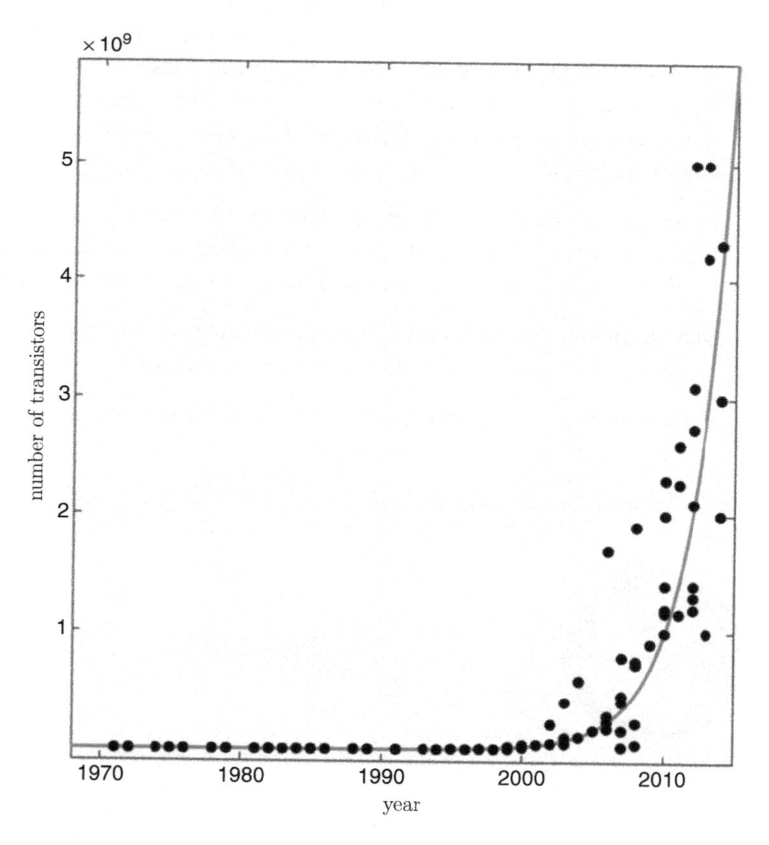

Fig. 3.16 As Moore proposed 50 years ago, the number of transistors in microprocessors versus the year they were invented follows an exponential pattern.

[12] One can find a modern reprinting of this paper in e.g., [57].

b) Formulate and minimize a Least Squares cost function for appropriate weights, and fit your model to the data in the original data space. The data shown here is located in the file *transistor_counts.csv*.

Exercises 3.8 Ohm's law and linear regression

Ohm's law, proposed by the German physicist Georg Simon Ohm following a series of experiments made by him in the 1820s, connects the magnitude of the current in a galvanic circuit to the sum of all the exciting forces in the circuit, as well as the length of the circuit. Although he did not publish any account of his experimental results, it is easy to verify his law using a simple experimental setup, shown in the left panel of Fig. 3.17, that is very similar to what he then utilized (the data in this figure is taken from [56]). The spirit lamp heats up the circuit, generating an electromotive force which creates a current in the coil deflecting the needle of the compass. The tangent of the deflection angle is directly proportional to the magnitude of the current passing through the circuit. The magnitude of this current, denoted by I, varies depending on the length of the wire used to close the circuit (dashed curve). In the right panel of Fig. 3.17 we plot the readings of the current I (in terms of the tangent of the deflection angle) when the circuit is closed with a wire of length x (in cm), for five different values of x.

a) Suggest a suitable nonlinear transformation of the original data to fit (located in the file *ohms_data.csv*) so that the transformed input/output data is related linearly. *Hint: to produce a linear relationship you will likely end up having to transform the output.*

b) Formulate a proper Least Squares cost function using your transformed data and minimize it to recover ideal parameters for your model.

c) Fit your proposed model to the data and display it in the original data space.

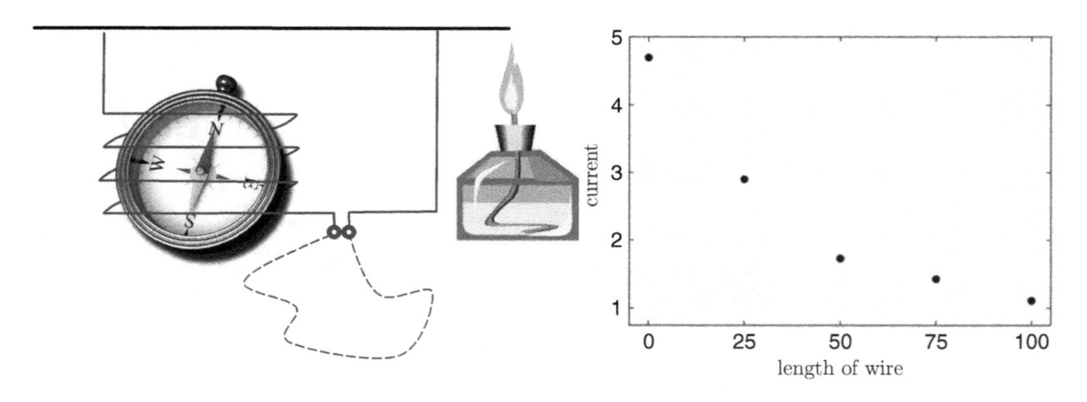

Fig. 3.17 (left panel) Experimental setup for verification of Ohm's law. Black and brown wires are made up of constantan and copper, respectively. (right panel) Current measurements for five different lengths of closing wire.

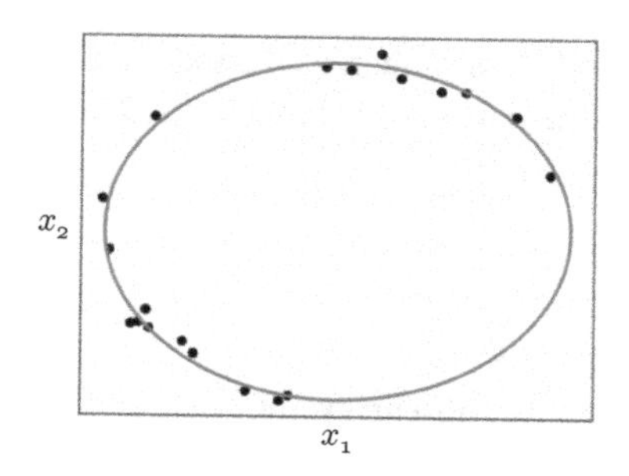

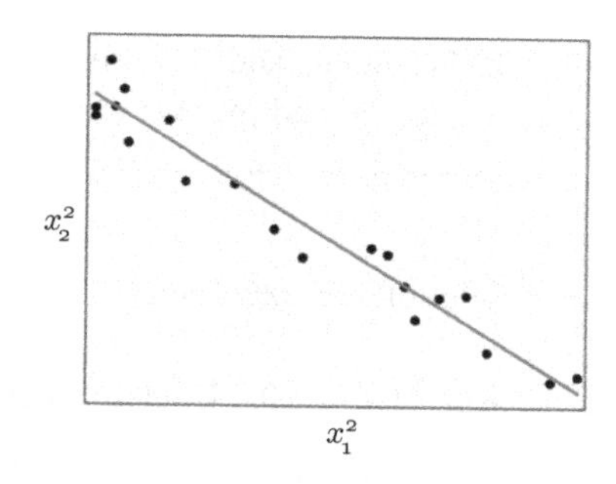

Fig. 3.18 Simulated observation data for the location of the asteroid Pallas on its orbital plane. The ellipsoidal curve fit to the data approximates the true orbit of Pallas. (right panel) Fitting an ellipsoid to the data in the original data space is equivalent to fitting a line to the data in a new space where both dimensions are squared.

Exercises 3.9 Determining the orbit of celestial bodies

One of the first recorded uses of regression via the Least Squares approach was made by Carl Frederich Gauss, a German mathematician, physicist, and all round polymath, who was interested in calculating the orbit of the asteroid Pallas by leveraging a dataset of recorded observations. Although Gauss solved the problem using ascension and declination data observed from the earth (see [61] and references therein), here we modify the problem so that the simulated data shown in the left panel of Fig. 3.18 simulates Cartesian coordinates of the location of the asteroid on its orbital plane. With this assumption, and according to Kepler's laws of planetary motion, we need to fit an ellipse to a series of observation points in order to recover the true orbit.

In this instance the data comes in the form of $P = 20$ noisy coordinates $\left\{\left(x_{1,p}, x_{2,p}\right)\right\}_{p=1}^{P}$ taken from an ellipsoid with the standard form of

$$\left(\frac{x_{1,p}}{\nu_1}\right)^2 + \left(\frac{x_{2,p}}{\nu_2}\right)^2 \approx 1 \quad \text{for all } p = 1 \ldots P, \tag{3.33}$$

where ν_1 and ν_2 are tunable parameters. By making the substitutions $w_1 = \left(\frac{1}{\nu_1}\right)^2$ and $w_2 = \left(\frac{1}{\nu_2}\right)^2$ this can be phrased equivalently as a set of approximate linear equations

$$x_{1,p}^2 w_1 + x_{2,p}^2 w_2 \approx 1 \quad \text{for all } p = 1 \ldots P. \tag{3.34}$$

a) Reformulate the equations shown above using vector notation as

$$\mathbf{f}_p^T \mathbf{w} \approx y_p \quad \text{for all } p = 1 \ldots P \tag{3.35}$$

by determining the appropriate $\mathbf{f}_p$ and y_p where $\mathbf{w} = \begin{bmatrix} w_1 & w_2 \end{bmatrix}^T$.

b) Formulate and solve the associated Least Squares cost function to recover the proper weights $\mathbf{w}$ and plot the ellipse with the data shown in the left panel of Fig. 3.18 located in the file *asteroid_data.csv.*

Section 3.3 exercises

Exercises 3.10 Logistic regression as a linear system

In this exercise you will explore particular circumstances that allow one to transform the nonlinear system of equations in (3.24) into a system which is linear in the parameters b and w. In order to do this recall that a function f has an *inverse* at t if another function f^{-1} exists such that $f^{-1}(f(t)) = t$. For example, the exponential function $f(t) = e^t$ has the inverse $f^{-1}(t) = \log(t)$ for every t since we always have $f^{-1}(f(t)) = \log(e^t) = t$.

a) Show that the logistic sigmoid has an inverse for each t where $0 < t < 1$ of the form $\sigma^{-1}(t) = \log\left(\frac{t}{1-t}\right)$ and check that indeed $\sigma^{-1}(\sigma(t)) = t$ for all such values of t.

b) Suppose for a given dataset $\left\{(x_p, y_p)\right\}_{p=1}^{P}$ that $0 < y_p < 1$ for all p. Apply the sigmoid inverse to the system shown in Equation (3.24) to derive the equivalent set of linear equations

$$b + x_p w \approx \log\left(\frac{y_p}{1 - y_p}\right) \qquad p = 1, \ldots, P. \tag{3.36}$$

Since the equations in (3.36) are now linear in both b and w we may solve for these parameters by simply checking the first order condition for optimality.

c) Using the dataset *bacteria_data.csv* solve the Least Squares cost function based on the linear system of equations from part **b)** and plot the data, along with the logistic sigmoid fit to the data as shown in Fig. 3.19.

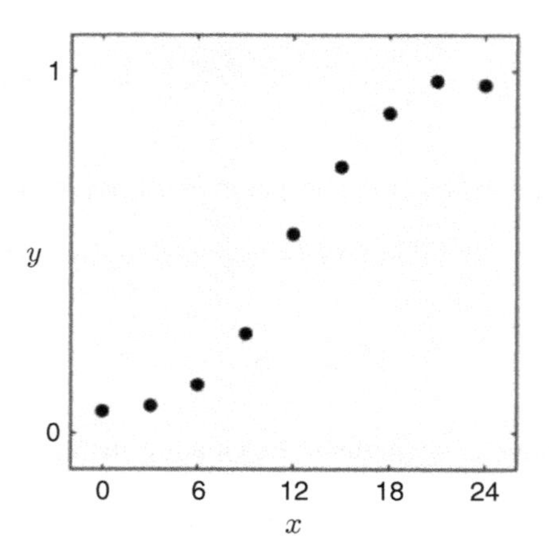

Fig. 3.19 The normalized cell concentration of Lactobacillus delbrueckii in a constrained laboratory environment over the period of 24 hours. Data in this figure is taken from [48].

Exercises 3.11 Code up gradient descent for logistic regression

In this exercise you will reproduce the gradient descent paths shown in Fig. 3.11.

a) Verify that the gradient descent step shown in Equation (3.27) is correct.

Note that this gradient can be written more compactly by denoting $\sigma_p^{k-1} = \sigma\left(\tilde{\mathbf{x}}_p^T \tilde{\mathbf{w}}^{k-1}\right)$, $r_p^{k-1} = 2\left(\sigma_p^{k-1} - y_p\right)\sigma_p^{k-1}\left(1 - \sigma_p^{k-1}\right)$ for all $p = 1, ..., P$, and $\mathbf{r}^{k-1} = \left[\begin{array}{cccc} r_1^{k-1} & r_2^{k-1} & \cdots & r_P^{k-1} \end{array}\right]^T$, and stacking the $\widetilde{\mathbf{x}}_p$ column-wise into the matrix $\widetilde{\mathbf{X}}$. Then the gradient can be written as $\nabla g\left(\tilde{\mathbf{w}}^{k-1}\right) = \widetilde{\mathbf{X}}\mathbf{r}^{k-1}$. For programming languages like Python and MATLAB/OCTAVE that have especially efficient implementations of matrix/vector operations this can be much more efficient than explicitly summing over the P points as in Equation (3.27).

b) The surface in this figure was generated via the wrapper *nonconvex_logistic_growth* with the dataset *bacteria_data.csv*, and inside the wrapper you must complete a short gradient descent function to produce the descent paths called

$$[\text{in, out}] = \text{grad\_descent}\left(\widetilde{\mathbf{X}}, \mathbf{y}, \tilde{\mathbf{w}}^0\right), \tag{3.37}$$

where "in" and "out" contain the gradient steps $\tilde{\mathbf{w}}^k = \tilde{\mathbf{w}}^{k-1} - \alpha_k \nabla g\left(\tilde{\mathbf{w}}^{k-1}\right)$ taken and corresponding objective value $g\left(\tilde{\mathbf{w}}^k\right)$ respectively, $\widetilde{\mathbf{X}}$ is the input data matrix, $\mathbf{y}$ the output values, and $\tilde{\mathbf{w}}^0$ the initial point.

Almost all of this function has already been constructed for you. For example, the step length is fixed at $\alpha_k = 10^{-2}$ for all iterations, etc., and you must only enter the gradient of the associated cost function. Pressing "run" in the editor will run gradient descent and will reproduce Fig. 3.11.

Exercises 3.12 A general sinusoid model nonlinear in its parameters

Recall the periodic sinusoidal regression discussed in Example 3.2. There we chose a model $b + \sin\left(2\pi x_p w\right) \approx y_p$ that fit the given data which was linear in the weights b and w, and we saw that the corresponding Least Squares cost function was therefore convex. This allowed us to solve for the optimal values for these weights in closed form via the first order system, with complete assurance that they represent a global minimum of the associated Least Squares cost function. In this exercise you will investigate how a simple change to this model leads to a comparably much more challenging optimization problem to solve.

Figure 3.20 shows a set of $P = 75$ data points $\left\{\left(x_p, y_p\right)\right\}_{p=1}^P$ generated via the model

$$w_1 \sin\left(2\pi x_p w_2\right) + \epsilon = y_p \quad \text{for all } p = 1 \ldots P, \tag{3.38}$$

where $\epsilon > 0$ is a small amount of noise. This dataset may be located in the file *extended_sinusoid_data.csv*. Unlike the previous instance here the model is nonlinear

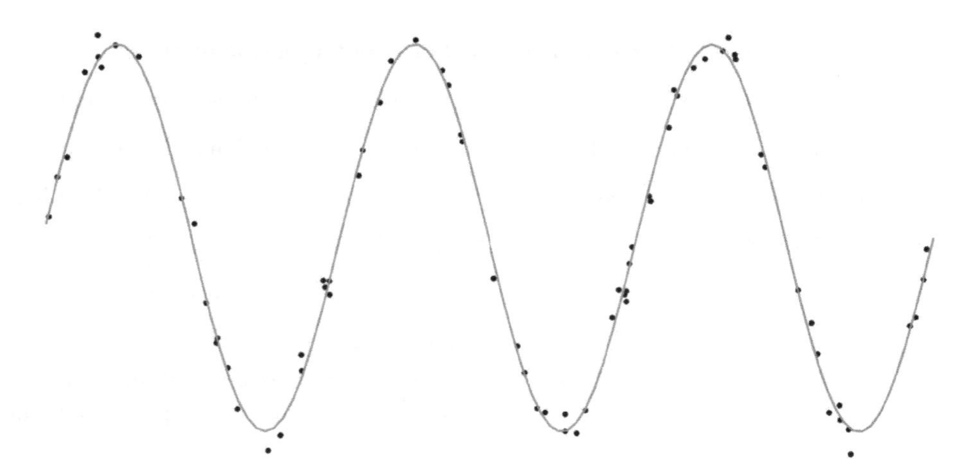

Fig. 3.20 A set of $P = 75$ periodic data points along with the underlying sinusoidal model used to generate the data in magenta.

in both the weights. Here w_1 controls the amplitude (i.e., stretches the model in the vertical direction) and w_2 controls the frequency (i.e., how quickly the sinusoid completes a single period) of the sinusoidal model.

We can then attempt to recover optimal weights of a representative curve for this data-set by minimizing the associated Least Squares cost function with respect to the dataset

$$g(\mathbf{w}) = \sum_{p=1}^{P} \left(w_1 \sin\left(2\pi x_p w_2\right) - y_p\right)^2. \tag{3.39}$$

a) Plot the surface of g over the region defined by $-3 \leq w_1, w_2 \leq 3$.

b) Discuss the approach you would take to find the best possible stationary point of this function, along with any potential difficulties you foresee in doing so.

Exercises 3.13 Code up gradient descent for ℓ_2 regularized logistic regression

In this exercise you will reproduce Fig. 3.13 by coding up gradient descent to minimize the regularized logistic regression Least Squares cost function shown in Equation (3.29).

a) Verify that the gradient of the cost function can be written as

$$\nabla g\left(\tilde{\mathbf{w}}\right) = 2\sum_{p=1}^{P} \left(\sigma\left(\tilde{\mathbf{x}}_p^T \tilde{\mathbf{w}}\right) - y_p\right) \sigma\left(\tilde{\mathbf{x}}_p^T \tilde{\mathbf{w}}\right) \left(1 - \sigma\left(\tilde{\mathbf{x}}_p^T \tilde{\mathbf{w}}\right)\right) \tilde{\mathbf{x}}_p + 2\lambda \begin{bmatrix} 0 \\ \mathbf{w} \end{bmatrix}. \tag{3.40}$$

b) The surface in this figure was generated via the wrapper *l2reg_nonconvex_logistic_growth* with the dataset *bacteria_data.csv*, and inside the wrapper you must complete a short gradient descent function to produce the descent paths called

$$[\text{in, out}] = \text{grad\_descent}\left(\widetilde{\mathbf{X}}, \mathbf{y}, \tilde{\mathbf{w}}^0\right), \tag{3.41}$$

where "in" and "out" contain the gradient steps $\tilde{\mathbf{w}}^k = \tilde{\mathbf{w}}^{k-1} - \alpha_k \nabla g\left(\tilde{\mathbf{w}}^{k-1}\right)$ taken and corresponding objective value $g\left(\tilde{\mathbf{w}}^k\right)$ respectively, $\widetilde{\mathbf{X}}$ is the input data matrix whose pth column is the input data $\tilde{\mathbf{x}}_p$, $\mathbf{y}$ the output values stacked into a column vector, and $\tilde{\mathbf{w}}^0$ the initial point.

Almost all of this function has already been constructed for you. For example, the step length is fixed at $\alpha_k = 10^{-2}$ for all iterations, etc., and you must only enter the gradient of the associated cost function. Pressing "run" in the editor will run gradient descent and will reproduce Fig. 3.13.

Exercises 3.14 The ℓ_2 regularized Newton's method

Recall from Section 2.2.4 that when applied to minimizing non-convex cost functions, Newton's method can climb to local maxima (or even diverge) due to the concave shape of the quadratic second order Taylor series approximation at concave points of the cost function (see Fig. 2.11). One very common way of dealing with this issue, which we explore formally in this exercise, is to add an ℓ_2 regularizer (centered at each step) to the quadratic approximation used by Newton's method in order to ensure that it is convex at each step. This is also commonly done when applying Newton's method to convex functions as well since, as we will see, the addition of a regularizer increases the eigenvalues of the Hessian and therefore helps avoid numerical problems associated with solving linear systems with zero (or near-zero) eigenvalues.

a) At the kth iteration of the regularized Newton's method we add an ℓ_2 regularizer centered at $\mathbf{w}^{k-1}$, i.e., $\frac{\lambda}{2}\left\|\mathbf{w} - \mathbf{w}^{k-1}\right\|_2^2$ where $\lambda \geq 0$, to the second order Taylor series approximation in (2.18), giving

$$\begin{aligned} h(\mathbf{w}) = g\left(\mathbf{w}^{k-1}\right) + \nabla g\left(\mathbf{w}^{k-1}\right)^T\left(\mathbf{w} - \mathbf{w}^{k-1}\right) + \frac{1}{2}\left(\mathbf{w} - \mathbf{w}^{k-1}\right)^T \\ \nabla^2 g\left(\mathbf{w}^{k-1}\right)\left(\mathbf{w} - \mathbf{w}^{k-1}\right) + \frac{\lambda}{2}\left\|\mathbf{w} - \mathbf{w}^{k-1}\right\|_2^2. \end{aligned} \tag{3.42}$$

Show that the first order condition for optimality leads to the following adjusted Newton's system for a stationary point of the above quadratic:

$$\left[\nabla^2 g\left(\mathbf{w}^{k-1}\right) + \lambda \mathbf{I}_{N\times N}\right]\mathbf{w} = \left[\nabla^2 g\left(\mathbf{w}^{k-1}\right) + \lambda \mathbf{I}_{N\times N}\right]\mathbf{w}^{k-1} - \nabla g\left(\mathbf{w}^{k-1}\right). \tag{3.43}$$

b) Show that the eigenvalues of $\nabla^2 g\left(\mathbf{w}^{k-1}\right) + \lambda \mathbf{I}_{N\times N}$ in the system above can all be made to be positive by setting λ large enough. What is the smallest value of λ that will make this happen? This is typically the value used in practice. *Hint: see Exercise* 2.9.

c) Using the value of λ determined in part **b)**, conclude that the ℓ_2 regularized second order Taylor series approximation centered at $\mathbf{w}^{k-1}$ in (3.42) is convex. *Hint: see Exercise 2.11.*

For a non-convex function, λ is typically adjusted at each step so that it just forces the eigenvalues of $\nabla^2 g\left(\mathbf{w}^{k-1}\right) + \lambda \mathbf{I}_{N\times N}$ to be all positive. In the case of a convex cost, since the eigenvalues of $\nabla^2 g\left(\mathbf{w}^{k-1}\right)$ are always nonnegative (via the second order definition of convexity) *any* positive value of λ will force the eigenvalues of $\nabla^2 g\left(\mathbf{w}^{k-1}\right) + \lambda \mathbf{I}_{N\times N}$ to be all positive. Therefore often for convex functions λ is set fixed for all iterations at some small value like $\lambda = 10^{-3}$ or $\lambda = 10^{-4}$.

4 Classification

In this chapter we discuss the problem of classification, where we look to distinguish between different types of distinct things. Beyond the crucial role such an ability plays in contributing to what we might consider as "intelligence," modern applications of classification arise in a wide range of fields including computer vision, speech processing, and digital marketing (see e.g., Sections 1.2.2 and 4.6). We begin by introducing the fundamental model for two class classification: the *perceptron*. As described pictorially in Fig. 1.10, the perceptron works by finding a line/hyperplane (or more generally a curve/surface) that separates two classes of data. We then describe two equally effective approximations to the basic perceptron known as the softmax and margin perceptrons, followed by a description of popular perspectives on these approximations where they are commonly referred to as *logistic regression* and *support vector machines*, respectively. In Section 4.4 we see how the two class framework can be easily generalized to deal with multiclass classification problems that have arbitrary numbers of distinct classes. Finally, we end the chapter by discussing knowledge-driven feature design methods for classification. This includes a description of basic histogram-based features commonly used for text, image, and speech classification problems.

4.1 The perceptron cost functions

In the most basic instance of a classification problem our data consists of just two classes. Common examples of two class classification problems include face detection, with classes consisting of facial versus non-facial images, textual sentiment analysis where classes consist of written product reviews ascribing a positive or negative opinion, and automatic diagnosis of medical conditions where classes consist of medical data corresponding to patients who either do or do not have a specific malady (see Sections 1.2.2 and 4.6 for further descriptions of these problems). In this section we introduce the most foundational tool for two class classification, the *perceptron*, as well as a popular variation called the *margin perceptron*. Both tools are commonly used and perform similarly in practice, as we discuss further in Section 4.1.7.

4.1.1 The basic perceptron model

Recall from the previous chapter that in a linear regression setting, given a training set of P continuous-valued input/output data points $\left\{\left(\mathbf{x}_p, y_p\right)\right\}_{p=1}^{P}$, we aim to learn a hyperplane $b + \mathbf{x}^T\mathbf{w}$ with parameters b and $\mathbf{w}$ such that

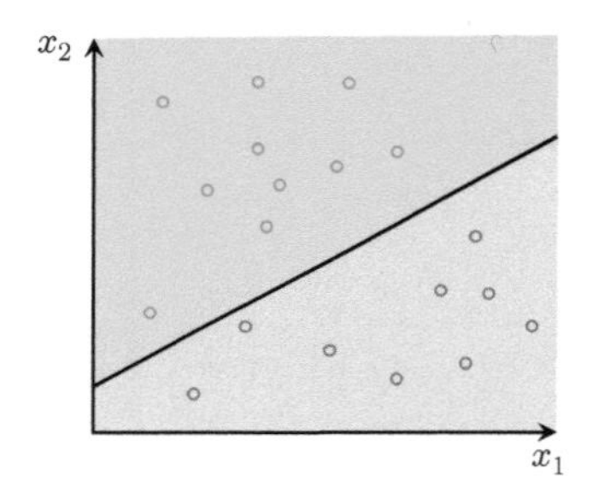

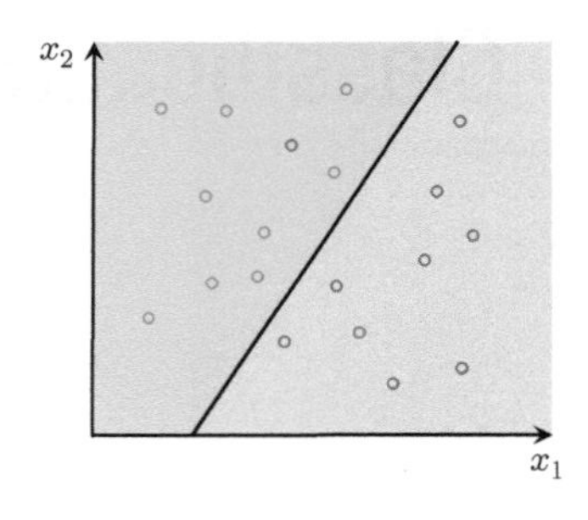

Fig. 4.1 With linear classification we aim to learn a hyperplane $b + \mathbf{x}^T\mathbf{w} = 0$ (shown here in black) to separate feature representations of the two classes, colored red (class "+1") and blue (class "−1"), by dividing the feature space into a red half-space where $b + \mathbf{x}^T\mathbf{w} > 0$, and a blue half-space where $b + \mathbf{x}^T\mathbf{w} < 0$. (left panel) A linearly separable dataset where it is possible to learn a hyperplane to perfectly separate the two classes. (right panel) A dataset with two overlapping classes. Although the distribution of data does not allow for perfect linear separation, we can still find a hyperplane that minimizes the number of misclassified points that end up in the wrong half-space.

$$b + \mathbf{x}_p^T\mathbf{w} \approx y_p \tag{4.1}$$

holds for $p = 1, \ldots, P$. In the case of linear classification a disparate yet simple motivation leads to the pursuit of a different sort of ideal hyperplane. As opposed to linear regression, where our aim is to *represent* a dataset, with classification our goal is to *separate* two distinct classes of the input/output data with a learned hyperplane. In other words, we want to learn a hyperplane $b + \mathbf{x}^T\mathbf{w} = 0$ that separates the two classes of points as much as possible, with one class lying "above" the hyperplane in the half-space given by $b + \mathbf{x}^T\mathbf{w} > 0$ and the other "below" it in the half-space $b + \mathbf{x}^T\mathbf{w} < 0$, as illustrated in Fig. 4.1.

More formally, with two class classification we still have a training set of P input/output data points $\left\{\left(\mathbf{x}_p, y_p\right)\right\}_{p=1}^{P}$ where each input $\mathbf{x}_p$ is N-dimensional (with each entry representing an input feature, just as with regression). However, the output data no longer takes on continuous but two discrete values or *labels* indicating class membership, i.e., points belonging to each class are assigned a distinct label. While one can choose any two values for this purpose, we will see that the values ± 1 are particularly useful and therefore will assume that $y_p \in \{-1, +1\}$ for $p = 1, \ldots, P$.

We aim to learn the parameters b and $\mathbf{w}$ of a hyperplane, so that the first class (where $y_p = +1$) lies largely *above* the hyperplane in the half-space defined by $b + \mathbf{x}^T\mathbf{w} > 0$, and the second class (where $y_p = -1$) lies mostly *below*[1] it in the half-space defined by $b + \mathbf{x}^T\mathbf{w} < 0$. If a given hyperplane places the point $\mathbf{x}_p$ on its correct side (or we say that it correctly classifies the point), then we have precisely that

[1] The choice of which class we assume lies "above" and "below" the hyperplane is arbitrary, i.e., if we instead suppose that those points with label $y_p = -1$ lie above and those with label $y_p = +1$ lie below, similar calculations can be made which lead to the perceptron cost function in Equation (4.5).

$$\begin{array}{ll} b + \mathbf{x}_p^T \mathbf{w} > 0 & \text{if } y_p = +1 \\ b + \mathbf{x}_p^T \mathbf{w} < 0 & \text{if } y_p = -1. \end{array} \quad (4.2)$$

Because we have chosen the labels ± 1 we can express (4.2) compactly by multiplying the two expressions by minus their respective label value $-y_p$, giving one equivalent expression

$$-y_p \left(b + \mathbf{x}_p^T \mathbf{w} \right) < 0. \quad (4.3)$$

By taking the maximum of this quantity and zero we can then write this condition, which states that a hyperplane correctly classifies the point $\mathbf{x}_p$, equivalently as

$$\max \left(0, -y_p \left(b + \mathbf{x}_p^T \mathbf{w} \right) \right) = 0. \quad (4.4)$$

Note that the expression $\max \left(0, -y_p \left(b + \mathbf{x}_p^T \mathbf{w} \right) \right)$ returns zero if $\mathbf{x}_p$ is classified correctly, but it returns a *positive* value if the point is classified incorrectly. This is useful not only because it characterizes the sort of hyperplane we wish to have, but more importantly by simply summing this expression over all the points we have the non-negative cost function

$$g_1 (b, \mathbf{w}) = \sum_{p=1}^{P} \max \left(0, -y_p \left(b + \mathbf{x}_p^T \mathbf{w} \right) \right), \quad (4.5)$$

referred to as the *perceptron* or *max* cost function.[2] Solving the minimization problem

$$\underset{b, \mathbf{w}}{\text{minimize}} \sum_{p=1}^{P} \max \left(0, -y_p \left(b + \mathbf{x}_p^T \mathbf{w} \right) \right), \quad (4.6)$$

then determines the optimal parameters for our separating hyperplane. However, while this problem is fine in principle, there are two readily apparent technical issues regarding the minimization itself. First, one minimum of g_1 always presents itself at the trivial and undesirable values $b = 0$ and $\mathbf{w} = \mathbf{0}_{N \times 1}$ (which indeed gives $g_1 = 0$). Secondly, note that while g_1 is continuous (and it is in fact convex) it is not everywhere differentiable (see Fig. 4.2), thus prohibiting the use of gradient descent and Newton's method.[3] One simple work-around for both of these issues is to make a particular smooth approximation to the perceptron function, which we discuss next.

4.1.2 The softmax cost function

One popular way of approximating the perceptron cost is to replace the non-differentiable "max" function $\max(s_1, s_2)$ (which returns the maximum of the two scalar inputs s_1 and s_2) in (4.5) with the smooth *softmax function* defined as

$$\text{soft}(s_1, s_2) = \log \left(e^{s_1} + e^{s_2} \right). \quad (4.7)$$

[2] The perceptron is also referred to as the *hinge* (as it is shaped like a hinge, see Fig. 4.2 for an illustration) or *rectified linear unit*.

[3] While specialized algorithms can be used to tune the perceptron (see e.g., [19]) differentiable approximations (that permit the use of gradient descent and/or Newton's method) are typically preferred over these options due to their superior efficacy and speed.

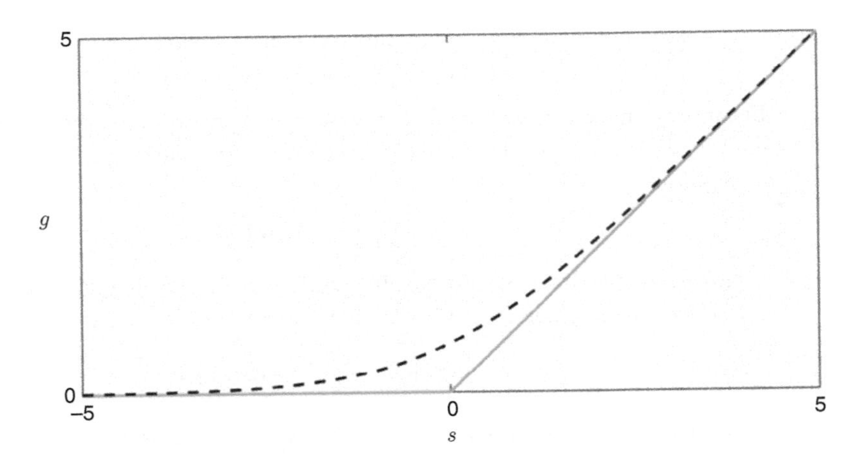

Fig. 4.2 Plots of the non-differentiable perceptron or hinge cost $g(s) = \max(0, s)$ (shown in green) as well as its smooth softmax approximation $g(s) = \text{soft}(0, s) = \log(1 + e^s)$ (shown in dashed black).

That $\text{soft}(s_1, s_2) \approx \max(s_1, s_2)$, or in words that the softmax approximates the max function, can be verified formally[4] and intuited visually in the particular example shown in Fig. 4.2.

Replacing the "max" function in the pth summand of g_1 in (4.5) with its softmax approximation,

$$\text{soft}\left(0, -y_p\left(b + \mathbf{x}_p^T\mathbf{w}\right)\right) = \log\left(1 + e^{-y_p\left(b+\mathbf{x}_p^T\mathbf{w}\right)}\right), \tag{4.8}$$

we have a smooth approximation of the perceptron cost given by

$$g_2(b, \mathbf{w}) = \sum_{p=1}^{P} \log\left(1 + e^{-y_p\left(b+\mathbf{x}_p^T\mathbf{w}\right)}\right), \tag{4.9}$$

which we will refer to as the *softmax cost function.* Note that this cost function does not have a trivial minimum at $b = 0$ and $\mathbf{w} = \mathbf{0}_{N\times 1}$ as was the case with the original perceptron. It also has the benefit of being smooth and hence we may apply gradient descent or Newton's method for its minimization as detailed in Section 2.2, the latter

[4] The fact that the softmax function provides a good approximation to the max function can be shown formally by the following simple argument. Suppose momentarily that $s_1 \leq s_2$, so that $\max(s_1, s_2) = s_2$. Therefore $\max(s_1, s_2)$ can be written as $\max(s_1, s_2) = s_1 + (s_2 - s_1)$, or equivalently as $\max(s_1, s_2) = \log(e^{s_1}) + \log(e^{s_2 - s_1})$ since $s = \log(e^s)$ for any s. Written in this way we can see that $\log(e^{s_1}) + \log(1 + e^{s_2 - s_1}) = \log(e^{s_1} + e^{s_2}) = \text{soft}(s_1, s_2)$ is always larger than $\max(s_1, s_2)$ but not by much, especially when $e^{s_2 - s_1} \gg 1$. Since the same argument can be made if $s_1 \geq s_2$ we can say generally that $\text{soft}(s_1, s_2) \approx \max(s_1, s_2)$.

Note also that the softmax approximation to the max function applies more generally for C inputs, as

$$\max(s_1, \ldots, s_C) \approx \text{soft}(s_1, \ldots, s_C) = \log\left(\sum_{c=1}^{C} e^{s_c}\right).$$

of which we may safely use as the softmax cost is indeed convex (see Exercise 4.2). Formally, the softmax minimization problem is written as

$$\underset{b,\,\mathbf{w}}{\text{minimize}} \sum_{p=1}^{P} \log\left(1+e^{-y_p\left(b+\mathbf{x}_p^T\mathbf{w}\right)}\right). \tag{4.10}$$

This approximation to the perceptron cost is very commonly used in practice, most often referred to as the *logistic regression* for classification (see Section 4.2) or *log-loss support vector machines* (see Section 4.3). Due to its immense popularity as the logistic regression, we will at times refer to the minimization of the softmax cost as the learning of the softmax or logistic regression classifier.

Example 4.1 Optimization of the softmax cost

Using the compact notation $\tilde{\mathbf{x}}_p = \begin{bmatrix} 1 \\ \mathbf{x}_p \end{bmatrix}$ and $\tilde{\mathbf{w}} = \begin{bmatrix} b \\ \mathbf{w} \end{bmatrix}$ we can rewrite the softmax cost function in (4.9) more conveniently as

$$g_2\left(\tilde{\mathbf{w}}\right) = \sum_{p=1}^{P} \log\left(1+e^{-y_p\tilde{\mathbf{x}}_p^T\tilde{\mathbf{w}}}\right). \tag{4.11}$$

Using the chain rule[5] we can then compute the gradient, and setting it equal to zero we check the first order condition (see Section 2.1.2),

$$\nabla g_2\left(\tilde{\mathbf{w}}\right) = -\sum_{p=1}^{P} \sigma\left(-y_p\tilde{\mathbf{x}}_p^T\tilde{\mathbf{w}}\right) y_p\tilde{\mathbf{x}}_p = \mathbf{0}_{(N+1)\times 1}. \tag{4.12}$$

Note here that $\sigma\left(-t\right) = \frac{1}{1+e^t}$ denotes the logistic sigmoid function[6] evaluated at $-t$ (see Section 3.3.1). However (4.12) is an unwieldy and highly nonlinear system of

[5] To see how to employ the chain rule let us briefly rewrite the pth summand in (4.11) explicitly as a composition of functions

$$\log\left(1+e^{-y_p\tilde{\mathbf{x}}_p^T\tilde{\mathbf{w}}}\right) = f\left(r\left(s\left(\tilde{\mathbf{w}}\right)\right)\right),$$

where $f\left(r\right) = \log\left(r\right)$, $r\left(s\right) = 1+e^{-s}$, and $s\left(\tilde{\mathbf{w}}\right) = y_p\tilde{\mathbf{x}}_p^T\tilde{\mathbf{w}}$. To compute the derivative of this with respect to a single entry $\tilde{w}_n$ the chain rule gives

$$\frac{\partial}{\partial \tilde{w}_n} f\left(r\left(s\left(\tilde{\mathbf{w}}\right)\right)\right) = \frac{df}{dr}\cdot\frac{dr}{ds}\cdot\frac{\partial}{\partial \tilde{w}_n} s\left(\tilde{\mathbf{w}}\right) = \frac{1}{r}\left(-e^{-s}\right) y_p\tilde{x}_{n,p} = \frac{1}{1+e^{-y_p\tilde{\mathbf{x}}_p^T\tilde{\mathbf{w}}}}\left(-e^{-y_p\tilde{\mathbf{x}}_p^T\tilde{\mathbf{w}}}\right) y_p\tilde{x}_{n,p},$$

which can be written more compactly as $-\sigma\left(-y_p\tilde{\mathbf{x}}_p^T\tilde{\mathbf{w}}\right) y_p\tilde{x}_{p,n}$ using the fact that $\frac{1}{1+e^{-t}}\left(-e^{-t}\right) = \frac{1}{1+e^{-t}}\cdot\frac{-1}{e^t} = \frac{-1}{1+e^t} = -\sigma\left(-t\right)$, where $\sigma\left(t\right)$ is the logistic sigmoid function. By combining the result for all entries in $\tilde{\mathbf{w}}$ and summing over all P summands we then get the gradient as shown in (4.12).

[6] Writing the derivative in this way also helps avoid numerical problems associated with using the exponential function on a modern computer. This is due to the exponential "overflowing" with large exponents, like e.g., e^{1000}, as these numbers are too large to store explicitly on the computer and so are represented symbolically as ∞. This becomes a problem when dividing two exponentials like e.g., $\frac{e^{1000}}{1+e^{1000}}$ which, although basically equal to the value 1, is thought of by the computer to be a NaN (not a

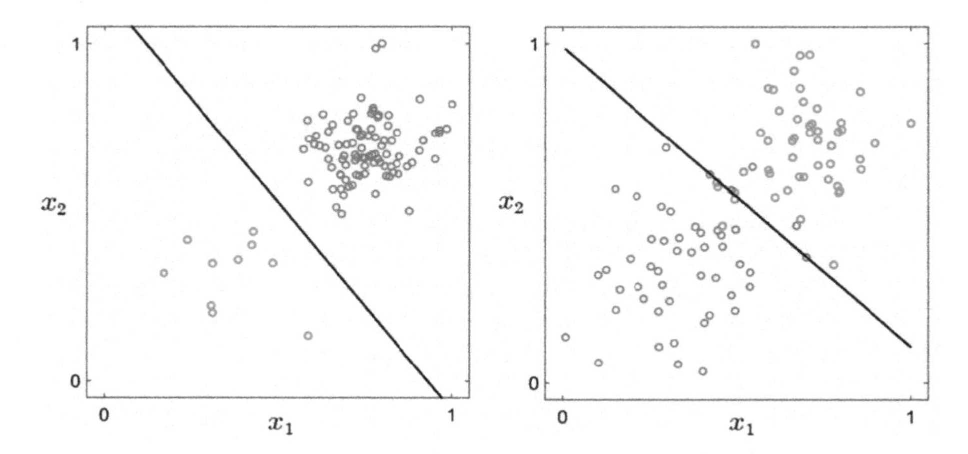

Fig. 4.3 (left panel) A two-dimensional toy dataset with linearly separable classes consisting of $P = 100$ points in total (90 in the "+1" class and 10 in the "−1" class), along with the softmax classifier learned using gradient descent. (right panel) A two-dimensional toy dataset with overlapping classes consisting of $P = 100$ points in total (50 points in each class), with the softmax classifier learned again using gradient descent. In both cases the learned classifier does a good job separating the two classes.

$N + 1$ equations which must be solved numerically by applying e.g., gradient descent or Newton's method. By again employing the chain rule, and noting that we always have that $y_p^2 = 1$ since $y_p \in \{-1, +1\}$, one may additionally compute the Hessian of the softmax as the following sum of weighted outer product matrices (see Exercise 2.10):

$$\nabla^2 g_2\left(\tilde{\mathbf{w}}\right) = \sum_{p=1}^{P} \sigma\left(-y_p \tilde{\mathbf{x}}_p^T \tilde{\mathbf{w}}\right)\left(1 - \sigma\left(-y_p \tilde{\mathbf{x}}_p^T \tilde{\mathbf{w}}\right)\right) \tilde{\mathbf{x}}_p \tilde{\mathbf{x}}_p^T. \tag{4.13}$$

Figure 4.3 illustrates the classification of two toy datasets, one linearly separable (left panel) and the other non-separable or overlapping (right panel), using the softmax classifier. In both cases a gradient descent scheme is used to learn the hyperplanes' parameters.

4.1.3 The margin perceptron

Here we discuss an often used variation of the original perceptron, called the margin perceptron, that is once again based on analyzing the geometry of the classification problem

number) as it thinks $\frac{e^{1000}}{1+e^{1000}} = \frac{\infty}{\infty}$ which is undefined. By writing each summand of the gradient such that it has an exponential in its denominator only we avoid the problem of dividing two overflowing exponentials. The overflowing exponential issue is discussed further in the exercises, as it is also something to keep in mind when both choosing an initial point for gradient descent/Newton's method as well as recording the value of the softmax cost at each iteration.

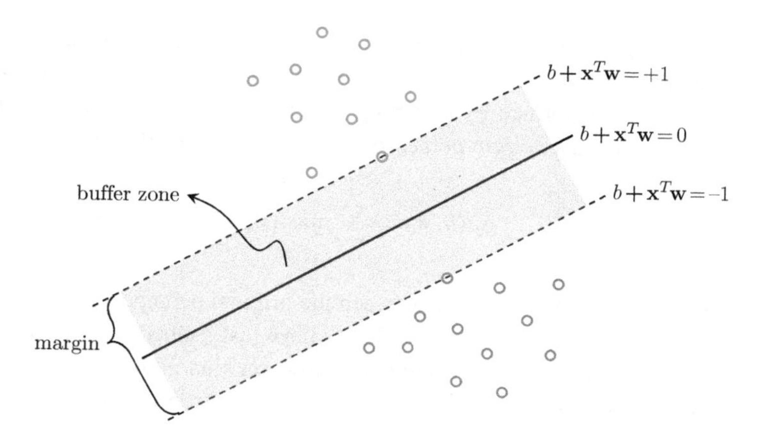

Fig. 4.4 For linearly separable data the width of the buffer zone confined between two evenly spaced translates of a separating hyperplane that just touches each respective class, defines the margin of that separating hyperplane.

where a line (or hyperplane in higher dimensions) is used to separate two classes of data. Due to the great similarity between the two perceptron concepts, what follows closely mirrors Sections 4.1.1 and 4.1.2.

Suppose for a moment that we are dealing with a two class dataset that is linearly separable with a known hyperplane $b+\mathbf{x}^T\mathbf{w}=0$ passing evenly between the two classes as illustrated in Fig. 4.4. This separating hyperplane creates a buffer zone between the two classes confined between two evenly shifted versions of itself: one version that lies *above* the separator and just touches the class having labels $y_p = +1$ taking the form $b+\mathbf{x}^T\mathbf{w} = +1$, and one lying below it just touching the class with labels $y_p = -1$ taking the form $b+\mathbf{x}^T\mathbf{w} = -1$. The width of this buffer zone is commonly referred to as the *margin* of such a hyperplane.[7]

The fact that all points in the "+1" class lie on or above $b+\mathbf{x}^T\mathbf{w} = +1$, and all points in the "−1" class lie on or below $b+\mathbf{x}^T\mathbf{w} = -1$ can be written formally as the following conditions:

$$\begin{aligned} b+\mathbf{x}_p^T\mathbf{w} &\geq 1 \qquad \text{if } y_p = +1 \\ b+\mathbf{x}_p^T\mathbf{w} &\leq -1 \qquad \text{if } y_p = -1. \end{aligned} \tag{4.14}$$

We can combine these conditions into a single statement by multiplying each by their respective label values, giving the single inequality $y_p\left(b+\mathbf{x}_p^T\mathbf{w}\right) \geq 1$, which can be equivalently written as

$$\max\left(0,\, 1 - y_p\left(b+\mathbf{x}_p^T\mathbf{w}\right)\right) = 0. \tag{4.15}$$

[7] The translations above and below the separating hyperplane are more generally defined as $b+\mathbf{x}^T\mathbf{w} = +\beta$ and $b+\mathbf{x}^T\mathbf{w} = -\beta$ respectively, where $\beta > 0$. However, by dividing off β in both equations and reassigning the variables as $\mathbf{w} \longleftarrow \frac{\mathbf{w}}{\beta}$ and $b \longleftarrow \frac{b}{\beta}$, we can leave out the redundant parameter β and have the two translations as stated, $b+\mathbf{x}^T\mathbf{w} = \pm 1$.

Dropping the assumption that we know the parameters of the hyperplane we can propose, as we did in devising the perceptron cost in (4.5), to *learn* them by minimizing the cost function formed by summing the criterion in (4.15) over all points in the dataset. Referred to as a *margin perceptron* or *hinge cost* this function takes the form

$$g_3(b, \mathbf{w}) = \sum_{p=1}^{P} \max\left(0,\ 1 - y_p\left(b + \mathbf{x}_p^T \mathbf{w}\right)\right). \tag{4.16}$$

Note the striking similarity between the original perceptron cost in (4.5) and the margin perceptron cost in (4.16): naively we have just "added a 1" to the nonzero input of the "max" function in each summand. However this additional "1" prevents the issue of a trivial zero solution with the original perceptron discussed in Section 4.1.1, which simply does not arise here.

If the data is indeed linearly separable, any hyperplane passing between the two classes will have a parameter pair $(b, \mathbf{w})$ where $g_3(b, \mathbf{w}) = 0$. However, the margin perceptron is still a valid cost function even if the data is not linearly separable. The only difference is that with such a dataset we cannot make the criteria in (4.14) hold for all points in the dataset. Thus a violation for the pth point adds the positive value of $1 - y_p\left(b + \mathbf{x}_p^T \mathbf{w}\right)$ to the cost function in (4.16).

Regardless of whether the two classes are linearly separable or not, by minimizing the margin perceptron cost stated formally as

$$\underset{b,\,\mathbf{w}}{\text{minimize}} \sum_{p=1}^{P} \max\left(0,\ 1 - y_p\left(b + \mathbf{x}_p^T \mathbf{w}\right)\right), \tag{4.17}$$

we can learn the parameters for the margin perceptron classifier. However, like the original perceptron, the margin cost is still not everywhere differentiable due to presence of the "max" function. Again it is common practice to make simple differentiable approximations to this cost so that descent methods, such as gradient descent and Newton's method, may be employed.

4.1.4 Differentiable approximations to the margin perceptron

To produce a differentiable approximation to the margin perceptron cost in (4.16) we can of course employ the softmax function first introduced in Section 4.1.2, replacing each summand's $\max(\cdot)$ function with $\text{soft}(\cdot)$. Specifically, taking the softmax approximation of the pth summand of (4.16) gives $\text{soft}\left(0,\ 1 - y_p\left(b + \mathbf{x}_p^T \mathbf{w}\right)\right) \approx \max\left(0,\ 1 - y_p\left(b + \mathbf{x}_p^T \mathbf{w}\right)\right)$, where

$$\text{soft}\left(0,\ 1 - y_p\left(b + \mathbf{x}_p^T \mathbf{w}\right)\right) = \log\left(1 + e^{1 - y_p\left(b + \mathbf{x}_p^T \mathbf{w}\right)}\right). \tag{4.18}$$

Summing over $p = 1, \ldots, P$ we can produce a cost function that approximates the margin perceptron, with the added benefit of differentiability. However, note that, as illustrated in Fig. 4.7, in fact the softmax approximation of the original perceptron summand $\text{soft}\left(0, -y_p\left(b + \mathbf{x}_p^T\mathbf{w}\right)\right) = \log\left(1 + e^{-y_p\left(b+\mathbf{x}_p^T\mathbf{w}\right)}\right)$ provides, generally speaking, just as good approximation of the margin perceptron.[8] Therefore the original softmax cost in (4.9) also provides a useful differentiable approximation to the margin perceptron as well!

Another perhaps more straightforward way of making a differentiable approximation to the margin perceptron cost is simply to square each of its summands, giving the *squared margin perceptron* cost function

$$g_4(b, \mathbf{w}) = \sum_{p=1}^{P} \max^2\left(0,\ 1 - y_p\left(b + \mathbf{x}_p^T\mathbf{w}\right)\right), \tag{4.19}$$

where $\max^2(s_1, s_2)$ is a brief way of writing $(\max(s_1, s_2))^2$. Note that when the two classes are linearly separable, solutions to the corresponding minimization problem,

$$\underset{b,\,\mathbf{w}}{\text{minimize}} \sum_{p=1}^{P} \max^2\left(0,\ 1 - y_p\left(b + \mathbf{x}_p^T\mathbf{w}\right)\right), \tag{4.20}$$

are precisely those of the original problem in (4.17). Moreover its differentiability permits easily computed gradient for use in gradient descent and Newton's method.

Example 4.2 Optimization of the squared margin perceptron

Using the compact notation $\tilde{\mathbf{x}}_p = \begin{bmatrix} 1 \\ \mathbf{x}_p \end{bmatrix}$ and $\tilde{\mathbf{w}} = \begin{bmatrix} b \\ \mathbf{w} \end{bmatrix}$ we can compute the gradient of the squared margin perceptron cost using the chain rule, and form the first order system of $N + 1$ equations

$$\nabla g_4\left(\tilde{\mathbf{w}}\right) = -2\sum_{p=1}^{P} \max\left(0,\ 1 - y_p\tilde{\mathbf{x}}_p^T\tilde{\mathbf{w}}\right) y_p\tilde{\mathbf{x}}_p = \mathbf{0}_{(N+1)\times 1}. \tag{4.21}$$

Because once again it is impossible to solve this system for $\tilde{\mathbf{w}}$ in closed form, a solution must be found iteratively by applying gradient descent. Since g_4 is convex (see Exercise 4.6) it is also possible to apply Newton's method, with the Hessian easily computable (noting that we always have that $y_p^2 = 1$ since $y_p \in \{-1, +1\}$) as[9]

$$\nabla^2 g_4\left(\tilde{\mathbf{w}}\right) = 2\sum_{p\in\Omega_{\tilde{\mathbf{w}}}} \tilde{\mathbf{x}}_p\tilde{\mathbf{x}}_p^T, \tag{4.22}$$

where $\Omega_{\tilde{\mathbf{w}}}$ is the index set defined as $\Omega_{\tilde{\mathbf{w}}} = \left\{p \mid 1 - y_p\tilde{\mathbf{x}}_p^T\tilde{\mathbf{w}} > 0\right\}$.

[8] As shown in Fig. 4.7 while the function $\text{soft}(0, 1 - t)$ better approximates $\max(0, 1 - t)$ for values of $t \leq 0$, $\text{soft}(0, -t)$ provides a better approximation for $t > 0$.

[9] This is actually a "generalized" Hessian since the "max" function in the gradient is not everywhere differentiable. Nevertheless, it still makes a highly effective Newton's method for the squared margin cost (see e.g., [27]).

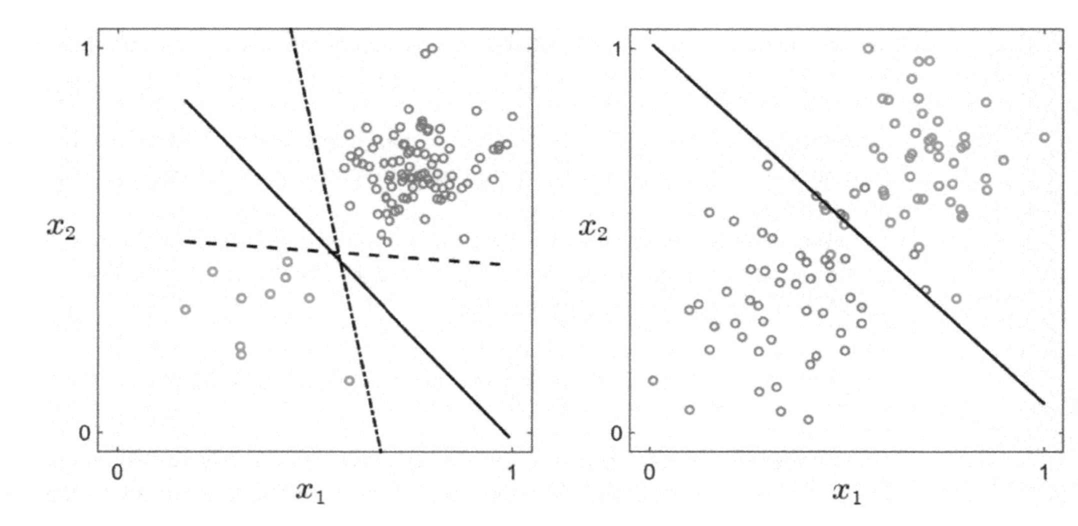

Fig. 4.5 Classification of two toy datasets, first shown in Fig. 4.3, using gradient descent for minimizing the squared margin perceptron cost function. Initializing gradient descent with three different starting points results in three different classifiers for the linearly separable dataset in the left panel, each perfectly separating the two classes.

In Fig. 4.5 we show the resulting linear classifiers learned by minimizing the squared margin perceptron cost for the two toy datasets first shown in Fig. 4.3. As is the case with the dataset in the left panel of this figure, when the two classes of data are linearly separable there are infinitely many distinct separating hyperplanes, and correspondingly infinitely many distinct minima of g_4. Thus on such a dataset initializing gradient descent or Newton's method with a random starting point means we may reach a different solution at each run, along with a distinct separating hyperplane.

4.1.5 The accuracy of a learned classifier

From our discussion of the original perceptron in Section 4.1.1, note that given any parameter pair $(b, \mathbf{w})$ (learned by any of the cost functions described in this section) we can determine whether a point $\mathbf{x}_p$ is classified correctly or not via the following simple evaluation:

$$\text{sign}\left(-y_p\left(b+\mathbf{x}_p^T\mathbf{w}\right)\right)=\begin{cases}+1 & \text{if } \mathbf{x}_p \text{ incorrectly classified}\\ -1 & \text{if } \mathbf{x}_p \text{ correctly classified,}\end{cases} \tag{4.23}$$

where sign $(\cdot)$ takes the mathematical sign of the input. Also note that by taking the maximum of this value and 0,

$$\max\left(0,\,\text{sign}\left(-y_p\left(b+\mathbf{x}_p^T\mathbf{w}\right)\right)\right)=\begin{cases}+1 & \text{if } \mathbf{x}_p \text{ incorrectly classified}\\ 0 & \text{if } \mathbf{x}_p \text{ correctly classified,}\end{cases} \tag{4.24}$$

we can count the precise number of misclassified points for a given set of parameters $(b, \mathbf{w})$ by summing (4.24) over all p. This observation naturally leads to a fundamental *counting cost* function, which precisely counts the number of points from the training data classified incorrectly as

$$g_0(b, \mathbf{w}) = \sum_{p=1}^{P} \max\left(0, \text{ sign}\left(-y_p\left(b + \mathbf{x}_p^T \mathbf{w}\right)\right)\right). \tag{4.25}$$

By plugging in any learned weight pair $(b^\star, \mathbf{w}^\star)$, the value of this cost function provides a metric for evaluating the performance of the associated linear classifier, i.e., the number of misclassifications for the given weight pair. This can be used to define the *accuracy* of a classifier with the weights $(b^\star, \mathbf{w}^\star)$ on the training data as

$$\text{accuracy} = 1 - \frac{g_0(b^\star, \mathbf{w}^\star)}{P}. \tag{4.26}$$

This metric ranges from 0 to 1, with an ideal classification corresponding to an accuracy of 1 or 100%. If possible it is also a good idea to compute the accuracy of a learned classifier on a set of new testing data, i.e., data that was not used to learn the model itself, in order to provide some assurance that the learned model will perform well on future data points. This is explored further in Chapter 6 in the context of *cross-validation*.

4.1.6 Predicting the value of new input data

As illustrated in Fig. 4.6, to predict the label y_{new} of a new point $\mathbf{x}_{\text{new}}$ we simply check which side of the learned hyperplane it lies on as

$$y_{\text{new}} = \text{sign}\left(b^\star + \mathbf{x}_{\text{new}}^T \mathbf{w}^\star\right), \tag{4.27}$$

where this hyperplane has parameters $(b^\star, \mathbf{w}^\star)$ learned over the current dataset via any of the cost functions described in this section. In other words, if the new point lies above the learned hyperplane $(b^\star + \mathbf{x}_{\text{new}}^T \mathbf{w}^\star > 0)$ it is given the label $y_{\text{new}} = 1$, and likewise if

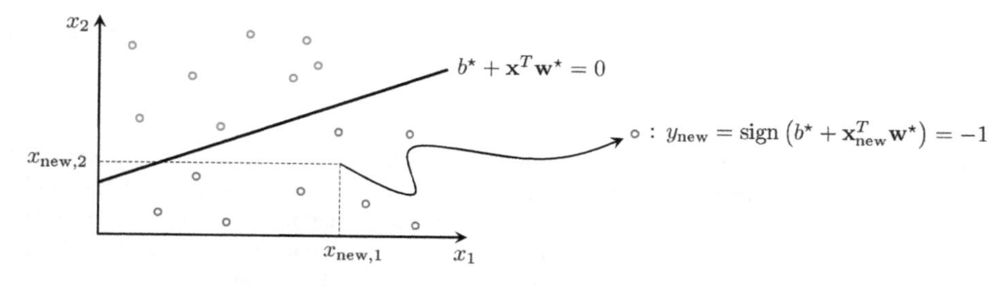

Fig. 4.6 Once a hyperplane has been learned to the current dataset with optimal parameters $\left(b^\star, \mathbf{w}^\star\right)$, the label y_{new} of a new point $\mathbf{x}_{\text{new}}$ can be determined by simply checking which side of the boundary it lies on. In the illustration shown here $\mathbf{x}_{\text{new}}$ lies below the learned hyperplane $(b^\star + \mathbf{x}_{\text{new}}^T \mathbf{w}^\star < 0)$ and so is given the label $y_{\text{new}} = \text{sign}\left(b^\star + \mathbf{x}_{\text{new}}^T \mathbf{w}^\star\right) = -1$.

the point lies below the boundary ($b^{\star} + \mathbf{x}_{\text{new}}^T \mathbf{w}^{\star} < 0$) it receives the label $y_{\text{new}} = -1$. If on the off chance the point lies on the boundary itself (i.e., $b^{\star} + \mathbf{x}_{\text{new}}^T \mathbf{w}^{\star} = 0$) then $\mathbf{x}_{\text{new}}$ may be assigned to either class.

4.1.7 Which cost function produces the best results?

In terms of accuracy, which (differentiable) cost function works the best in practice, the softmax or squared margin perceptron? Nothing we have seen so far seems to indicate one cost function's superiority over the other. In fact, the various geometric derivations given so far have shown how both are intimately related to the original perceptron cost in (4.5). Therefore it should come as little surprise that while they can differ from dataset to dataset in terms of their performance, in practice both differentiable costs typically produce very similar results.

> The softmax and squared margin costs perform similarly well in practice.

Thus one should feel comfortable using either one or, if resources allow, apply both and keep the higher performer on a case by case basis. Figure 4.7 shows a visual comparison of all classification cost functions we have discussed so far.

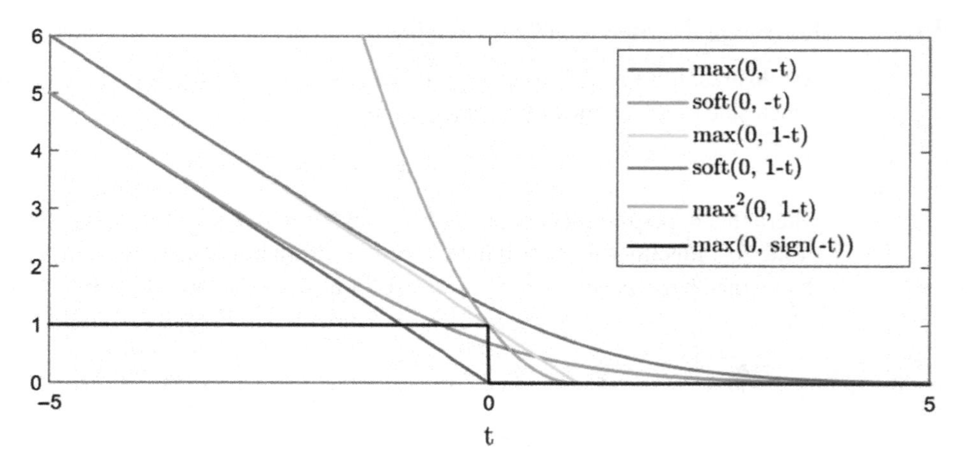

Fig. 4.7 Comparison of various classification cost functions. For visualization purposes we show here only one summand of each cost function plotted versus $t = b + \mathbf{x}_p^T \mathbf{w}$ with the label y_p assumed to be 1. The softmax cost (red) is a smooth approximation to the non-differentiable perceptron or hinge cost (blue), which can be thought of itself as a continuous surrogate for the discontinuous counting loss (black). The margin cost (yellow) is a shifted version of the basic perceptron, and is non-differentiable at its corner point. The squared margin cost (green) resolves this issue by taking its square (as does the softmax cost). Note that all the cost functions (except for the counting cost) are convex.

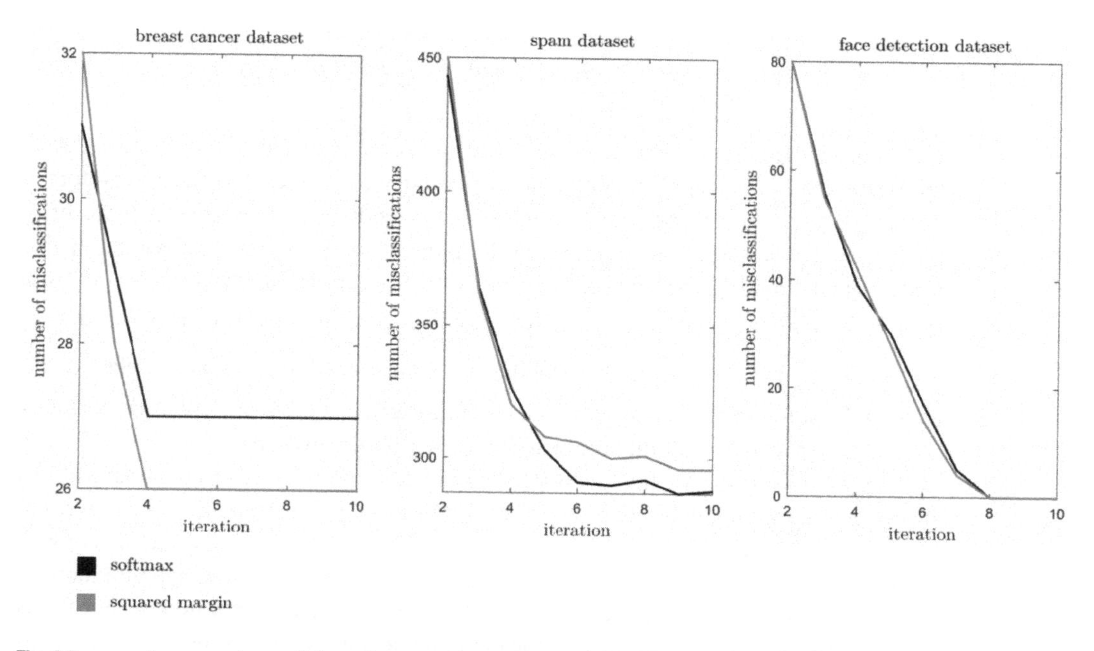

Fig. 4.8 A comparison of the softmax and margin costs on three real training datasets. Shown in each panel is the number of misclassifications per iteration of Newton's method (only ten iterations were required for convergence in all instances) applied to minimizing the softmax cost (shown in black) and squared margin cost (shown in magenta) over (left panel) breast cancer, (middle panel) spam email, and (right panel) face detection datasets respectively. While the performance of each cost function differs from case to case, generally they perform similarly well.

Example 4.3 Real dataset comparison of the softmax and squared margin costs

In Fig. 4.8 we illustrate the similar efficacy of the softmax and squared margin costs on three real training datasets. For each dataset we show the number of misclassifications resulting from the use of ten iterations of Newton's method (as only ten iterations were required for the method to converge in all cases), by evaluating the counting cost in (4.25) at each iteration, to minimize both cost functions over the data.

The left, middle, and right panels of the figure display these results on breast cancer (consisting of $P = 569$), spam email (with $P = 4601$ points), and face detection datasets (where $P = 10\,000$) respectively. While their performance differs from case to case the softmax and margin costs perform similarly well in these examples. For more information about the datasets used here see Exercise 4.9, as well as Examples 4.9 and 4.10.

4.1.8 The connection between the perceptron and counting costs

Note that with the cost function defined in (4.25) as our true desired criterion for linear classification, we could have begun our discussion by trying to minimize it formally as

$$\underset{b,\,\mathbf{w}}{\text{minimize}} \sum_{p=1}^{P} \max\left(0,\ \text{sign}\left(-y_p\left(b+\mathbf{x}_p^T\mathbf{w}\right)\right)\right). \tag{4.28}$$

Unfortunately this problem is not only non-convex but is highly *discontinuous* due to the presence of the "sign" function in each summand of the objective. Therefore it is extremely difficult to attempt to minimize it directly. However, note that the original perceptron cost derived in (4.5) can be thought of simply as a relaxation of this fundamental counting cost, where we remove the discontinuous "sign" function from each summand (or in other words, approximate $\text{sign}\left(-y_p\left(b+\mathbf{x}_p^T\mathbf{w}\right)\right)$ linearly as $-y_p\left(b+\mathbf{x}_p^T\mathbf{w}\right)$). Thus while the original perceptron cost, as well as its relatives including the softmax[10] and margin costs, are intimately related to this counting cost they are still *approximations* of the true criterion we wish to minimize.

In Fig. 4.9 we illustrate this point by showing both the number of misclassifications and objective value of gradient descent applied to minimizing the softmax cost over the toy datasets shown first in Fig. 4.3. Specifically, we show results from three runs of gradient descent applied to both the linearly separable (top panels) and overlapping (bottom panels) datasets. In the left panels of Fig. 4.9 we show the number of misclassifications per iteration calculated by evaluating the counting cost in (4.25), while in the right panels we show the corresponding softmax cost values from each run per iteration. In other words, the left and right panels show the value of the counting cost function from (4.25) and the softmax cost from (4.9) per iteration of gradient descent, respectively.

Comparing the left and right panels for each dataset note that, in both instances, the per iteration counting and softmax values do not perfectly match. Further note how with the second dataset, shown in the lower two panels, the counting cost value actually fluctuates (by a small amount) as we increase the number of iterations while the corresponding softmax cost value continues to fall. Both of these phenomena are caused by the fact that we are directly minimizing an approximation of the counting cost, and not the counting cost itself. While neither effect is ideal, they are examples of the tradeoff we must accept for working with cost functions we can actually minimize properly in practice.

4.2 The logistic regression perspective on the softmax cost

This section describes a common way of both deriving and thinking about the softmax cost function first introduced in Section 4.1.2. Here we will see how the softmax cost naturally arises as a direct approximation of the fundamental counting cost discussed in Section 4.1.5. However the major benefit of this new perspective is in adding a useful geometric viewpoint,[11] that of regression/surface-fitting, to the classification framework in general, and the softmax cost in particular.

[10] We will also see in Section 4.2 how the softmax cost can be thought of as a direct approximation of the counting cost.

[11] Logistic regression can also be interpreted from a *probabilistic* perspective (see Exercise 4.12).

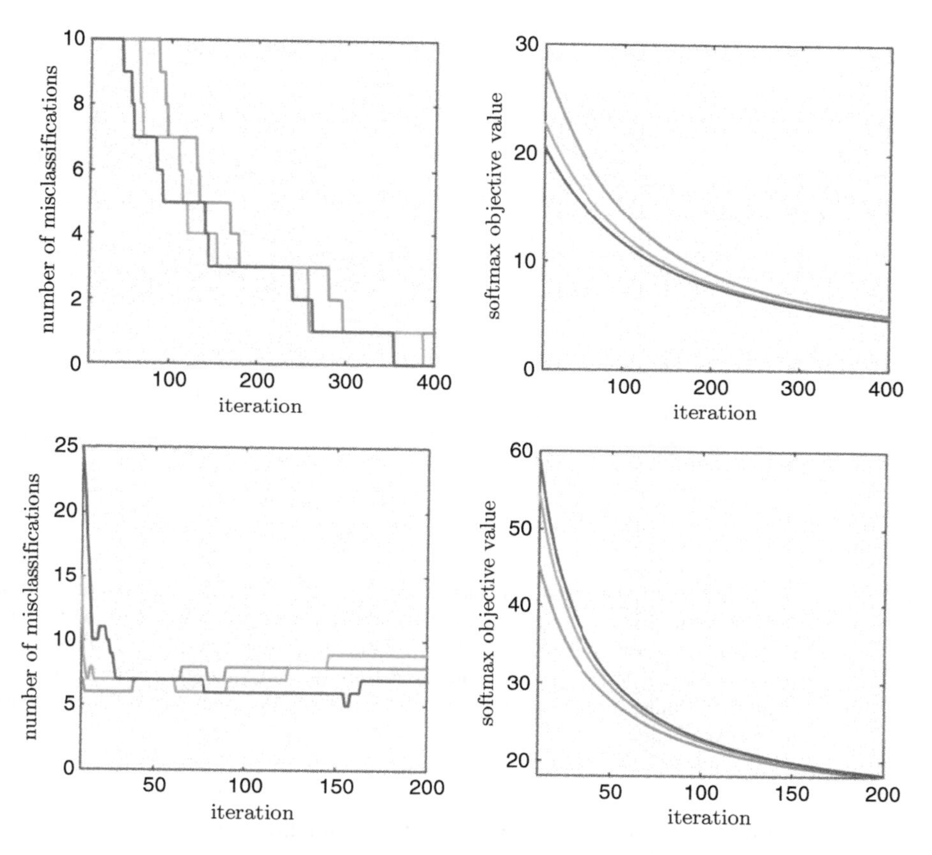

Fig. 4.9 The number of misclassifications (left panels) and objective value (right panels) plotted versus number of iterations of three runs of gradient descent applied to minimizing the softmax cost over two toy datasets, one linearly separable (top panels) and the other overlapping (bottom panels), both shown originally in Fig. 4.3.

4.2.1 Step functions and classification

Two class classification can be fruitfully considered as a particular instance of regression or surface-fitting, wherein the output of a dataset of P points $\left\{\left(\mathbf{x}_p, y_p\right)\right\}_{p=1}^{P}$ is no longer continuous but takes on two fixed values, $y_p \in \{-1, +1\}$, corresponding to the two classes. As illustrated in Fig. 4.10, an ideal *data generating function* for classification (i.e., a function that can be assumed to generate the data we receive) is a discontinuous step function (shown in yellow). When the step function is viewed "from above", as also illustrated in this figure, we return to viewing classification from the "separator" point of view described in the previous section, and the linear boundary separating the two classes is defined exactly by the hyperplane where the step function transitions from its lower to higher step, defined by

$$b + \mathbf{x}^T \mathbf{w} = 0. \tag{4.29}$$

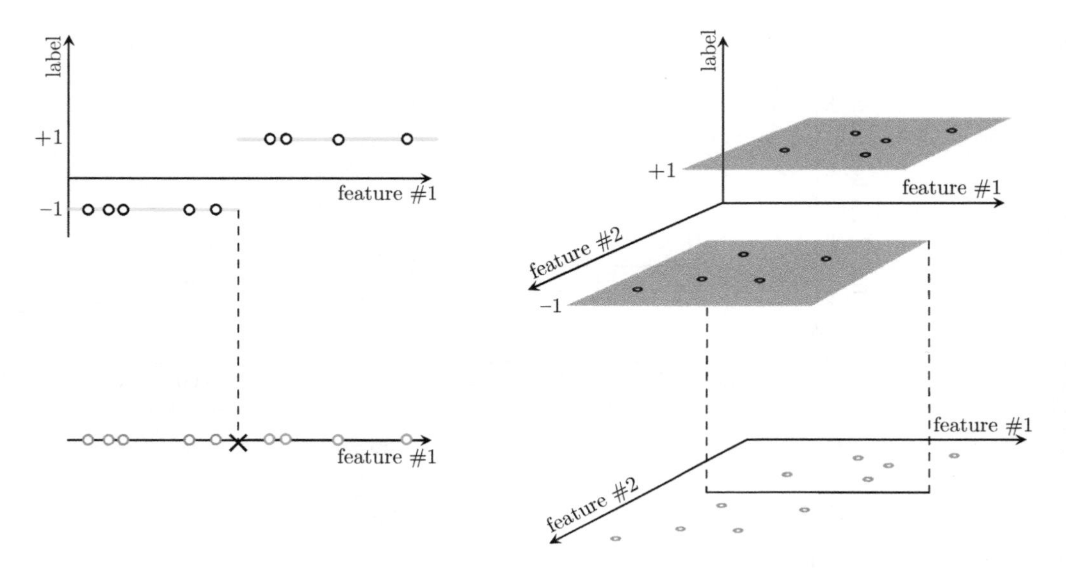

Fig. 4.10 Classification from a regression/surface-fitting perspective for 1-dimensional (left panel) and 2-dimensional (right panel) toy datasets. This surface-fitting view is equivalent to the "separator" perspective described in Section 4.1, where the separating hyperplane is precisely where the step function (shown here in yellow) transitions from its lower to higher step. In the separator view the actual y value (or label) is represented by coloring the points red or blue to denote their respective classes.

With this, the equation for any step function taking values on $\{-1, +1\}$ can be written explicitly as

$$\operatorname{sign}\left(b + \mathbf{x}^T\mathbf{w}\right) = \begin{cases} +1 & \text{if } b + \mathbf{x}^T\mathbf{w} > 0 \\ -1 & \text{if } b + \mathbf{x}^T\mathbf{w} < 0. \end{cases} \tag{4.30}$$

We ideally would like to find a set of parameters $(b, \mathbf{w})$ for a hyperplane so that data points having label $y_p = +1$ lie on the top step, and those having label $y_p = -1$ lie on the bottom step. To say then that a particular parameter choice places a point $\mathbf{x}_p$ on its correct step means that $\operatorname{sign}\left(b + \mathbf{x}_p^T\mathbf{w}\right) = y_p$, and because $y_p \in \{-1, +1\}$ this can be written equivalently as

$$\operatorname{sign}\left(y_p\left(b + \mathbf{x}_p^T\mathbf{w}\right)\right) = 1. \tag{4.31}$$

In what follows we will make a smooth approximation to the step function, in particular deriving a smoothed equivalent of the criterion in (4.31) for the parameters of a desired hyperplane. This will quickly lead us to the minimization of the softmax cost function first described in Section 4.1.2 in order to properly fit a smoothed step function to our labeled data.

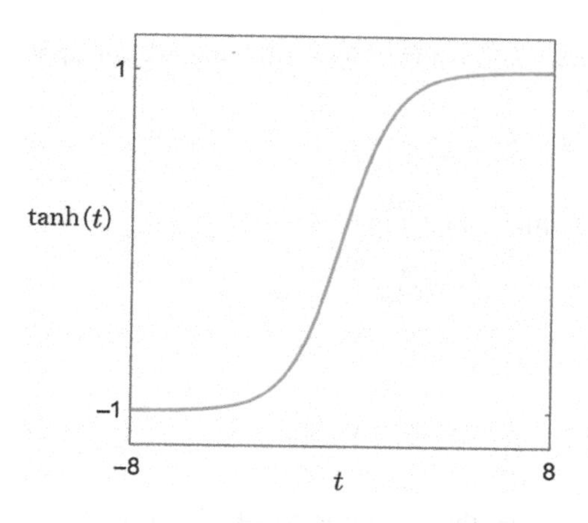

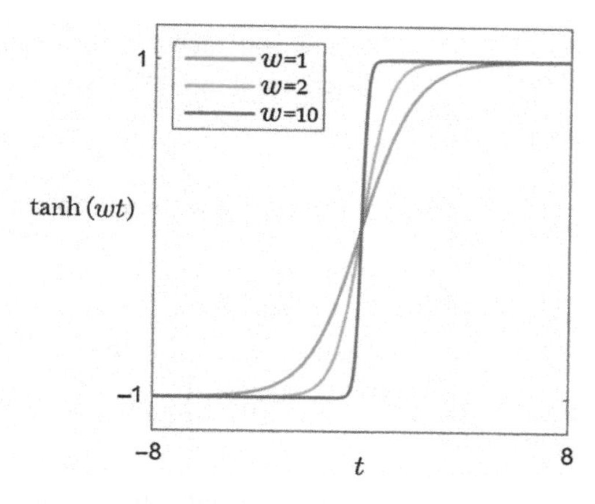

Fig. 4.11 (left panel) Plot of the tanh function defined as $\tanh(t) = 2\sigma(t) - 1$. (right panel) By increasing the weight w of the function $\tanh(wt)$ from $w = 1$ (red) to $w = 2$ (green) and finally to $w = 10$ (blue), it becomes an increasingly good approximator of a step function taking on values -1 and $+1$.

4.2.2 Convex logistic regression

We have actually already seen an excellent smooth approximator of a step function, i.e., the sigmoid function

$$\sigma\left(b + \mathbf{x}^T\mathbf{w}\right) = \frac{1}{1 + e^{-\left(b+\mathbf{x}^T\mathbf{w}\right)}}, \tag{4.32}$$

introduced in Section 3.3.1 in its original context as a model for population growth. As shown in Fig. 3.10, by adjusting the parameters $(b, \mathbf{w})$ the sigmoid can be made to approximate a step function taking on the values $\{0, 1\}$. By simply multiplying the sigmoid by 2 and then subtracting 1 we can stretch it so that it approximates a step function taking on the values $\{-1, +1\}$. This stretched sigmoid is referred to as the "tanh" function

$$\tanh\left(b + \mathbf{x}^T\mathbf{w}\right) = 2\sigma\left(b + \mathbf{x}^T\mathbf{w}\right) - 1. \tag{4.33}$$

As shown in the left panel of Fig. 4.11, the tanh function retains the desired property of the sigmoid by being a fine approximator to the step function, this time one that takes on values $\{-1, +1\}$.

Thus we have for any pair $(b, \mathbf{w})$ that any desired step function of the form given in (4.30) may be roughly approximated as $\operatorname{sign}\left(b + \mathbf{x}^T\mathbf{w}\right) \approx \tanh\left(b + \mathbf{x}^T\mathbf{w}\right)$, or in other words

$$\tanh\left(b + \mathbf{x}^T\mathbf{w}\right) \approx \begin{cases} +1 & \text{if } b + \mathbf{x}^T\mathbf{w} > 0 \\ -1 & \text{if } b + \mathbf{x}^T\mathbf{w} \leq 0. \end{cases} \tag{4.34}$$

To make this approximation finer we can, as illustrated in the right panel of Fig. 4.11, multiply the input argument of the tanh by a large positive constant.

Now with tanh as a smooth approximation of the "sign" function, we can approximate the criterion in (4.31) as

$$\tanh\left(y_p\left(b+\mathbf{x}_p^T\mathbf{w}\right)\right)\approx 1, \tag{4.35}$$

which can be written, using the definition of tanh in (4.33), as

$$1+e^{-y_p\left(b+\mathbf{x}_p^T\mathbf{w}\right)}\approx 1. \tag{4.36}$$

Taking the log of both sides[12] then leads to

$$\log\left(1+e^{-y_p\left(b+\mathbf{x}_p^T\mathbf{w}\right)}\right)\approx 0. \tag{4.37}$$

Since we want a hyperplane that forces the condition in (4.37) to hold for all $p = 1,\ldots,P$, a reasonable way of learning associated parameters is to simply minimize the sum of these expressions over the entire dataset as

$$\underset{b,\,\mathbf{w}}{\text{minimize}}\ \sum_{p=1}^{P}\log\left(1+e^{-y_p\left(b+\mathbf{x}_p^T\mathbf{w}\right)}\right). \tag{4.38}$$

This is precisely the minimization of the softmax cost first introduced in Section 4.1.2 as a smooth approximation to the original perceptron. Here, however, our interpretation has changed: we think of the minimization of the softmax cost in the current section in the context of logistic regression surface-fitting (where the output takes on only the values ± 1), determining ideal parameters for a smoothed step function to fit to our labeled data.

Through the perspective of logistic regression, we can think of classification simultaneously as:

(1) finding a hyperplane that best separates the data; and

(2) finding a step-like surface that best places the positive and negative classes on its top and bottom steps, respectively.

In Fig. 4.12 we show an example of both the resulting linear separator and surface fit corresponding to minimizing the softmax cost via Newton's method as described in Example 4.1 on a toy dataset first shown in the left panel of Fig. 4.3. The resulting

[12] Without taking the log on both sides one can deduce instead a desired approximation $e^{-y_p\left(b+\mathbf{x}_p^T\mathbf{w}\right)}\approx 0$ to hold, leading to the analogous conclusion that we should minimize the cost function $\sum_{p=1}^{P} e^{-y_p\left(b+\mathbf{x}_p^T\mathbf{w}\right)}$. This approximation, however, is less useful for classification as it is much more sensitive to the presence of *outliers* in the data (see Exercise 4.11). Regardless, it is used for instance as the objective function in a greedy classification method called *boosting* [34].

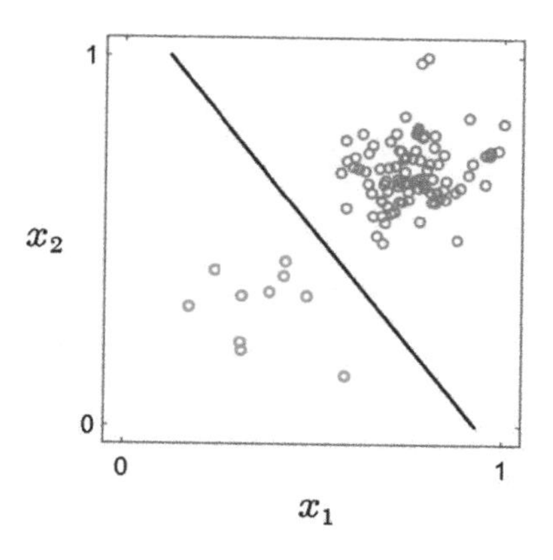

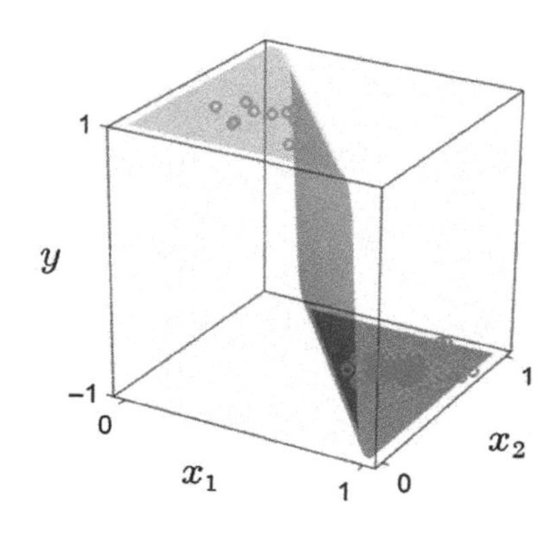

Fig. 4.12 Minimizing the softmax cost in (4.38) gives an optimal weight pair $(b^\star, \mathbf{w}^\star)$ that define the linear separator $b^\star + \mathbf{x}^T\mathbf{w}^\star = 0$ shown in the left panel (in black), as well as the surface $y(\mathbf{x}) = \tanh\left(b^\star + \mathbf{x}^T\mathbf{w}^\star\right)$ shown in the right panel (in gray).

parameters found $(b^\star, \mathbf{w}^\star)$ define both the linear separator $b^\star + \mathbf{x}^T\mathbf{w}^\star = 0$, as well as the surface $y(\mathbf{x}) = \tanh\left(b^\star + \mathbf{x}^T\mathbf{w}^\star\right)$.

4.3 The support vector machine perspective on the margin perceptron

In deriving the margin perceptron in Section 4.1.3 we introduced the concept of a margin for a hyperplane as the width of the buffer zone it creates between two linearly separable classes. We now extend this idea to its natural conclusion, leading to the so-called support vector machine (SVM) classifier. While an intriguing notion in the ideal case where data is perfectly separable, we will see by the end of this section that practically speaking the SVM classifier is a margin perceptron with the addition of an ℓ_2 regularizer (ℓ_2 regularization was first introduced in Section 3.3.2).

4.3.1 A quest for the hyperplane with maximum margin

As discussed in Section 4.1.3, when two classes of data are linearly separable, infinitely many hyperplanes could be drawn to separate the data. In Fig. 4.5 we displayed three such hyperplanes for a given synthetic dataset, each derived by starting the gradient descent procedure for minimizing the squared margin perceptron cost with a different initialization. Given that all these three classifiers (as well as any other separating hyperplane derived from this procedure) would perfectly classify the data, is there one that we can say is the "best" of all possible separating hyperplanes? One reasonable standard for judging the quality of these hyperplanes is via their margin lengths, that is the distance between the evenly spaced translates that just touch each class. The larger this distance is, the intuitively better the associated hyperplane separates the entire space, given the particular distribution of the data. This idea is illustrated in Fig. 4.13.

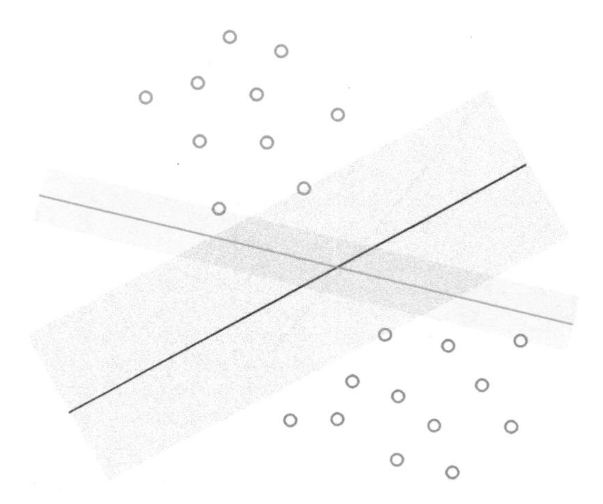

Fig. 4.13 Of the infinitely many hyperplanes that exist between two classes of linearly separable data the one with maximum margin does an intuitively better job than the rest at distinguishing between classes because it more equitably partitions the entire space based on how the data is distributed. In this illustration two separators are shown along with their respective margins. While both perfectly distinguish between the two classes the green separator (with smaller margin) divides up the space in a rather awkward fashion given how the data is distributed, and will therefore tend to more easily misclassify future data points. On the other hand, the black separator (having a larger margin) divides up the space more evenly with respect to the given data, and will tend to classify future points more accurately.

To find the separating hyperplane with maximum margin, first recall from Section 4.1.3 that the margin of a hyperplane $b + \mathbf{x}^T\mathbf{w} = 0$ is the width of the buffer zone confined between two symmetric translations of itself, written conveniently as $b + \mathbf{x}^T\mathbf{w} = \pm 1$, each just touching one of the two classes. As shown in Fig. 4.14, the margin can be determined by calculating the distance between any two points (one from each translated hyperplane) both lying on the normal vector $\mathbf{w}$. Denoting by $\mathbf{x}_1$ and $\mathbf{x}_2$ the points on vector $\mathbf{w}$ belonging to the *upper* and *lower* translated hyperplanes, respectively, the margin is computed simply as the length of the line segment connecting $\mathbf{x}_1$ and $\mathbf{x}_2$, i.e., $\|\mathbf{x}_1 - \mathbf{x}_2\|_2$.

The margin can be written much more conveniently by taking the difference of the two translates evaluated at $\mathbf{x}_1$ and $\mathbf{x}_2$ respectively, as

$$\left(b + \mathbf{x}_1^T\mathbf{w}\right) - \left(b + \mathbf{x}_2^T\mathbf{w}\right) = \left(\mathbf{x}_1 - \mathbf{x}_2\right)^T \mathbf{w} = 2. \tag{4.39}$$

Using the inner product rule (see Appendix A) and the fact that the two vectors $\mathbf{x}_1 - \mathbf{x}_2$ and $\mathbf{w}$ are parallel to each other, we can solve for the margin directly in terms of $\mathbf{w}$, as

$$\|\mathbf{x}_1 - \mathbf{x}_2\|_2 = \frac{2}{\|\mathbf{w}\|_2}. \tag{4.40}$$

Therefore finding the separating hyperplane with maximum margin is equivalent to finding the one with the smallest possible normal vector $\mathbf{w}$.

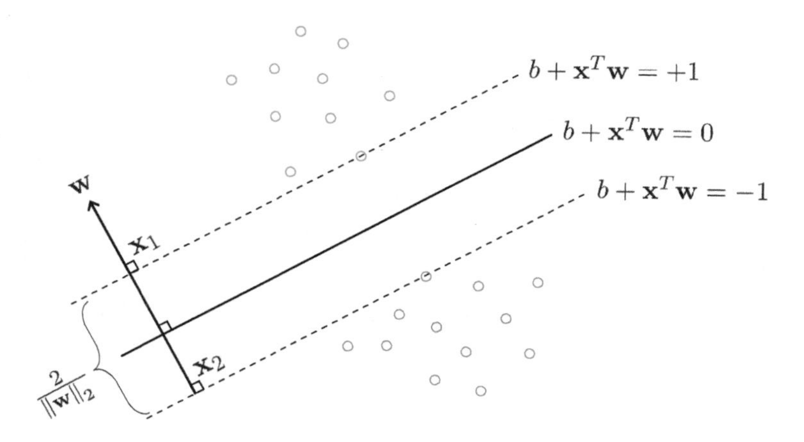

Fig. 4.14 The margin of a separating hyperplane can be calculated by measuring the distance between the two points of intersection of the normal vector $\mathbf{w}$ and the two equidistant translations of the hyperplane. This distance can be shown to have the value of $\frac{2}{\|\mathbf{w}\|_2}$ (see text for further details).

4.3.2 The hard-margin SVM problem

In order to find a separating hyperplane for the data with minimum length normal vector we can simply combine this with our desire to minimize $\|\mathbf{w}\|_2^2$ subject to the constraint that the hyperplane perfectly separates the data (given by the margin criterion in (4.14)). This gives the so-called *hard-margin SVM* constrained optimization problem

$$\begin{aligned} &\underset{b,\,\mathbf{w}}{\text{minimize}} \quad \|\mathbf{w}\|_2^2 \\ &\text{subject to} \quad \max\left(0,\, 1 - y_p\left(b + \mathbf{x}_p^T\mathbf{w}\right)\right) = 0, \quad p = 1, \ldots, P. \end{aligned} \tag{4.41}$$

Unlike the minimization problems we have seen so far, here we have a set of constraints on the permissible values of $(b, \mathbf{w})$ that guarantee that the hyperplane we recover separates the data perfectly. Problems of this sort can be solved using a variety of optimization techniques (see e.g., [23, 24, 50]) that we do not discuss here.

Figure 4.15 shows the SVM hyperplane learned for a toy dataset along with the buffer zone confined between the separating hyperplane's translates. The points from each class lying on either boundary of the buffer zone are called *support vectors*, hence the name "support vector machines," and are highlighted in green.

4.3.3 The soft-margin SVM problem

Because a priori we can never be entirely sure in practice that our data is perfectly linearly separable, the hard-margin SVM problem in (4.41) is of mostly theoretical interest. This is because if the data is not perfectly separable by a hyperplane, the hard-margin problem in (4.41) is "ill-defined," meaning that it has no solution (as the constraints can never be satisfied). As a result the hard-margin SVM problem, which again was designed on the assumption of perfect linear separability between the two classes, is not

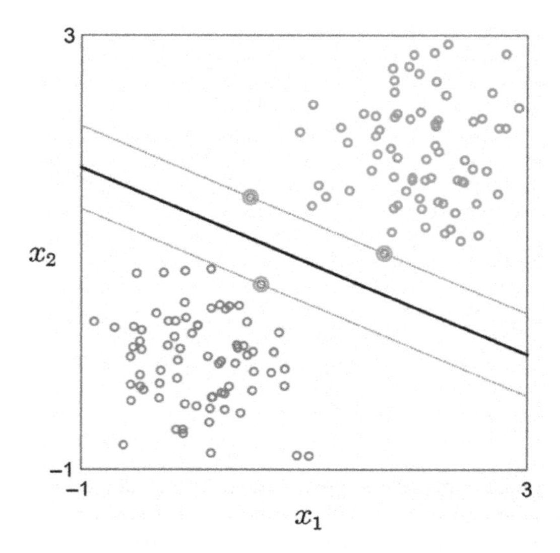

Fig. 4.15 A linearly separable toy dataset consisting of $P = 150$ points in total (75 per class) with the SVM classifier (in black) learned by solving the hard-margin SVM problem. Also shown are the buffer zone boundaries (dotted) and support vectors (highlighted in green).

commonly used in practice. Instead, its constraints are typically "relaxed" in order to allow for possible violations of linear separability. To relax the constraints[13] we make them part of a single cost function, which includes the original objective $\|\mathbf{w}\|_2^2$ as well, so that they are not all forced to hold exactly. This gives the *soft-margin SVM* cost function

$$g\left(b, \mathbf{w}\right) = \sum_{p=1}^{P} \max\left(0,\ 1 - y_p\left(b + \mathbf{x}_p^T\mathbf{w}\right)\right) + \lambda \left\|\mathbf{w}\right\|_2^2, \tag{4.42}$$

where the parameter $\lambda \geq 0$ controls the trade-off between how well we satisfy the original constraints in (4.41) while seeking a large margin classifier. The smaller we set λ the more pressure we put on satisfying the constraints of the original problem, and the less emphasis we put on the recovered hyperplane having a large margin (and vice versa). While λ is often set to a small value in practice, we discuss methods for automatically choosing the value of λ in Chapter 7. Formally, minimization of the soft-margin SVM cost function is written as

$$\underset{b,\,\mathbf{w}}{\text{minimize}} \sum_{p=1}^{P} \max\left(0,\ 1 - y_p\left(b + \mathbf{x}_p^T\mathbf{w}\right)\right) + \lambda \left\|\mathbf{w}\right\|_2^2. \tag{4.43}$$

[13] Generally speaking any relaxed version of the SVM, which allows for violations of perfect linear separability of the data, is referred to as a *soft-margin SVM* problem. While there is another popular relaxation of the basic SVM problem used in practice (see e.g., [22, 23]) it has no theoretical or practical advantage over the one presented here [21, 27].

Looking closely at the soft-margin cost we can see that, practically speaking, it is just the margin perceptron cost given in (4.16) with the addition of an ℓ_2 regularizer (as described in Section 3.3.2).

> Practically speaking, the soft-margin SVM cost is just an ℓ_2 regularized form of the margin perceptron cost.

As with the original margin perceptron cost described in Section 4.1, differentiable approximations of the same sort we have seen before (e.g., squaring the "max" function or using the softmax approximation) are typically used in place of the margin perceptron component of the soft-margin SVM cost function. For example, using the softmax approximation (see Section 4.1.2) the soft-margin SVM cost may be written as

$$g\left(b,\mathbf{w}\right)=\sum_{p=1}^{P}\log\left(1+e^{-y_p\left(b+\mathbf{x}_p^T\mathbf{w}\right)}\right)+\lambda\left\|\mathbf{w}\right\|_2^2. \tag{4.44}$$

With this approximation the soft-margin SVM cost is sometimes referred to as *log-loss SVM* (see e.g., [21]). However, note that, using the softmax approximation, we can also think of log-loss SVM as an ℓ_2 regularized form of logistic regression. ℓ_2 regularization, first described in Section 3.3 in the context of nonlinear regression, can be analogously applied to classification cost functions as well.

4.3.4 Support vector machines and logistic regression

While the motives for formally deriving the SVM and logistic regression classifiers differ significantly, due to the fact that their cost functions are so similar (or the same if the softmax cost is employed for SVM as in (4.44)) both perform similarly well in practice (as first discussed in Section 4.1.7). Unsurprisingly, as we will see later in Chapters 5 through 7, both classifiers can be extended (using so-called "kernels" and "feed-foward neural networks") in precisely the same manner to perform nonlinear classification.

> While the motives for formally deriving the SVM and logistic regression classifiers differ, due to their similar cost functions (which in fact can be entirely similar if the softmax cost is employed for SVM) both perform similarly well in practice.

4.4 Multiclass classification

In practice many classification problems have more than two classes we wish to distinguish, e.g., face recognition, hand gesture recognition, recognition of spoken phrases or

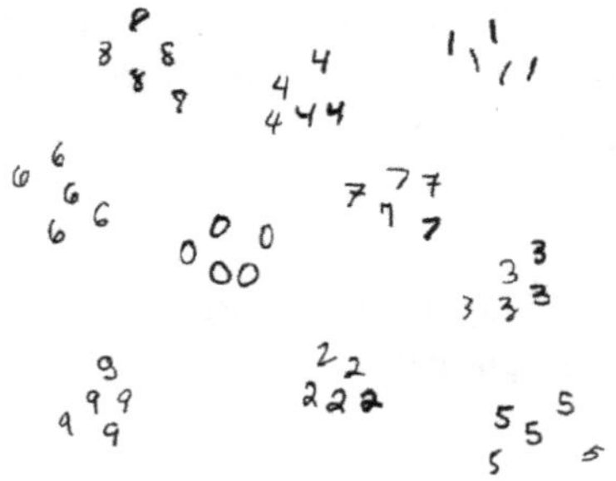

Fig. 4.16 Various handwritten digits in a feature space. Handwritten digit recognition is a common multiclass classification problem. The goal here is to determine regions in the feature space where current (and future) instances of each type of handwritten digit are present.

words, etc. Such a multiclass dataset $\left\{\left(\mathbf{x}_p, y_p\right)\right\}_{p=1}^{P}$ consists of C distinct classes of data, where each label y_p now takes on a value between 1 and C, i.e., $y_p \in \{1, 2, \ldots, C\}$. In this section we discuss two popular generalizations of the two class framework, namely, *one-versus-all* and *multiclass softmax classification* (sometimes referred to as *softmax regression*). Each scheme learns C two class linear separators to deal with the multiclass setting, differing only in how these linear separators are learned. Both methods are commonly used and perform similarly in practice, as we discuss further in Section 4.4.4.

Example 4.4 Handwritten digit recognition

Recognizing handwritten digits is a popular multiclass classification problem commonly built into the software of mobile banking applications, as well as more traditional automated teller machines, to give users e.g., the ability to automatically deposit paper checks. Here each class of data consists of (images of) several handwritten versions of a single digit in the range $0-9$, giving a total of ten classes. Using the methods discussed in this section, as well as their nonlinear extensions described in Section 6.3, we aim to learn a separator that distinguishes each of the ten classes from each other (as illustrated in Fig. 4.16). You can perform this task on a large dataset of handwritten digits by completing Exercise 4.16.

4.4.1 One-versus-all multiclass classification

Because it has only two sides, a single linear separator is fundamentally insufficient as a mechanism for differentiating between more than two classes of data. To overcome this shortcoming when dealing with $C > 2$ classes we can instead learn C linear classifiers (one per class), each distinguishing one class from the rest of the data. We illustrate this idea for a particular toy dataset with $C = 3$ classes in Fig. 4.17. By properly fusing these C learned linear separators, we can then form a classification rule for the entire dataset. This approach is called one-versus-all (OvA) classification.

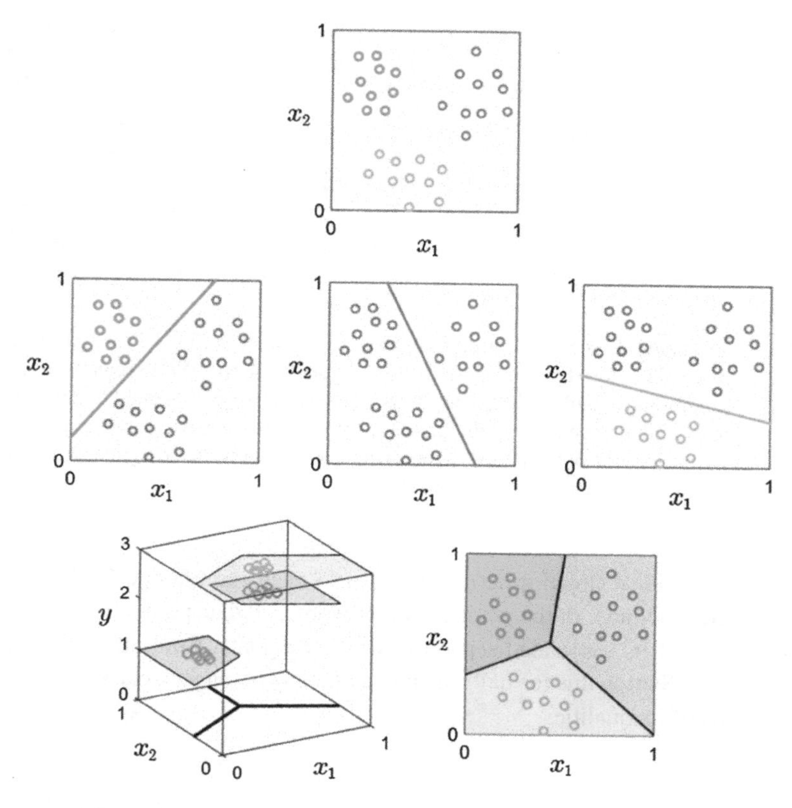

Fig. 4.17 One-versus-all multiclass scheme applied to (top panel) a toy classification dataset with $C = 3$ classes consisting of $P = 30$ data points in total (10 per class). (middle panels) The three classifiers learned to distinguish each class from the rest of the data. In each panel we have temporarily colored all data points not in the primary class gray for visualization purposes. (bottom panels) By properly fusing these $C = 3$ individual classifiers we determine a classification rule for the entire space, allowing us to predict the label value of every point. These predictions are illustrated as the colored regions shown from "the side" and "from above" in the left and right panels respectively.

Beginning, we first learn C individual linear separators in the manner described in previous sections (using any desired cost function and minimization technique). In learning the cth classifier we treat all points not in class c as a single "not-c" class by lumping them all together. To learn a two class classifier we then assign temporarily labels to the P training points: points in classes c and "not-c" are assigned temporary labels $+1$ and -1, respectively. With these temporary labels we can then learn a linear classifier distinguishing the points in class c from all other classes. This is illustrated in the middle panels of Fig. 4.17 for a $C = 3$ class dataset.

Having done this for all C classes we then have C linear separators of the form

$$b_c + \mathbf{x}^T \mathbf{w}_c = 0, \quad c = 1, \ldots, C. \tag{4.45}$$

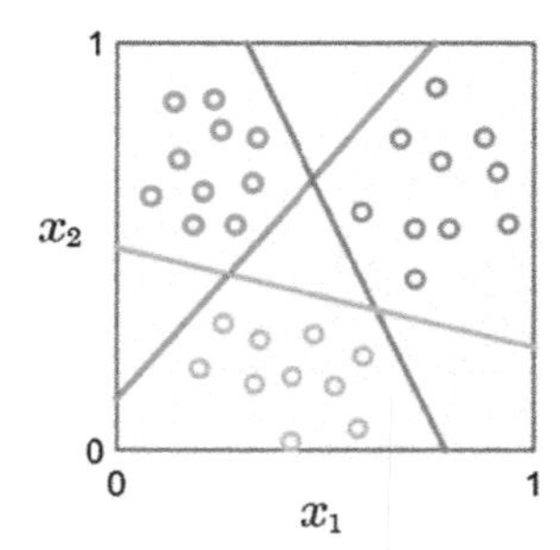

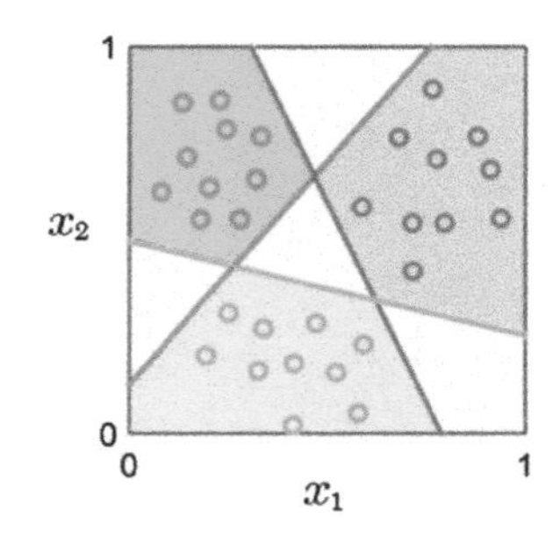

Fig. 4.18 (left panel) Linear separators from the middle panel of Fig. 4.17. (right panel) Regions of the space are colored according to the set of rules in (4.46). White regions do not satisfy these conditions, meaning that points in these areas cannot be assigned to any class/color. In the case shown here those points lying in the three white regions between any two classes are positive with respect to both classes' hyperplanes, while the white triangular region in the middle is negative with respect to all three classifiers.

In the ideal situation shown in Fig. 4.17 each classifier perfectly separates its class from the remainder of the points. In other words, all data points from class c lie on the *positive side* of its associated separator, while the points from other classes lie on its *negative side*. Stating this formally, a known point $\mathbf{x}_p$ belongs to class c if it satisfies the following set of inequalities

$$\begin{aligned} &b_c + \mathbf{x}_p^T \mathbf{w}_c > 0 \\ &b_j + \mathbf{x}_p^T \mathbf{w}_j < 0 \quad j = 1, \ldots, C, \; j \neq c. \end{aligned} \tag{4.46}$$

While this correctly describes the labels of the current set of points in an ideal scenario, using this criterion more generally to assign labels to other points in the space would be a very poor idea, as illustrated in Fig. 4.18 where we show the result of using the set of rules in (4.46) to assign labels to all points $\mathbf{x}$ in the feature space of our toy dataset from Fig. 4.17. As can be seen in the figure there are entire regions of the space for which the inequalities in (4.46) do not simultaneously hold, meaning that points in these regions cannot be assigned a class at all. These regions, left uncolored in the figure, include those areas lying on the positive side of more than one classifier (the three white regions lying between each pair of classes), and those lying on the negative side of all the classifiers (the triangular region in the middle of all three).

However, by generalizing the criteria in (4.46) we can in fact produce a useful rule that assigns labels to every point in the entire space. For a point $\mathbf{x}$ the rule is generalized by determining not the classifier that provides a positive evaluation $b_c + \mathbf{x}^T \mathbf{w}_c > 0$ (if there even is one such classifier), but by assigning $\mathbf{x}$ the label according to whichever classifier produces the largest evaluation (even if this evaluation is negative). In other words, we generalize (4.46) by assigning the label y to a point $\mathbf{x}$ by taking

$$y = \underset{j=1,\ldots,C}{\operatorname{argmax}} \; b_j + \mathbf{x}^T \mathbf{w}_j. \tag{4.47}$$

This criterion, which we refer to as the *fusion rule*,[14] was used to assign labels[15] to the entire space of the toy dataset shown in the bottom panel of Fig. 4.17. Although devised in the context of an ideal scenario where the classes are not overlapping, the fusion rule is effective in dealing with overlapping multiclass datasets as well (see Example 4.5). As we will see in Section 4.4.2, the fusion rule is also the basis for the second multiclass method described here, multiclass softmax classification.

To perform one-versus-all classification on a dataset with C classes:

(1) Learn C individual classifiers using any approach (e.g., logistic regression, support vector machines, etc.), each distinguishing one class from the remainder of the data.

(2) Combine the learned classifiers using the fusion rule in (4.47) to make final assignments.

Example 4.5 OvA classification for overlapping data

In Fig. 4.19 we show the results of applying the OvA framework to a toy dataset with $C = 4$ overlapping classes. In this example we use the logistic regression classifier (i.e., softmax cost) and Newton's method for minimization, as described in Section 4.1.2. After learning each of the four individual classifiers (shown in the middle panels) they are fused using the rule in (4.47) to form the final partitioning of the space as shown in the bottom panels of this figure.

4.4.2 Multiclass softmax classification

As we have just seen, in the OvA framework we learn C linear classifiers separately and fuse them afterwards to create a final assignment rule for the entire space. A popular

[14] One might smartly suggest that we should first normalize the learned hyperplanes by the length of their respective normal vectors as $\frac{b_j+\mathbf{x}^T\mathbf{w}_j}{\|\mathbf{w}_j\|_2}$ prior to fusing them as in (4.47) in order to put all the classifiers "on equal footing." Or, in other words, so that no classifier is given an unwanted advantage or disadvantage in fusing due to the size of its learned weight pair $(b_j, \mathbf{w}_j)$, as this size is arbitrary (since the hyperplane $b + \mathbf{x}^T\mathbf{w} = 0$ remains unchanged when multiplied by a positive scalar γ as $\gamma \cdot (b + \mathbf{x}^T\mathbf{w}) = \gamma \cdot 0 = 0$. While this is rarely done in practice it is certainly justified, and one should feel free to normalize each hyperplane in practice prior to employing the fusion rule if desired.

[15] Note that while the boundary resulting from the fusion rule is always piecewise-linear, as in the toy examples shown here, the fusion rule itself does *not* explicitly define this boundary, i.e., it does not provide us with a nice formula for it (although one may work out a somewhat convoluted formula describing the boundary in general). This is perfectly fine since remember that our goal is not to find a formula for some separating boundary, but rather a reliable rule for accurately predicting labels (which the fusion rule provides). In fact the piecewise-linear boundaries shown in the figures of this section were drawn *implicitly* by labeling (and appropriately coloring) every point in the region shown using the fusion rule.

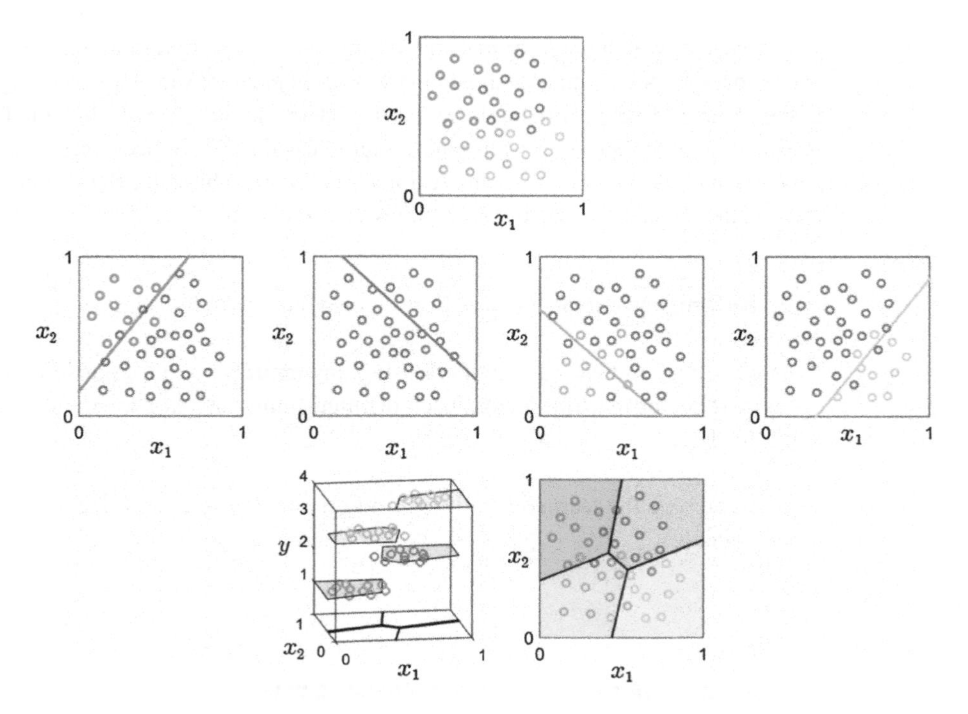

Fig. 4.19 One-versus-all multiclass scheme applied to (top panel) a toy classification dataset with $C = 4$ classes consisting of $P = 40$ data points in total (10 per class). (middle panels) The four classifiers learned to distinguish each class from the rest of the data. (bottom panels) Having determined proper linear separators for each class, we use the fusion rule in (4.47) to form the final partitioning of the space. The left and right panels illustrate the predicted labels (shown as colored regions) from both "the side" and "from above." These regions implicitly define the piecewise linear boundary shown in the right panel.

alternative, referred to as *multiclass softmax classification*, determines the C classifiers jointly by learning all of their parameters together using a cost function based on the fusion rule in (4.47). According to the fusion rule if we want a point $\mathbf{x}_p$ belonging to class c (i.e., $y_p = c$) to be classified correctly we must have that

$$c = \underset{j=1,\ldots,C}{\operatorname{argmax}} \left(b_j + \mathbf{x}_p^T \mathbf{w}_j\right). \tag{4.48}$$

This means that we must have that

$$b_c + \mathbf{x}_p^T \mathbf{w}_c = \max_{j=1,\ldots,C} \left(b_j + \mathbf{x}_p^T \mathbf{w}_j\right), \tag{4.49}$$

or equivalently

$$\max_{j=1,\ldots,C} \left(b_j + \mathbf{x}_p^T \mathbf{w}_j\right) - \left(b_c + \mathbf{x}_p^T \mathbf{w}_c\right) = 0. \tag{4.50}$$

Indeed we would like to tune the weights so that (4.50) holds for all points in the dataset (with their respective class label). Because the quantity on the left hand side of (4.50) is always nonnegative, and is exactly zero if the point $\mathbf{x}_p$ is classified correctly, it makes sense to form a cost function using this criterion that we then minimize in order to determine proper weights. Summing the expression in (4.50) over all P points in the dataset, denoting Ω_c the index set of points belonging to class c, we have a nonnegative cost function

$$g\left(b_1,\ldots,b_C,\mathbf{w}_1,\ldots,\mathbf{w}_C\right)=\sum_{c=1}^{C}\sum_{p\in\Omega_c}\left[\max_{j=1,\ldots,C}\left(b_j+\mathbf{x}_p^T\mathbf{w}_j\right)-\left(b_c+\mathbf{x}_p^T\mathbf{w}_c\right)\right]. \tag{4.51}$$

Note that there are only P summands in this sum, one for each point in the dataset. However, the problem here, which we also encountered when deriving the original perceptron cost function for two class classification in Section 4.1, is that the max function is continuous but not differentiable and that the trivial solution ($b_j = 0$ and $\mathbf{w}_j = \mathbf{0}_{N\times 1}$ for all j) successfully minimizes the cost. One useful work-around approach we saw there for dealing with this issue, which we will employ here as well, is to approximate $\max_{j=1,\ldots,C}\left(b_j+\mathbf{x}_p^T\mathbf{w}_j\right)$ using the smooth *softmax* function.

Recall from Section 4.1.2 that the softmax function of C scalar inputs $s_1,\ldots,s_C$, written as soft $(s_1,\ldots,s_C)$, is defined as

$$\text{soft}\left(s_1,\ldots,s_C\right)=\log\left(\sum_{j=1}^{C}e^{s_j}\right), \tag{4.52}$$

and provides a good approximation to max $(s_1,\ldots,s_C)$ for a wide range of input values. Substituting the softmax function in (4.51) we have a smooth approximation to the original cost, given as

$$g\left(b_1,\ldots,b_C,\mathbf{w}_1,\ldots,\mathbf{w}_C\right)=\sum_{c=1}^{C}\sum_{p\in\Omega_c}\left[\log\left(\sum_{j=1}^{C}e^{b_j+\mathbf{x}_p^T\mathbf{w}_j}\right)-\left(b_c+\mathbf{x}_p^T\mathbf{w}_c\right)\right]. \tag{4.53}$$

Using the facts that $s=\log\left(e^s\right)$ and that $\log\left(\frac{s}{t}\right)=\log\left(s\right)-\log\left(t\right)$ and $\frac{e^a}{e^b}=e^{a-b}$, the above may be written equivalently as

$$g\left(b_1,\ldots,b_C,\mathbf{w}_1,\ldots,\mathbf{w}_C\right)=\sum_{c=1}^{C}\sum_{p\in\Omega_c}\log\left(1+\sum_{\substack{j=1\\ j\neq c}}^{C}e^{\left(b_j-b_c\right)+\mathbf{x}_p^T\left(\mathbf{w}_j-\mathbf{w}_c\right)}\right). \tag{4.54}$$

This is referred to as the *multiclass softmax cost function*, or because the softmax cost for two class classification can be interpreted through the lens of surface fitting as logistic regression (as we saw in Section 4.2), for similar reasons multiclass softmax classification is often referred to as *softmax regression*.[16] When $C = 2$ one can show that this cost

[16] When thought about in this way the multiclass softmax cost is commonly written as

$$g\left(b_1,\ldots,b_C,\mathbf{w}_1,\ldots,\mathbf{w}_C\right)=-\sum_{c=1}^{C}\sum_{p\in\Omega_c}\log\left(\frac{e^{b_c+\mathbf{x}_p^T\mathbf{w}_c}}{\sum_{j=1}^{C}e^{b_j+\mathbf{x}_p^T\mathbf{w}_j}}\right), \tag{4.55}$$

which is also equivalent to (4.53).

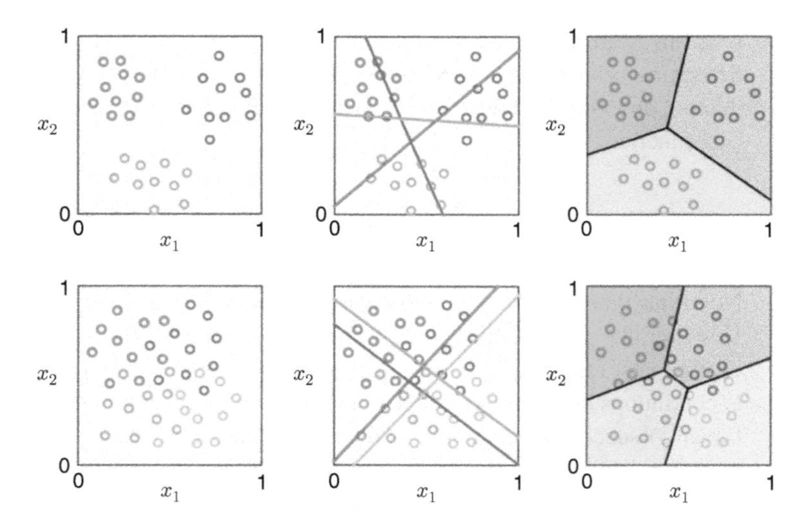

Fig. 4.20 (top left panel) Toy dataset from Fig. 4.17 with $C = 3$ classes. (top middle panel) Individual linear classifiers learned by the multiclass softmax scheme. (top right panel) Final partitioning of the feature space resulting from the application of the fusion rule in (4.47). (bottom left panel) Toy dataset from Fig. 4.19 with $C = 4$ classes. (bottom middle panel) Individual linear classifiers learned by the multiclass softmax scheme. (bottom right panel) Final partitioning of the feature space.

function reduces to the two class softmax cost originally given in (4.9). Furthermore, because the multiclass softmax cost function is convex[17] we can apply either gradient descent or Newton's method to minimize it and recover optimal weights for all C classifiers simultaneously.

In the top and bottom panels of Fig. 4.20 we show multiclass softmax classification applied to the toy datasets previously shown in the context of OvA in Fig. 4.17 and 4.19, respectively. Note that unlike the OvA separators shown in the middle panel of Fig. 4.17, the linear classifiers learned by the multiclass softmax scheme do not individually create perfect separation between one class and the remainder of the data. However, when combined according to the fusion rule in (4.47), they still perfectly partition the three classes of data. Also note that similar to OvA, the multiclass softmax scheme still produces a very good classification of the data even with overlapping classes. In both instances shown in Fig. 4.20 we used gradient descent for minimization of the multiclass softmax cost function, as detailed in Example 4.6.

Example 4.6 Optimization of the multiclass softmax cost

To calculate the gradient of the multiclass softmax cost in (4.54), we first rewrite it more compactly as

[17] This is perhaps most easily verified by noting that it is the composition of linear terms $b_j + \mathbf{x}_p^T \mathbf{x}_j$ with the convex nondecreasing softmax function. Such a composition is always guaranteed to be convex [24].

$$g\left(\tilde{\mathbf{w}}_1,\ldots,\tilde{\mathbf{w}}_C\right)=\sum_{c=1}^{C}\sum_{p\in\Omega_c}\log\left(1+\sum_{\substack{j=1\\j\neq c}}^{C}e^{\tilde{\mathbf{x}}_p^T\left(\tilde{\mathbf{w}}_j-\tilde{\mathbf{w}}_c\right)}\right),\tag{4.56}$$

where we have used the compact notation $\tilde{\mathbf{w}}_j=\begin{bmatrix}b_j\\\mathbf{w}_j\end{bmatrix}$ and $\tilde{\mathbf{x}}_p=\begin{bmatrix}1\\\mathbf{x}_p\end{bmatrix}$ for all $c=1,\ldots,C$ and $p=1,\ldots,P$. In this form, the gradient of g with respect to $\tilde{\mathbf{w}}_c$ may be computed[18] as

$$\nabla_{\tilde{\mathbf{w}}_c}g=\sum_{p=1}^{P}\left(\frac{1}{1+\sum\limits_{\substack{j=1\\j\neq c}}^{C}e^{\tilde{\mathbf{x}}_p^T\left(\tilde{\mathbf{w}}_j-\tilde{\mathbf{w}}_c\right)}}-\mathbb{1}_{p\in\Omega_c}\right)\tilde{\mathbf{x}}_p,\tag{4.57}$$

for $c=1,\ldots,C$, where $\mathbb{1}_{p\in\Omega_c}=\begin{cases}1 & \text{if } p\in\Omega_c\\0 & \text{else}\end{cases}$ is an indicator function on the set Ω_c. Concatenating all individual classifiers' parameters into a single weight vector $\tilde{\mathbf{w}}_{\text{all}}$ as

$$\tilde{\mathbf{w}}_{\text{all}}=\begin{bmatrix}\tilde{\mathbf{w}}_1\\\tilde{\mathbf{w}}_2\\\vdots\\\tilde{\mathbf{w}}_C\end{bmatrix},\tag{4.58}$$

the gradient of g with respect to $\tilde{\mathbf{w}}_{\text{all}}$ is formed by stacking block-wise gradients found in (4.57) into

$$\nabla g=\begin{bmatrix}\nabla_{\tilde{\mathbf{w}}_1}g\\\nabla_{\tilde{\mathbf{w}}_2}g\\\vdots\\\nabla_{\tilde{\mathbf{w}}_C}g\end{bmatrix}.\tag{4.59}$$

4.4.3 The accuracy of a learned multiclass classifier

To calculate the accuracy of both the OvA and multiclass softmax classifiers we use the labeling mechanism in (4.47). That is, denoting $\left(b_j^\star,\mathbf{w}_j^\star\right)$ the learned parameters for the jth boundary, we assign the predicted label $\hat{y}_p$ to the pth point $\mathbf{x}_p$ as

$$\hat{y}_p=\underset{j=1\ldots C}{\operatorname{argmax}}\ b_j^\star+\mathbf{x}_p^T\mathbf{w}_j^\star.\tag{4.60}$$

[18] Writing the gradient in this way helps avoid potential numerical problems posed by the "overflowing" exponential problem described in footnote 6.

We then compare each predicted label to its true label using an indicator function

$$\mathcal{I}\left(y_p, \hat{y}_p\right) = \begin{cases} 1 & \text{if } y_p \neq \hat{y}_p \\ 0 & \text{if } y_p = \hat{y}_p, \end{cases} \tag{4.61}$$

which we use towards computing the accuracy of the multiclass classifier on our training set as

$$\text{accuracy} = 1 - \frac{1}{P}\sum_{p=1}^{P}\mathcal{I}\left(y_p, \hat{y}_p\right). \tag{4.62}$$

This quantity ranges between 1 when every point is classified correctly, and 0 when no point is correctly classified. When possible it is also recommended to compute the accuracy of the learned model on a new testing dataset (i.e., data not used to train the model) in order to provide some assurance that the learned model will perform well on future data points. This is explored further in Chapter 6 in the context of *cross-validation*.

4.4.4 Which multiclass classification scheme works best?

As we have now seen, both OvA and multiclass softmax approaches are built using the fusion rule given in Equation (4.47). While the multiclass softmax approach more directly aims at optimizing this criterion, both OvA and softmax multiclass perform similarly well in practice (see e.g., [70, 77] and references therein).

> One-versus-all (OvA) and multiclass softmax classifiers perform similarly well in practice, having both been built using the fusion rule in (4.47).

The two methods largely differ in how they are applied in practice as well as their computational burden. In learning each of the C linear separators individually the computation required for the OvA classifier is naturally parallelizable, as each linear separator can be learned independently of the rest. On the other hand, while both OvA and multiclass softmax may be naturally extended for use with nonlinear multiclass classification (as we will discuss in Chapter 6), the multiclass softmax scheme provides a more commonly used framework for performing nonlinear multiclass classification using neural networks.

4.5 Knowledge-driven feature design for classification

Often with classification we observe not linear separability between classes but some sort of nonlinear separability. As with regression (detailed in Section 3.2), here we formulate *feature transformations* of the input data to capture this nonlinearity and use

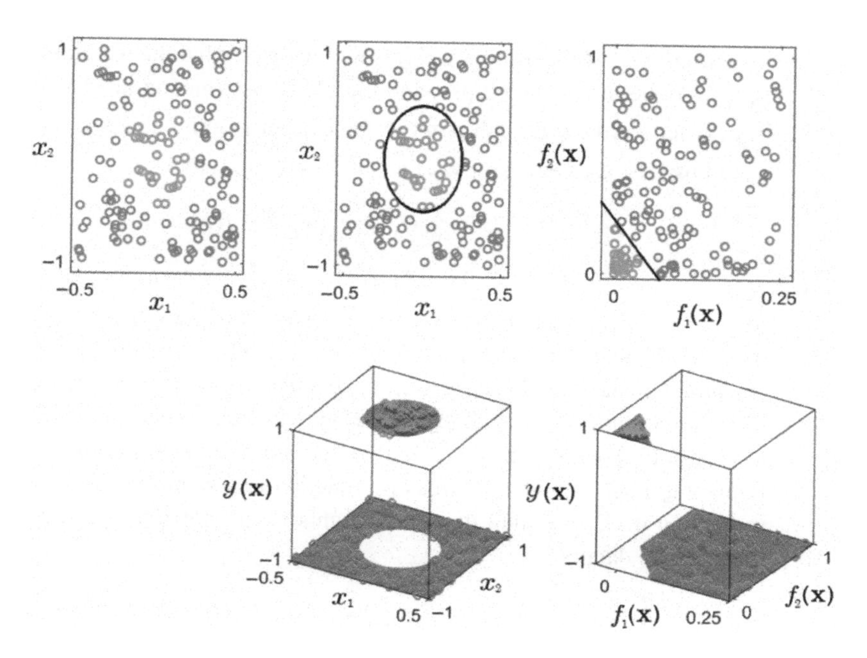

Fig. 4.21 (top left panel) A toy classification dataset where the two classes are separable via an elliptical boundary. (top middle panel) A proper learned boundary given as $1 + x_1^2 w_1^\star + x_2^2 w_2^\star = 0$ can perfectly separate the two classes. (top right panel) Finding this elliptical boundary in the original feature space is equivalent to finding a line to separate the data in the transformed space where both input features have undergone a feature transformation $\mathbf{x} = \begin{bmatrix} x_1 & x_2 \end{bmatrix}^T \longrightarrow \begin{bmatrix} f_1(\mathbf{x}) & f_2(\mathbf{x}) \end{bmatrix}^T = \begin{bmatrix} x_1^2 & x_2^2 \end{bmatrix}^T$. (bottom panels) The estimated data generating function (in gray) corresponding to each learned boundary, which is a "step function" in the transformed space.

these to construct an *estimated data generating function* (i.e., a function that appears to generate the data at hand). In very rare instances, when the dimension of the input data is low and the distribution of data is "nice," we can visualize the data and determine features by inspecting the data itself. We begin this brief section with such an example in order to practice the concept of feature design in a simple setting, and conclude by making some general points about feature design for classification. In the section following this one we then give a high level overview of common features used for classification problems involving high dimensional text, image, and audio data.

Example 4.7 Data separable by an ellipse

In the top left panel of Fig. 4.21 we show a toy dataset where, by visual inspection, it appears that a nonlinear elliptical boundary can perfectly separate the two classes of data. Recall that the equation of a standard ellipse (i.e., one aligned with the horizontal and vertical axes and centered at the origin) can be written as $1 + x_1^2 w_1 + x_2^2 w_2 = 0$, where w_1 and w_2 determine how far the ellipse stretches in the x_1 and x_2 directions, respectively.

Here we would like to find weights w_1 and w_2 so that the red class (which have label $y_p = +1$) lie inside the ellipse, and the blue class (having label $y_p = -1$) lie outside it. In other words, if the pth point of the dataset is written as $\mathbf{x}_p = \begin{bmatrix} x_{1,p} & x_{2,p} \end{bmatrix}^T$ we would like the weight vector $\mathbf{w} = \begin{bmatrix} w_1 & w_2 \end{bmatrix}^T$ to satisfy

$$\begin{aligned} 1 + x_{1,p}^2 w_1 + x_{2,p}^2 w_2 < 0 \text{ if } y_p = +1 \\ 1 + x_{1,p}^2 w_1 + x_{2,p}^2 w_2 > 0 \text{ if } y_p = -1, \end{aligned} \tag{4.63}$$

for $p = 1, \ldots, P$. Note that these equations are *linear* in their weights, and so we can interpret the above as precisely a linear separation criterion for the original perceptron (as in (4.2)) where the bias has been fixed at $b = 1$. In other words, denoting the feature transformations $f_1(\mathbf{x}) = x_1^2$ and $f_2(\mathbf{x}) = x_2^2$, we can combine the above two conditions as $y_p\left(1 + f_1\left(\mathbf{x}_p\right) w_1 + f_2\left(\mathbf{x}_p\right) w_2\right) < 0$ or $\max\left(0, y_p\left(1 + f_1\left(\mathbf{x}_p\right) w_1 + f_2\left(\mathbf{x}_p\right) w_2\right)\right) = 0$. Replacing $\max(\cdot)$ with a softmax $(\cdot)$ function and summing over p (as first described in Section 4.1.2) we may tune the weights by minimizing the softmax cost over the transformed data as

$$\underset{b,\mathbf{w}}{\text{minimize}} \sum_{p=1}^{P} \log\left(1 + e^{y_p\left(1 + f_1\left(\mathbf{x}_p\right) w_1 + f_2\left(\mathbf{x}_p\right) w_2\right)}\right). \tag{4.64}$$

This can be minimized precisely as shown in Section 4.1.2, i.e., by using gradient descent or Newton's method. Shown in the top middle and top right panels of Fig. 4.21 are the corresponding learned boundary given by

$$1 + f_1(\mathbf{x}) w_1^\star + f_2(\mathbf{x}) w_2^\star = 0, \tag{4.65}$$

whose weights were tuned by minimizing (4.64) via gradient descent, forming an ellipse in the original feature space and a line in the transformed feature space.

Finally note, as displayed in the bottom panels of Fig. 4.21, that the data generating function, that is a function determined by our chosen features as one which generates the given dataset, is not a "step function" in the original feature space because the boundary between the upper and lower sections is nonlinear. However, it is in fact a step function in the transformed feature space since the boundary there is linear. Since every point above[19] or below the learned linear boundary is declared to be of class $+1$ or -1 respectively, the estimated data generating function is given by simply taking the sign of the boundary as

$$y(\mathbf{x}) = \text{sign}\left(1 + f_1(\mathbf{x}) w_1^\star + f_2(\mathbf{x}) w_2^\star\right). \tag{4.66}$$

4.5.1 General conclusions

As shown in the previous example, a general characteristic of well-designed feature transformations is that they produce good nonlinear separation in the original feature

[19] Note that the linear separator in this case has negative slope, and we refer to the half-space to its left as the area "above" the separator.

space while simultaneously producing good linear separation in the transformed feature space.[20]

Properly designed features for linear classification provide good *nonlinear* separation in the original feature space and, simultaneously, good *linear* separation in the transformed feature space.

For any given dataset of arbitrary input dimension, N, if we determine a set of feature transformations $f_1, \ldots, f_M$ so that the boundary given by

$$b + \sum_{m=1}^{M} f_m(\mathbf{x})\, w_m = 0 \tag{4.67}$$

provides proper separation in the original space, it simultaneously splits the data equally well as a hyperplane in the transformed feature space whose M coordinate axes are given by $f_1(\mathbf{x}), \ldots, f_M(\mathbf{x})$. This is in complete analogy to the case of regression where, as we saw in Section 3.2, proper features produce a nonlinear fit in the original feature space and a corresponding linear fit in the transformed feature space. The corresponding estimated data generating function in general is then given by

$$y(\mathbf{x}) = \text{sign}\left(b + \sum_{m=1}^{M} f_m(\mathbf{x})\, w_m\right), \tag{4.68}$$

which produces the sort of generalized step function we saw in the previous example.

Rarely, however, can we design perfect features using our knowledge of a dataset. In many applications data is too high dimensional to visualize or to perfectly understand through some scientific framework. Even in the instance where the data can be visualized, as with the example dataset shown in Fig. 4.22, determining a precise functional form for each feature transformation by visual inspection can be extremely difficult. Later in Chapter 6 we describe a set of tools for the automatic design, or *learning*, of feature transformations directly from the data which can ameliorate this problem.

4.6 Histogram features for real data types

Unlike the synthetic dataset described in Example 4.7, more often than not real instances of classification data cannot be visualized due to the high dimensionality. Because of this, knowledge can rarely be used to define features algebraically for real data, i.e., by proposing a specific functional form for a set of feature transformations (as was

[20] Technically speaking there is one subtle yet important caveat to the use of the word "good" in this statement, in that we do not want to "overfit" the data (an issue we discuss at length in Chapter 6). However, for now this issue will not concern us.

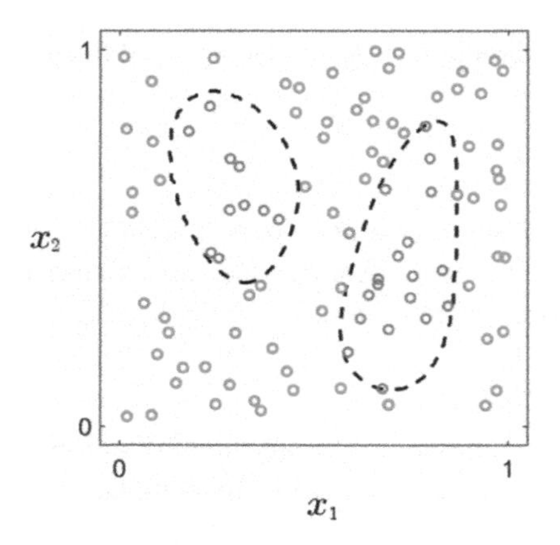

Fig. 4.22 A toy classification dataset where determining proper feature transformations by visual inspection is challenging. The two ovoid boundaries in dashed black are the boundaries between the top and bottom "steps" (i.e., the regions taking on values 1 and -1 respectively) of the data generating function.

done with the toy dataset in Example 4.7). Instead, due to our weaker level of understanding of real data, feature transformations often consist of discrete processing steps which aim at ensuring that instances within a single class are "similar" while those from different classes are "dissimilar." These processing steps are still feature transformations $f_1(\mathbf{x}), \ldots, f_M(\mathbf{x})$ of the input data, but again they are not so easily expressed algebraically.

Feature transformations for real data often consist of discrete processing steps which aim at ensuring that instances within a single class are "similar" while those from different classes are "dissimilar." While they are not always easily expressed as closed form mathematical equations, these processing steps are, when taken as a whole, still feature transformations $f_1(\mathbf{x}), \ldots, f_M(\mathbf{x})$ of the input data. Thus, properly designed instances will (as discussed in Section 4.5.1) produce good nonlinear separation in the original space and equally good linear separation in the transformed feature space.

In this section we briefly overview methods of knowledge-driven feature design for naturally high dimensional text, image, and audio data types, all of which are based on the same core concept for representing data: the *histogram*. A histogram is just a simple way of summarizing/representing the contents of an array of numbers as a vector showing how many times each number appears in the array. Although each of the aforementioned data types differs substantially in nature, we will see how the notion of a histogram-based feature makes sense in each context. While histogram features are not guaranteed to produce perfect separation, their simplicity and all around solid performance make them quite popular in practice.

Lastly, note that the discussion in this section is only aimed at giving the reader a high level, intuitive understanding of how common knowledge-driven feature design methods work. The interested reader is encouraged to consult specialized texts (referenced throughout this section) on each subject for further study.

4.6.1 Histogram features for text data

Many popular uses of classification, including spam detection and sentiment analysis (see Examples 4.8 and 4.9), are based on text data (e.g., online articles, emails, social-media updates, etc.). However with text data, the initial input (i.e., the document itself) requires a significant amount of preprocessing and transformation prior to further feature design and classification. The most basic yet widely used feature of a document for regression/classification tasks is called a Bag of Words (BoW) histogram or feature vector. Here we introduce the BoW histogram and discuss its strengths, weaknesses, and common extensions.

A BoW feature vector of a document is a simple histogram count of the different words it contains with respect to a single corpus or collection of documents (each count of an individual word is a feature, and taken together gives a feature vector), minus those nondistinctive words that do not characterize the document. To illustrate this idea let us build a BoW representation for the following corpus of two documents each containing a single sentence:

$$\begin{array}{l} \text{1) dogs are the best.} \\ \text{2) cats are the worst.} \end{array} \tag{4.69}$$

To make the BoW representation of these documents we begin by *parsing* them, creating representative vectors (histograms) $\mathbf{x}_1$ and $\mathbf{x}_2$ which contain the number of times each word appears in each document. For the two documents in (4.69) these vectors take the form

$$\mathbf{x}_1 = \frac{1}{\sqrt{2}}\begin{bmatrix} 1 \\ 0 \\ 1 \\ 0 \end{bmatrix}\begin{pmatrix} \text{best} \\ \text{cat} \\ \text{dog} \\ \text{worst} \end{pmatrix} \quad \mathbf{x}_2 = \frac{1}{\sqrt{2}}\begin{bmatrix} 0 \\ 1 \\ 0 \\ 1 \end{bmatrix}\begin{pmatrix} \text{best} \\ \text{cat} \\ \text{dog} \\ \text{worst} \end{pmatrix}. \tag{4.70}$$

Note that uninformative words such as "are" and "the", typically referred to as *stop words*, are not included in the representation. Further note that we count the singular "dog" and "cat" in place of their plural which appeared in the actual documents in (4.69). This preprocessing step is commonly called *stemming*, where related words with a common stem or root are reduced to and then represented by their common root. For instance, the words "learn," "learning," "learned," and "learner," in the final BoW feature vector are represented by and counted as "learn." Additionally, each BoW vector is normalized to have unit length.

Given that the BoW vector contains only non-negative entries and has unit length, the correlation between two BoW vectors $\mathbf{x}_1$ and $\mathbf{x}_2$ always ranges between $0 \leq \mathbf{x}_1^T\mathbf{x}_2 \leq 1$. When the correlation is zero (i.e., the vectors are perpendicular), as with the two vectors in (4.70), the two vectors are considered maximally different and will therefore (hopefully) belong to different classes. In the instances shown in (4.70) the fact that $\mathbf{x}_1^T\mathbf{x}_2 = 0$ makes sense: the two documents are completely different, containing entirely different words and polar opposite sentiment. On the other hand, the higher the correlation between two vectors the more similar the documents are purported to be, with highly correlated documents (hopefully) belonging to the same class. For example, the BoW vector of the document "I love dogs" would have positive correlation with $\mathbf{x}_1$, the document in (4.70) about dogs.

However, because the BoW vector is such a simple representation of a document, completely ignoring word order, punctuation, etc., it can only provide a gross summary of a document's contents and is thus not always distinguishing. For example, the two documents "dogs are better than cats" and "cats are better than dogs" would be considered the same document using BoW representation, even though they imply completely opposite relations. Nonetheless, the gross summary provided by BoW can be distinctive enough for many applications. Additionally, while more complex representations of documents (capturing word order, parts of speech, etc.,) may be employed they can often be unwieldy (see e.g., [54]).

Example 4.8 Sentiment analysis

Determining the aggregated feelings of a large base of customers, using text-based content like product reviews, tweets, and comments, is commonly referred to as *sentiment analysis* (as first discussed in Example 1.5). Classification models are often used to perform sentiment analysis, learning to identify consumer data of either positive or negative feelings.

For example, Fig. 4.23 shows BoW vector representations for two brief reviews of a controversial comedy movie, one with a positive opinion and the other with a negative one. The BoW vectors are rotated sideways in this figure so that the horizontal axis contains the common words between the two sentences (after stop word removal

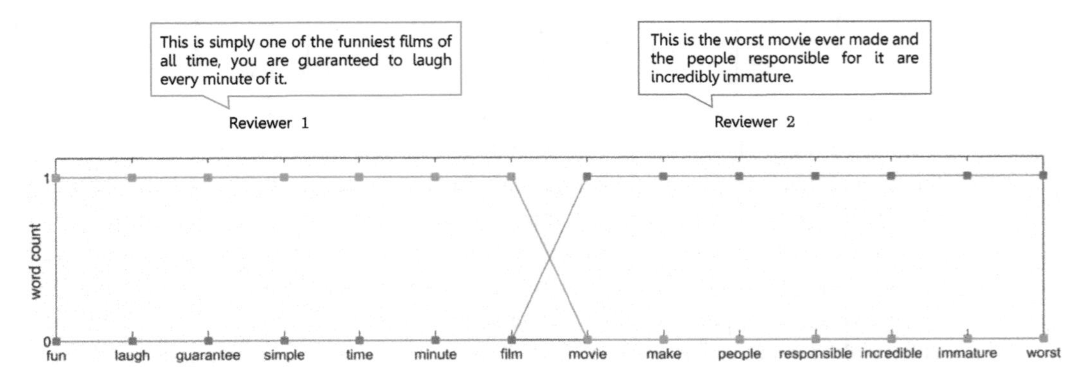

Fig. 4.23 BoW representation of two movie review excerpts, with words (after the removal of stop words and stemming) shared between the two reviews listed along the horizontal axis. The vastly different opinion of each review is reflected very well by the BoW histograms, which have zero correlation.

and stemming), and the vertical axis represents the count for each word (before normalization). The polar opposite sentiment of these two reviews is perfectly represented in their BoW representations, which as one can see are orthogonal (i.e., they have zero correlation).

Example 4.9 Spam detection

Spam detection is a standard text-based two class classification problem. Implemented in most email systems, spam detection automatically identifies unwanted messages (e.g., advertisements), referred to as spam, as distinct from the emails users want to see. Once trained, a spam detector can remove unwanted messages without user input, greatly improving a user's email experience. In many spam detectors the BoW feature vectors are formed with respect to a specific list of spam words (or phrases) including "free," "guarantee," "bargain," "act now," "all natural," etc., that are frequently seen in spam emails. Additionally features like the frequency of certain characters like ! and * are appended to the BoW feature, as are other spam-targeted features like the total number of capital letters in the email and the length of longest uninterrupted sequence of capital letters, as these features can further distinguish the two classes.

In Fig. 4.24 we show classification results on a spam email dataset consisting of BoW, character frequencies, and other spam-focused features (including those mentioned

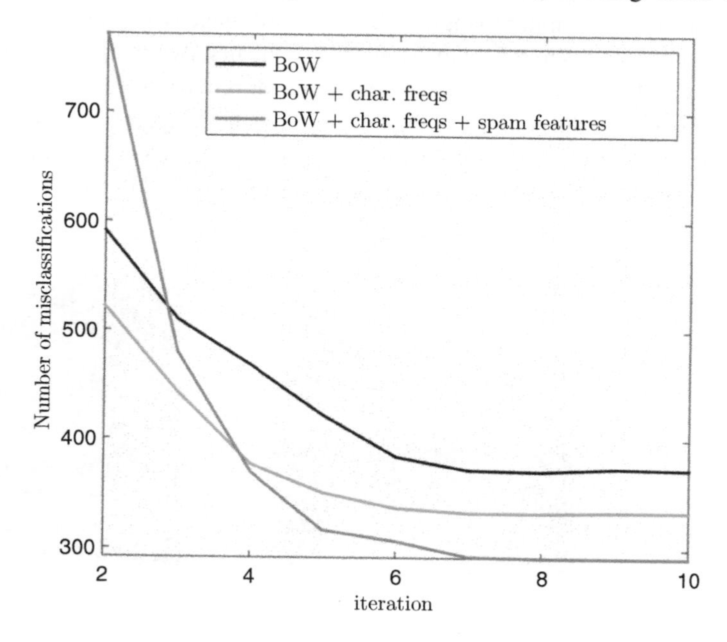

Fig. 4.24 Results of applying the softmax cost (using Newton's method) to distinguish spam from real email using BoW and additional features. The number of misclassifications per iteration of Newton's method is shown in the case of BoW features (in black), BoW and character frequencies (in green), and BoW, character frequencies, as well as spam-focused features (in magenta). In each case adding more distinguishing features (on top of the BoW vector) improves classification. Data in this figure is taken from [47].

previously) taken from 1813 spam and 2788 real email messages for a total of $P = 4601$ data points (this data is taken from [47]). Employing the softmax cost to learn the separator, the figure shows the number of misclassifications per iteration of Newton's method (using the counting cost in (4.25) at each iteration). More specifically these classification results are shown for the same dataset using only BoW features (in black), BoW and character frequencies (in green), and the BoW/character frequencies as well as spam-targeted features (in magenta) (see Exercise 4.20 for further details). Unsurprisingly the addition of character frequencies improves the classification, with the best performance occurring when the spam-focused features are used as well.

4.6.2 Histogram features for image data

To perform classification tasks on image data, like object detection (see Example 1.4), the raw input features are pixel values of an image itself. The pixel values of an 8-bit grayscale image are each just a single integer in the range of 0 (black) to 255 (white), as illustrated in Fig. 4.25. In other words, a grayscale image is just a matrix of integers ranging from 0 to 255. A color image is then just a set of three such grayscale matrices: one for each of the red, blue, and green channels.

Pixel values themselves are typically not discriminative enough to be useful for classification tasks. We illustrate why this is the case using a simple example in Fig. 4.26. Consider the three simple images of shapes shown in the left column of this figure. The first two are similar triangles while the third shape is a square, and we would like an ideal set of features to reflect the similarity of the first two images as well as their distinctness from the last image. However, due to the difference in their relative size, position in the image, and the contrast of the image itself (the image with the smaller triangle is darker toned overall), if we were to use raw pixel values to compare the images (by taking the difference between each image pair[21]) we would find that the square and larger triangle

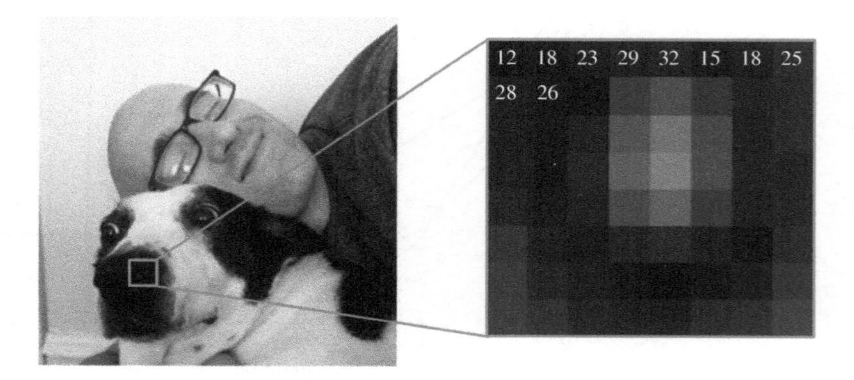

Fig. 4.25 An 8-bit grayscale image consists of pixels, each taking a value between 0 (black) and 255 (white). To visualize individual pixels, a small 8×8 block from the original image is enlarged on the right.

[21] This is to say that if we denote by $\mathbf{X}_i$ the ith image then we would find that $\|\mathbf{X}_1 - \mathbf{X}_3\|_F < \|\mathbf{X}_1 - \mathbf{X}_2\|_F$.

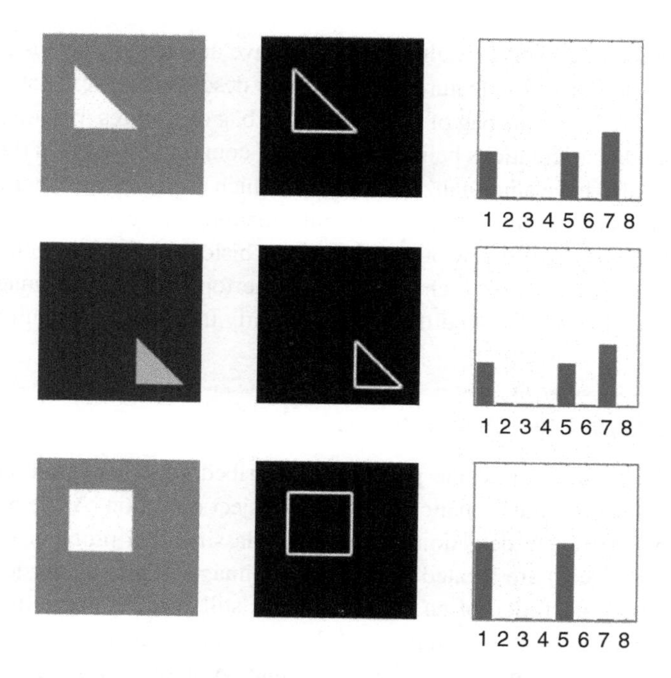

Fig. 4.26 (left column) Three images of simple shapes. While the triangles in the top two images are visually similar, this similarity is not reflected by comparing their raw pixel values. (middle column) Edge-detected versions of the original images, here using eight edge orientations, retain the distinguishing structural content while significantly reducing the amount of information in each image. (right column) By taking normalized histograms of the edge content we have a feature representation that captures the similarity of the two triangles quite well while distinguishing both from the square.

in the top image are more similar than the two triangles themselves. This is because the pixel values of the first and third image, due to their identical contrast and location of the triangle/square, are indeed more similar than those of the two triangle images.

In the middle and right columns of Fig. 4.26 we illustrate a two step procedure that generates the sort of discriminating feature transformation we are after. In the first part we shift perspective from the pixels themselves to the edge content at each pixel. As first detailed in Example 1.8, by taking edges instead of pixel values we significantly reduce the amount of information we must deal with in an image without destroying its identifying structures. In the middle column of the figure we show corresponding edge-detected images, in particular highlighting eight equally (angularly) spaced edge orientations, starting from 0 degrees (horizontal edges) with seven additional orientations at increments of 22.5 degrees, including 45 degrees (capturing the diagonal edges of the triangles) and 90 degrees (vertical edges). Clearly the edges retain distinguishing characteristics from each original image, while significantly reducing the amount of total information in each case.

We then make a normalized histogram of each image's edge content. That is, we make a vector consisting of the amount of each edge orientation found in the image and normalize the resulting vector to have unit length. This is completely analogous to the bag of words feature representation described for text data previously and is often referred to as the bag of visual words or bag of features method [1, 2], with the counting of edge orientations being the analog of counting "words" in the case of text data. Here we also have a normalized histogram which represents an image grossly while ignoring the location and ordering of its information. However, as shown in the right panel of the figure, unlike raw pixel values these histogram feature vectors capture characteristic information about each image, with the top two triangle images having very similar histograms and both differing significantly from that of the third image of the square.

Example 4.10 Object detection

Generalizations of the previously described edge histogram concept are widely used as feature transformations for visual object detection. As detailed in Example 1.4, the task of object detection is a popular classification problem where objects of interest (e.g., faces) are located in an example image. While the basic principles which led to the consideration of an edge histogram still hold, example images for such a task are significantly more complicated than the simple geometric shapes shown in Fig. 4.26. In particular, preserving local information at smaller scales of an image is considerably more important. Thus a natural way to extend the edge histogram feature is to compute it not over the entire image, but by breaking the image into relatively small patches and computing an edge histogram of each patch, then concatenating the results. In Fig. 4.27 we show a diagram of a common variation of this technique often used in

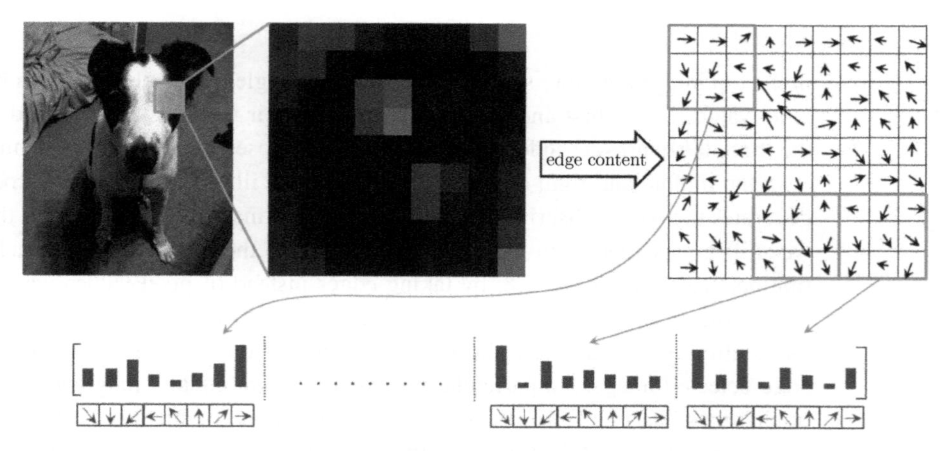

Fig. 4.27 A representation of the sort of generalized edge histogram feature transformation commonly used for object detection. An input image is broken down into small (here 9×9) blocks, and an edge histogram is computed on each of the smaller non-overlapping (here 3×3) patches that make up the block. The resulting histograms are then concatenated and normalized jointly, producing a feature vector for the entire block. Concatenating such block features by scanning the block window over the entire image gives the final feature vector.

Fig. 4.28 Example images taken from a large face detection dataset of (left panel) 3000 facial and (right panel) 7000 non-facial images (see text for further details). The facial images shown in this figure are taken from [2].

practice where we normalize neighboring histograms jointly in larger blocks (for further details see e.g., [3, 29, 65] and [4, 5] for extensions of this approach). Interestingly this sort of feature transformation can in fact be written out algebraically as a set of quadratic transformations of the input image [25].

To give a sense of just how much histogram-based features improve our ability to detect visual objects we now show the results of a simple experiment on a large face detection dataset. This data consists of 3000 cropped 28×28 (or dimension $N = 784$) images of faces (taken from [2]) and 7000 equal sized non-face images (taken from various images not containing faces), a sample of which is shown in Fig. 4.28.

We then compare the classification accuracy of the softmax classifier on this large training set of data using a) raw pixels and b) a popular histogram-based feature known as the histogram of oriented gradients (HoG) [29]. HoG features were extracted using the Vlfeat software library [67], providing a corresponding feature vector of each image in the dataset (of length $N = 496$). In Fig. 4.29 we show the resulting number of misclassifications per iteration of Newton's method applied to the raw pixel (black) and HoG feature (magenta) versions of data. While the raw images are not linearly separable, with over 300 misclassifications upon convergence of Newton's method, the HoG feature version of the data is perfectly separable by a hyperplane and presents zero misclassifications upon convergence.

4.6.3 Histogram features for audio data

Like images, raw audio signals are not discriminative enough to be used for audio-based classification tasks (e.g., speech recognition) and once again properly designed histogram-based features are used. In the case of an audio signal it is the histogram of its frequencies, otherwise known as its *spectrum,* that provides a robust summary of its contents. As illustrated in Fig. 4.30, the spectrum of an audio signal counts up

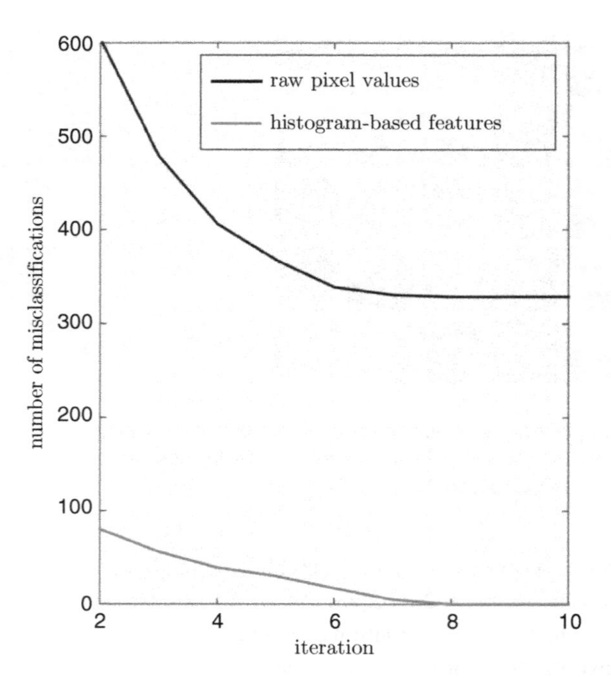

Fig. 4.29 An experiment comparing the classification efficacy of raw pixel versus histogram-based features for a large training set of face detection data (see text for further details). Employing the softmax classifier, the number of misclassifications per iteration of Newton's method is shown for both raw pixel data (in black) and histogram-based features (in magenta). While the raw data itself has overlapping classes, with a large number of misclassifications upon convergence of Newton's method, the histogram-based feature representation of the data is perfectly linearly separable with zero misclassifications upon convergence.

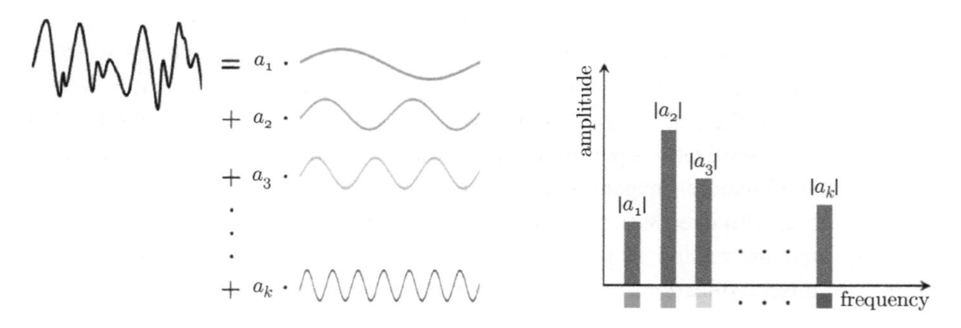

Fig. 4.30 A representation of an audio signal and its representation as a frequency histogram or spectrum. (left panel) A figurative audio signal can be decomposed as a linear combination of simple sinusoids with varying frequencies (or oscillations). (right panel) The frequency histogram then contains the strength of each sinusoid in the representation of the audio signal.

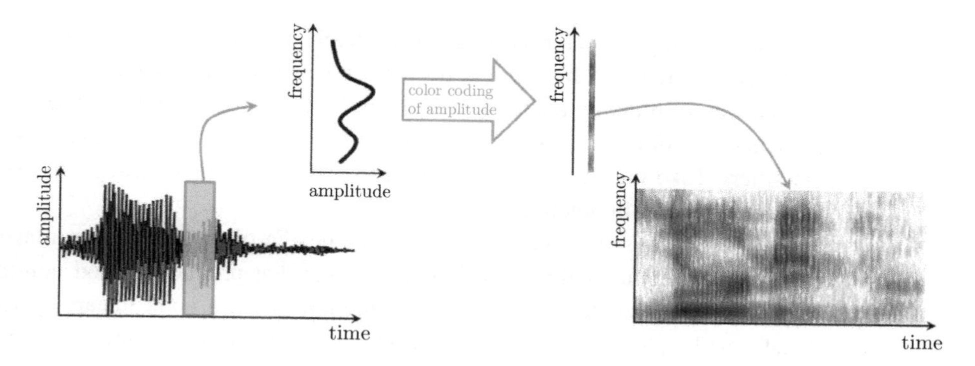

Fig. 4.31 A representation of histogram-based features for audio data. The original speech signal (shown on the left) is broken up into small (overlapping) windows whose frequency histograms are computed and stacked vertically to produce a spectrogram (shown on the right). Classification tasks like speech recognition are then performed using this feature representation, or a further refinement of it (see text for further details).

(in histogram fashion) the strength of each level of its frequency or oscillation. This is done by decomposing the speech signal over a basis of sine waves of ever increasing frequency, with the weights on each sinusoid representing the amount of that frequency in the original signal. Each oscillation level is analogous to an edge direction in the case of an image, or an individual word in the case of a BoW text feature.

Example 4.11 Speech recognition

In Example 4.10 we discussed how edge histograms computed on overlapping blocks of an image provide a useful feature transformation for object detection since they preserve characteristic local information. Likewise computing frequency histograms over overlapping windows of an audio signal (forming a "spectrogram" as illustrated in Fig. 4.31) produces a feature vector that preserves important local information as well, and is a common feature transformation used for speech recognition. Further processing of the windowed histograms, in order to e.g., emphasize the frequencies of sound best recognized by the human ear, are also commonly performed in practical implementations of this sort of feature transformation [40, 67].

4.7 Summary

In Section 4.1 we first described the fundamental cost function associated with linear two class classification: the perceptron. We then saw how to derive two convex and differentiable relatives of the basic perceptron, the softmax and squared margin perceptron cost functions. These two costs are often used in practice and, given their close resemblance, typically perform very similarly. We then saw in Sections 4.2 and 4.3 how these two cost functions can be derived classically, as logistic regression and soft-margin support vector machines respectively, with the logistic regression "surface-fitting perspective" of the softmax perceptron being of particular value as it provides a second way of thinking about classification. Next, in Section 4.4 we discussed two approaches to multiclass classification, the multiclass softmax and one-versus-all (OvA) classifiers. Like the two commonly used two class cost functions, these two methods perform similarly in practice as well.

We then discussed in Section 4.5 how the design of proper feature transformations corresponds geometrically with finding features that produce a good nonlinear separator in the original feature space and, simultaneously, a good linear separator in the transformed feature space. In the final section we described common histogram-based features for text, image, and audio data types and how understanding of each guides both their representation as well as practical feature design for common classification problems.

4.8 Exercises

Section 4.1 exercises

Exercises 4.1 The perceptron cost is convex

In this exercise you will show that the original perceptron cost given in Equation (4.5) is convex using two steps.

a) Use the zeroth order definition of convexity (described in Appendix D) to show that $\max\left(0,\, -y_p\left(b+\mathbf{x}_p^T\mathbf{w}\right)\right)$ is convex in both parameters $(b, \mathbf{w})$.

b) Use the zeroth order definition of convexity to show that if both $g(t)$ and $h(t)$ are convex, then so too is $g(t) + h(t)$. Use this to conclude that the perceptron cost is indeed convex.

Exercises 4.2 The softmax/logistic regression cost is convex

Show that the softmax/logistic regression cost function given in Equation (4.9) is convex by verifying that it satisfies the second order definition of convexity. *Hint: the Hessian, already given in Equation (4.13), is a weighted outer product matrix like the one described in Exercise 2.10.*

Exercises 4.3 Code up gradient descent for the softmax cost/logistic regression on a toy dataset

In this exercise you will code up gradient descent to minimize the softmax cost function on a toy dataset, reproducing the left panel of Fig. 4.3 in Example 4.1.

a) Verify the gradient of the softmax cost shown in Equation (4.12).

b) (optional) This gradient can be written more efficiently for programming languages like Python and MATLAB/OCTAVE that have especially good implementations of matrix/vector operations by writing it in matrix-vector form as

$$\nabla g(\tilde{\mathbf{w}}) = \tilde{\mathbf{X}}\mathbf{r}, \tag{4.71}$$

where $\tilde{\mathbf{X}}$ is the $(N+1) \times P$ matrix formed by stacking the P vectors $\tilde{\mathbf{x}}_p$ column-wise, and where $\mathbf{r}$ is a $P \times 1$ vector based on the form of the gradient shown in Equation (4.12). Verify that this can be done and determine $\mathbf{r}$.

c) Code up gradient descent to minimize the softmax cost, reproducing the left panel of Fig. 4.3. This figure is generated via the wrapper *softmax_grad_demo_hw* using the dataset *imbalanced_2class.csv*. You must complete a short gradient descent function located within the wrapper which takes the form

$$\tilde{\mathbf{w}} = \text{softmax\_grad}\left(\tilde{\mathbf{X}}, \mathbf{y}, \tilde{\mathbf{w}}^0, \text{alpha}\right). \tag{4.72}$$

Here $\tilde{\mathbf{w}}$ is the optimal weights learned via gradient descent, $\tilde{\mathbf{X}}$ is the input data matrix, $\mathbf{y}$ the output values, and $\tilde{\mathbf{w}}^0$ the initial point.

Almost all of this function has already been constructed for you. For example, the step length is given and fixed for all iterations, etc., and you must only enter the gradient of the associated cost function. All of the additional code necessary to generate the associated plot is already provided in the wrapper.

Exercises 4.4 Code up Newton's method to learn a softmax/logistic regression classifier on a toy dataset

In this exercise you will code up Newton's method to minimize the softmax/logistic regression cost function on a toy dataset, producing a plot similar to the right panel of Fig. 4.3 in Example 4.1.

a) Verify that the Hessian of the softmax given in (4.13) is correct.

b) (optional) The gradient and Hessian can be written more efficiently for programming languages like Python and MATLAB/OCTAVE that have especially good implementations of matrix/vector operations by writing them more compactly. In particular the gradient can be written compactly as discussed in part b) of Exercise 4.3, and likewise the Hessian can be written more compactly as

$$\nabla^2 g(\tilde{\mathbf{w}}) = \tilde{\mathbf{X}} \text{diag}(\mathbf{r}) \tilde{\mathbf{X}}^T, \tag{4.73}$$

where $\tilde{\mathbf{X}}$ is the $(N+1) \times P$ matrix formed by stacking P data vectors $\tilde{\mathbf{x}}_p$ column-wise, and where $\mathbf{r}$ is a $P \times 1$ vector based on the form of the Hessian shown in Equation (4.13). Verify that this can be done and determine $\mathbf{r}$. Note that for large datasets you do not want to explicitly form the matrix $\text{diag}(\mathbf{r})$, but compute $\tilde{\mathbf{X}}\text{diag}(\mathbf{r})$ by broadcasting the multiplication of each entry of $\mathbf{r}$ across the columns of $\tilde{\mathbf{X}}$.

c) Using the wrapper *softmax_Newton_demo_hw* code up Newton's method to minimize the softmax cost with the dataset *overlapping_2class.csv*. You must complete a short Newton's method function located within the wrapper,

$$\tilde{\mathbf{w}} = \text{softmax\_newton}\left(\tilde{\mathbf{X}}, \mathbf{y}, \tilde{\mathbf{w}}^0\right). \tag{4.74}$$

Here $\widetilde{\mathbf{w}}$ is the optimal weights learned via Newton's method, $\widetilde{\mathbf{X}}$ is the input data matrix, $\mathbf{y}$ the output values, and $\widetilde{\mathbf{w}}^0$ the initial point.

Almost all of this function has already been constructed for you and all you must do is enter the form of the Newton step of the associated cost function. All of the additional code necessary to generate the associated plot is already provided in the wrapper.

Exercises 4.5 The softmax cost and diverging weights with linearly separable data

Suppose that a two class dataset of P points is linearly separable, and that the pair of finite-valued parameters $(b, \mathbf{w})$ defines a separating hyperplane for the data.

a) Show while multiplying these weights by a positive constant $C > 1$, that as $(C \cdot b,\ C \cdot \mathbf{w})$ does not alter the equation of the separating hyperplane, the scaled parameters reduce the value of the softmax cost as $g\left(C \cdot b, C \cdot \mathbf{w}\right) < g\left(b, \mathbf{w}\right)$ where g is the softmax cost in (4.9). *Hint: remember from (4.3) that if the point $\mathbf{x}_p$ is classified correctly then* $-y_p\left(b + \mathbf{x}_p^T\mathbf{w}\right) < 0$.

b) Using part a) describe how, in minimizing the softmax cost over a linearly separable dataset, it is possible for the parameters to grow infinitely large. Why do you think this is a problem, practically speaking?

There are several simple ways to prevent this problem: one is to add a stopping condition that halts gradient descent/Newton's method if the parameters $(b,\ \mathbf{w})$ become larger than a preset maximum value. A second option is to add an ℓ_2 regularizer (see Section 3.3.2) to the softmax cost with a small penalty parameter λ, since adding the regularizer $\lambda \left\|\mathbf{w}\right\|_2^2$ will stop $\mathbf{w}$ from growing too large (since otherwise the value of the regularized softmax cost will grow to infinity).

Exercises 4.6 The margin cost function is convex

In this exercise you will show that the margin and squared margin cost functions are convex using two steps.

a) Use the zeroth order definition of convexity (described in Appendix D) to show that $\max\left(0,\ 1 - y_p\left(b + \mathbf{x}_p^T\mathbf{w}\right)\right)$ is convex in both parameters $(b, \mathbf{w})$. Do the same for the squared margin $\max^2\left(0,\ 1 - y_p\left(b + \mathbf{x}_p^T\mathbf{w}\right)\right)$.

b) Use the zeroth order definition of convexity to show that if both $g\left(t\right)$ and $h\left(t\right)$ are convex, then so too is $g\left(t\right) + h\left(t\right)$. Use this to conclude that the margin and squared margin perceptron costs are indeed convex.

Exercises 4.7 Code up gradient descent to learn a squared margin classifier

In this exercise you will code up gradient descent for minimizing the squared margin cost function discussed in Section 4.1.4.

a) Verify that the gradient of the squared margin cost is given as in Equation (4.21).

b) (optional) This gradient can be written more efficiently for programming languages like Python and MATLAB/OCTAVE that have especially good implementations of matrix/vector operations by writing it in matrix-vector form as

$$\nabla g\left(\widetilde{\mathbf{w}}\right) = -2\widetilde{\mathbf{X}}\text{diag}\left(\mathbf{y}\right)\mathbf{max}\left(\mathbf{0}_{P\times 1},\ \mathbf{1}_{P\times 1} - \text{diag}\left(\mathbf{y}\right)\widetilde{\mathbf{X}}^{T}\widetilde{\mathbf{w}}\right), \tag{4.75}$$

where **max** is the maximum function applied entrywise, $\widetilde{\mathbf{X}}$ is the $(N+1) \times P$ matrix formed by stacking P data vectors $\widetilde{\mathbf{x}}_p$ column-wise. Verify that this can be done. (Note that for large datasets you do not want to explicitly form the matrix $\text{diag}\left(\mathbf{y}\right)$, but compute $\widetilde{\mathbf{X}}\text{diag}\left(\mathbf{y}\right)$ by broadcasting the multiplication of each entry of $\mathbf{y}$ across the columns of $\widetilde{\mathbf{X}}$.)

c) Code up gradient descent to minimize the squared margin cost, reproducing the left panel of Fig. 4.3. This figure is generated via the wrapper *squared_margin_grad_demo_hw* using the dataset *imbalanced_2class.csv*. You must complete a short gradient descent function located within the wrapper which takes the form

$$\widetilde{\mathbf{w}} = \text{squared\_margin\_grad}\left(\widetilde{\mathbf{X}}, \mathbf{y}, \widetilde{\mathbf{w}}^{0}, \text{alpha}\right). \tag{4.76}$$

Here $\widetilde{\mathbf{w}}$ is the optimal weights learned via gradient descent, $\widetilde{\mathbf{X}}$ is the input data matrix, $\mathbf{y}$ the output values, and $\widetilde{\mathbf{w}}^{0}$ the initial point.

Almost all of this function has already been constructed for you. For example, the step length is given and fixed for all iterations, etc., and you must only enter the gradient of the associated cost function. All of the additional code necessary to generate the associated plot is already provided in the wrapper.

Exercises 4.8 Code up Newton's method to learn a squared margin classifier

In this exercise you will code up Newton's method to minimize the squared margin cost function on a toy dataset, producing a plot similar to the right panel of Fig. 4.5 in Example 4.2.

a) Code up Newton's method to minimize the squared margin cost. You may use the wrapper *squared_margin_Newton_demo_hw* with the dataset *overlapping_2class.csv*. You must complete a short Newton's method function located within the wrapper, which takes the form

$$\widetilde{\mathbf{w}} = \text{squared\_margin\_newton}\left(\widetilde{\mathbf{X}}, \mathbf{y}, \widetilde{\mathbf{w}}^{0}\right). \tag{4.77}$$

Here $\widetilde{\mathbf{w}}$ is the optimal weights learned via Newton's method, $\widetilde{\mathbf{X}}$ is the input data matrix, $\mathbf{y}$ the output values, and $\widetilde{\mathbf{w}}^{0}$ the initial point.

Almost all of this function has already been constructed for you and all you must do is enter the form of the Newton step. All of the additional code necessary to generate the associated plot is already provided in the wrapper.

Exercises 4.9 Perform classification on the breast cancer dataset

Compare the efficacy of the softmax and squared margin costs in distinguishing healthy from cancerous tissue using the entire breast cancer dataset as training data, located in *breast_cancer_dataset.csv*, first discussed in Example 4.3. This dataset consists of $P = 699$ data points, with each data point having nine medically valuable features (i.e., $N = 9$) which you may read about by reviewing the readme file *breast_cancer_readme.txt*. Note that for simplicity we have removed the sixth feature from the original version of this data, taken from [47], due to its absence from many of the data points.

To compare the two cost functions create a plot like the one shown in Fig. 4.8, which compares the number of misclassifications per iteration of Newton's method as applied to minimize each cost function over the data (note: depending on your initialization it could take between 10–20 iterations to achieve the results shown in this figure). As mentioned in footnote 6, you need to be careful not to overflow the exponential function used with the softmax cost here. In particular make sure to choose a small initial point for your Newton's method algorithm with the softmax cost.

Exercises 4.10 Perform classification on histogram-based features for face detection

Compare the efficacy of the softmax and squared margin costs in distinguishing face from non-face images using the histogram-based feature face detection training dataset, located in *feat_face_data.csv*, first discussed in Example 4.3 and later in Example 4.10. This set of training data consists of $P = 10\,000$ feature data points from 3000 face images (taken from [2]) and 7000 non-face images like those shown in Fig. 4.28. Here each data point is a histogram-based feature vector of length $N = 496$ taken from a corresponding 28×28 grayscale image.

To compare the two cost functions create a plot like the one shown in Fig. 4.8 which compares the number of misclassifications per iteration of Newton's method as applied to minimize each cost function over the data. However, in this case use gradient descent to minimize both cost functions. You may determine a fixed step size for each cost function by trial and error, or by simply using the "conservatively optimal" fixed step lengths shown in Table 8.1 (which are guaranteed to cause gradient descent to converge to a minimum in each instance).

As mentioned in footnote 6, you need to be careful here not to overflow the exponential function used with the softmax cost, in particular make sure to choose a small initial point. In calculating the value of the softmax cost at each iteration you may find it useful to include a conditional statement that deals with the possibility of e^s overflowing for large values of s, which will cause $\log(1 + e^s)$ to be returned as ∞ (as the computer will see it as $\log(1 + \infty)$), by simply returning s since for large values $s \approx \log(1 + e^s)$.

Section 4.2 exercises

Exercises 4.11 Alternative form of logistic regression

In this section we saw how the desire for having the following approximation for the pth data point $(\mathbf{x}_p, y_p)$:

$$\tanh\left(y_p\left(b + \mathbf{x}_p^T\mathbf{w}\right)\right) \approx 1, \tag{4.78}$$

led us to forming the softmax perceptron cost function $h_1(b, \mathbf{w}) = \sum_{p=1}^{P} \log\left(1 + e^{-y_p\left(b + \mathbf{x}_p^T \mathbf{w}\right)}\right)$.

a) Following a similar set of steps, show that Equation (4.78) can be used to arrive at the related cost function given by $h_2(b, \mathbf{w}) = \sum_{p=1}^{P} e^{-y_p\left(b + \mathbf{x}_p^T \mathbf{w}\right)}$.

b) Code up gradient descent to minimize both cost functions using the two-dimensional dataset shown in Fig. 4.32 (located in the data file *exp_vs_log_data.csv*). After performing gradient descent on each, the final separators provided by h_1 and h_2 are shown in black and magenta respectively.

Using the wrapper *exp_vs_log_demo_hw* you must complete two short gradient descent functions corresponding to h_1 and h_2 respectively:

$$\widetilde{\mathbf{w}} = \text{grad\_descent\_soft\_cost}\left(\widetilde{\mathbf{X}}, \mathbf{y}, \widetilde{\mathbf{w}}^0, \text{alpha}\right) \tag{4.79}$$

and

$$\widetilde{\mathbf{w}} = \text{grad\_descent\_exp\_cost}\left(\widetilde{\mathbf{X}}, \mathbf{y}, \widetilde{\mathbf{w}}^0, \text{alpha}\right). \tag{4.80}$$

Here $\widetilde{\mathbf{w}}$ is the optimal weights, $\widetilde{\mathbf{X}}$ is the input data matrix, $\mathbf{y}$ the output values, and $\widetilde{\mathbf{w}}^0$ the initial point.

Almost all of this function has already been constructed for you. For example, the step length is fixed for all iterations, etc., and you must only enter the gradient of each associated cost function. All of the additional code necessary to generate the associated plot is already provided in the wrapper.

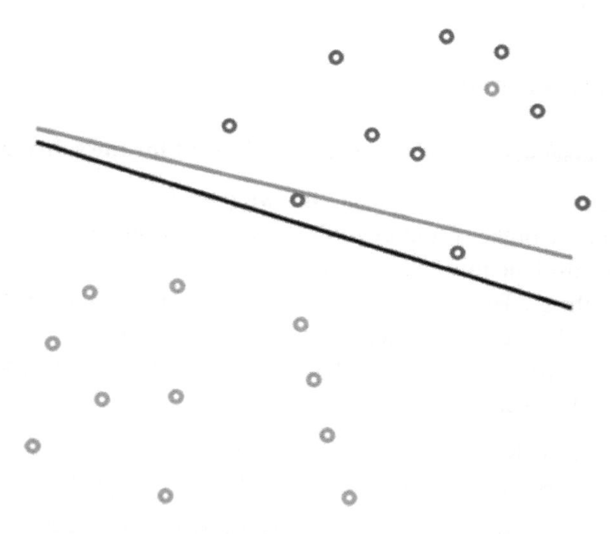

Fig. 4.32 A two-dimensional dataset used for Exercise 4.11. See text for details.

c) Compare the two separating hyperplanes found in the previous part of this exercise. Which cost function does a better job at separating the two classes of data? Why? *Hint: note the error contribution of the outlier to each cost function.*

Exercises 4.12 Probabilistic perspective of logistic regression

In the previous chapter (in Section 3.3.1) we first introduced logistic regression in the context of its original application: modeling population growth. We then followed this geometric perspective to re-derive the softmax cost function in the instance of classification.

In the classification setting, logistic regression may also be derived from a probabilistic perspective.[22] Doing so one comes to the following cost function for logistic regression:

$$h\left(b,\mathbf{w}\right) = -\sum_{p=1}^{P} \bar{y}_p \log \sigma\left(b + \mathbf{x}_p^T\mathbf{w}\right) + (1-\bar{y}_p)\log\left(1 - \sigma\left(b + \mathbf{x}_p^T\mathbf{w}\right)\right), \tag{4.81}$$

where the modified labels $\bar{y}_p$ are defined as

$$\bar{y}_p = \begin{cases} 0 & \text{if } y_p = -1 \\ 1 & \text{if } y_p = +1. \end{cases} \tag{4.82}$$

Show that the cost function $h\left(b,\mathbf{w}\right)$, also referred to as the *cross-entropy* cost for logistic regression, is equivalent to the softmax cost function $g\left(b,\mathbf{w}\right) = \sum_{p=1}^{P} \log\left(1 + e^{-y_p\left(b + \mathbf{x}_p^T\mathbf{w}\right)}\right)$.

Hint: this can be done in cases, i.e., suppose $y_p = +1$, show that the corresponding summand of the softmax cost becomes that of the cross-entropy cost when substituting $\bar{y}_p = 1$.

Section 4.3 exercises

Exercises 4.13 Code up gradient descent for the soft-margin SVM cost

Extend Exercise 4.7 by using the wrapper and dataset discussed there to test the performance of the soft-margin SVM classifier using the squared margin perceptron as the base cost function, i.e., an ℓ_2 regularized form of the squared margin cost function. How does the gradient change due to the addition of the regularizer? Input those changes into the gradient descent function described in that exercise and run the wrapper for values of $\lambda \in \left[10^{-2}, 10^{-1}, 1, 10\right]$. Describe the consequences of choosing each in terms of the final classification accuracy.

[22] Using labels $\bar{y}_p \in \{0, 1\}$ this is done by assuming a sigmoidal conditional probability for the point $\mathbf{x}_p$ to have label $\bar{y}_p = 1$ as $p\left(\mathbf{x}_p\right) = \sigma\left(b + \mathbf{x}_p^T\mathbf{w}\right) = \frac{1}{1+e^{-\left(b+\mathbf{x}_p^T\mathbf{w}\right)}}$. The cross-entropy cost in (4.81) is then found by maximizing the so-called log likelihood function associated to this choice of model (see e.g., [52] for further details).

Section 4.4 exercises

Exercises 4.14 One-versus-all classification

In this exercise you will reproduce the result of performing one-versus-all classification on the $C = 4$ dataset shown in Fig. 4.19.

a) Use the Newton's method subfunction produced in (4.83) to complete the one-versus-all wrapper *one_versus_all_demo_hw* to classify the $C = 4$ class dataset *four_class_data.csv* shown in Fig. 4.19. With your Newton's method module you must complete a short subfunction in this wrapper called

$$\widetilde{\mathbf{W}} = \text{learn\_separators}\left(\widetilde{\mathbf{X}}, \mathbf{y}\right), \tag{4.83}$$

that enacts the OvA framework, outputting learned weights for all C separators (i.e., this should call your Newton's method module C times, once for each individual two class classifier). Here $\widetilde{\mathbf{W}} = \left[\begin{array}{cccc} \widetilde{\mathbf{w}}_1 & \widetilde{\mathbf{w}}_2 & \cdots & \widetilde{\mathbf{w}}_C \end{array}\right]$ is an $(N+1) \times C$ matrix of weights, where $\widetilde{\mathbf{w}}_c$ is the compact weight/bias vector associated with the cth classifier, $\widetilde{\mathbf{X}}$ is the input data matrix, $\mathbf{y}$ the associated labels. All of the additional code necessary to generate the associated plot is already provided in the wrapper.

Exercises 4.15 Code up gradient descent for the multiclass softmax classifier

In this exercise you will code up gradient descent to minimize the multiclass softmax cost function on a toy dataset, reproducing the result shown in Fig. 4.20.

a) Confirm that the gradient of the multiclass softmax perceptron is given by Equation (4.57) for each class $c = 1, \ldots, C$.

b) Code up gradient descent to minimize the multiclass softmax perceptron, reproducing the result shown for the $C = 4$ class dataset shown in Fig. 4.20. This figure is generated via the wrapper *softmax_multiclass_grad_hw* and you must complete a short gradient descent function located within which takes the form

$$\widetilde{\mathbf{W}} = \text{softmax\_multiclass\_grad}\left(\widetilde{\mathbf{X}}, \mathbf{y}, \widetilde{\mathbf{W}}^0, \text{alpha}\right). \tag{4.84}$$

Here $\widetilde{\mathbf{W}} = \left[\begin{array}{cccc} \widetilde{\mathbf{w}}_1 & \widetilde{\mathbf{w}}_2 & \cdots & \widetilde{\mathbf{w}}_C \end{array}\right]$ is an $(N+1) \times C$ matrix of weights, where $\widetilde{\mathbf{w}}_c$ is the compact bias/weight vector associated with the cth classifier, $\widetilde{\mathbf{X}}$ is the input data matrix, $\mathbf{y}$ the associated labels, and $\widetilde{\mathbf{W}}^0$ the initialization for the weights. Almost all of this function has already been constructed for you. For example, the step length is fixed for all iterations, etc., and you must only enter the gradient of the associated cost function. All of the additional code necessary to generate the associated plot is already provided in the wrapper.

Exercises 4.16 Handwritten digit recognition

In this exercise you will perform $C = 10$ multiclass classification for handwritten digit recognition, as described in Example 4.4 , employing the OvA multiclass classification framework. Employ the softmax cost with gradient descent or Newton's method to solve each of the two-class subproblems.

a) Train your classifier on the training set located in *MNIST_training_data.csv*, that contains $P = 60\,000$ examples of handwritten digits $0-9$ (all examples are vectorized grayscale images of size 28×28 pixels). Report the accuracy of your trained model on this training set.

b) Using the weights learned from part a) report the accuracy of your model on a new test dataset of handwritten digits located in *MNIST_testing_data.csv*. This contains $P = 10\,000$ new examples of handwritten digits that were not used in the training of your model.

Exercises 4.17 Show the multiclass softmax reduces to two-class softmax when $C = 2$

Show that the multiclass softmax cost function given in (4.54) reduces to the two class softmax cost in (4.9) when $C = 2$.

Exercises 4.18 Calculating the Hessian of the multiclass softmax cost

Show that the Hessian of the multiclass softmax cost function can be computed blockwise as follows. For $s \neq c$ we have $\nabla_{\tilde{\mathbf{w}}_c \tilde{\mathbf{w}}_s} g = -\sum_{p=1}^{P} \frac{e^{\tilde{\mathbf{x}}_p^T \tilde{\mathbf{w}}_c + \tilde{\mathbf{x}}_p^T \tilde{\mathbf{w}}_s}}{\left(\sum_{d=1}^{C} e^{\tilde{\mathbf{x}}_p^T \tilde{\mathbf{w}}_d}\right)^2} \tilde{\mathbf{x}}_p \tilde{\mathbf{x}}_p^T$ and the second derivative block in $\tilde{\mathbf{w}}_c$ is given as $\nabla_{\tilde{\mathbf{w}}_c \tilde{\mathbf{w}}_c} g = \sum_{p=1}^{P} \frac{e^{\tilde{\mathbf{x}}_p^T \tilde{\mathbf{w}}_c}}{\sum_{d=1}^{C} e^{\tilde{\mathbf{x}}_p^T \tilde{\mathbf{w}}_d}} \left(1 - \frac{e^{\tilde{\mathbf{x}}_p^T \tilde{\mathbf{w}}_c}}{\sum_{d=1}^{C} e^{\tilde{\mathbf{x}}_p^T \tilde{\mathbf{w}}_d}}\right) \tilde{\mathbf{x}}_p \tilde{\mathbf{x}}_p^T$.

Section 4.5 exercises

Exercises 4.19 Learn a quadratic separator

Shown in the left panel of Fig. 4.33 are $P = 150$ data points which, by visual inspection, can be seen to be separable not by a line but by some quadratic boundary. In other words, points from each class all lie either above or below a quadratic of the form $f(x_1, x_2) = b + x_1^2 w_1 + x_2 w_2 = 0$ in the original feature space, i.e.,

$$\begin{aligned} b + x_{1,p}^2 w_1 + x_{2,p} w_2 &> 0 \text{ if } y_p = 1 \\ b + x_{1,p}^2 w_1 + x_{2,p} w_2 &< 0 \text{ if } y_p = -1. \end{aligned} \tag{4.85}$$

As illustrated in the right panel of the figure, this quadratic boundary is simultaneously a linear boundary in the *feature space* defined by the quadratic *feature transformation* or *mapping* of $(x_1, x_2) \longrightarrow (x_1^2, x_2)$.

Using any cost function and the dataset *quadratic_classification.csv*, reproduce the result shown in the figure by learning the proper parameters b and $\mathbf{w}$ for the quadratic boundary, and by plotting the data and its associated separator in both the original and transformed feature spaces.

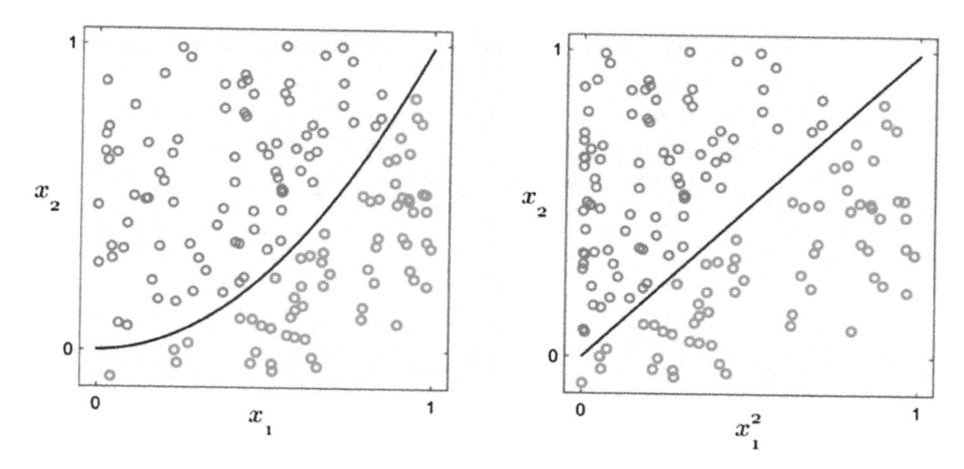

Fig. 4.33 Data separable via a quadratic boundary. (left panel) A quadratic boundary given as $b + x_1^2 w_1 + x_2 w_2 = 0$ can perfectly separate the two classes. (right panel) Finding the weights associated to this quadratic boundary in the original feature space is equivalent to finding a line to separate the data in the transformed feature space where the input has undergone a quadratic feature transformation $(x_1, x_2) \longrightarrow \left(x_1^2, x_2\right)$.

Section 4.6 exercises

Exercises 4.20 Perform spam detection using BoW and spam-specific features

Compare the efficacy of using various combinations of features to perform spam detection on a real dataset of emails, as described in Example 4.9. Your job is to reproduce as well as possible the final result (i.e., the final number of misclassifications) shown in Fig. 4.24, using only the squared margin cost and gradient descent (instead of the softmax cost and Newton's method as shown there). You may determine a fixed step size by trial and error or by using the "conservatively optimal" fixed step length shown in Table 8.1 (which is guaranteed to cause gradient descent to converge to a minimum).

Use the entire dataset, taken from [47] and consisting of features taken from 1813 spam and 2788 real email messages (for a total of $P = 4601$ data points), as your training data. The features for each data point include: 48 BoW features, six character frequency features, and three spam-targeted features (further details on these features can be found by reviewing the readme file *spambase_data_readme.txt*). This dataset may be found in *spambase_data.csv*. Note that you may find it useful to rescale the final two spam-targeted features by taking their natural log, as they are considerably larger than the other features.

Exercises 4.21 Comparing pixels and histogram-based features for face detection

In this exercise you will reproduce as well as possible the result shown in Fig. 4.29, using a cost function and descent algorithm of your choosing, which compares the classification efficacy of raw pixel features versus a set of standard histogram-based features on a large training set of face detection data (described in Example 4.10 and Exercise 4.10). Note that it may take between 10–20 Newton steps to achieve around the same number of misclassifications as shown in this figure depending on your initialization. The raw pixel features are located in *raw_face_data.csv* and the histogram-based features may be found in *feat_face_data.csv*.

Part II

Tools for fully data-driven machine learning

Overview of Part II

In Sections 3.2 and 4.5 we have discussed how understanding of regression and classification datasets can be used to forge useful features in particular instances. With regression we saw that by visualizing low-dimensional data we could form excellent features for particular datasets like e.g., data from Galileo's classic ramp experiment. Later, when discussing classification, we also saw how basic features can be designed for e.g., image data using our understanding of natural signals and the mammalian visual processing system. Unfortunately, due to our general ignorance regarding most types of phenomena in the universe, instances such as these are rare and we often have no knowledge on which to construct reasonable features at all. However, we can, as described in the next three chapters, automate the process of feature design itself by leveraging what we know strong features should accomplish for regression/classification tasks.

5 Automatic feature design for regression

As discussed in the end of Section 3.2, rarely can we design perfect or even strongly performing features for the general regression problem by completely relying on our understanding of a given dataset. In this chapter we describe tools for automatically designing proper features for the general regression problem, without the explicit incorporation of human knowledge gained from e.g., visualization of the data, philosophical reflection, or domain expertise.

We begin by introducing the tools used to perform regression in the ideal but extremely unrealistic scenario where we have complete and noiseless access to all possible input feature/output pairs of a regression phenomenon, i.e., a continuous function (as first discussed in Section 3.2). Here we will see how, in the case where we have such unfettered access to regression data, perfect features can be designed automatically by combining elements from a set of basic feature transformations. We then see how this process for building features translates, albeit imperfectly, to the general instance of regression where we have access to only noisy samples of a regression relationship. Following this we describe *cross-validation*, a crucial procedure to employing automatic feature design in practice. Finally we discuss several issues pertaining to the best choice of primary features for automatic feature design in practice.

5.1 Automatic feature design for the ideal regression scenario

In Fig. 5.1 we illustrate a prototypical dataset on which we perform regression, where our input feature and output have some sort of clear nonlinear relationship. Recall from Section 3.2 that at the heart of feature design for regression is the tacit assumption that the data we receive are in fact noisy samples of some underlying continuous function (shown in dashed black in Fig. 5.1). Our goal in solving the general regression problem is then, using the data at our disposal (which we may think of as noisy glimpses of the underlying function), to approximate this data-generating function as well as we can.

In this section we will assume the impossible: that we have complete access to a clean version of every input feature/output pair of a regression phenomenon, or in other words that our data completely traces out a continuous function $y(\mathbf{x})$. We do not assume that we know a functional form for $y(\mathbf{x})$, but in such an ideal scenario we will see how perfect features may be designed automatically to fit such data (regardless of the complexity or ambient dimension of $y(\mathbf{x})$) by combining different elements from a basis of primary features.

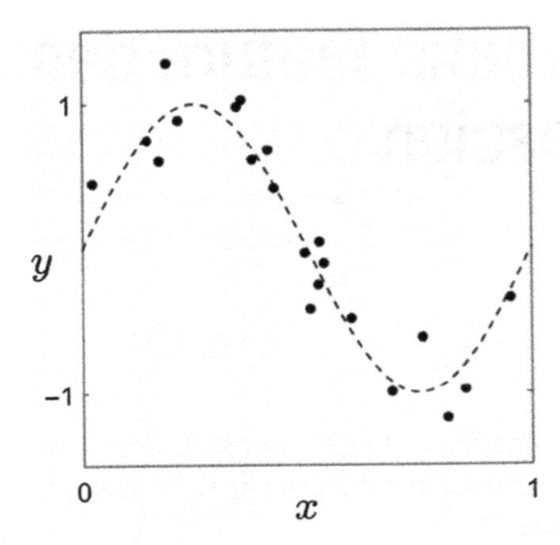

Fig. 5.1 A realistic dataset for regression made by taking noisy samples from the data generating function $y(x) = \sin(2\pi x)$ (shown in dashed black) over the unit interval.

The entire mathematical framework for the automatic design of features in this perfect regression scenario comes from the classic study of *continuous function approximation*,[1] which has been developed by mathematicians, physical scientists, and engineers over the past several centuries. Because of this we will use phrases like "function approximation" and "automatic feature design" synonymously in the description that follows.

5.1.1 Vector approximation

Recall the linear algebra fact that any vector $\mathbf{y}$ in $\mathbb{R}^P$, that is the set of all column vectors of length P with real entries, can be represented perfectly over any given basis of P linearly independent vectors. In other words, given a set of P linearly independent vectors $\left\{\mathbf{x}_p\right\}_{p=1}^{P}$ in $\mathbb{R}^P$, we can always express $\mathbf{y}$ precisely (i.e., without any error) as a linear combination of its elements,

$$\sum_{p=1}^{P} \mathbf{x}_p w_p = \mathbf{y}. \tag{5.1}$$

Now let us suppose that we only have access to a subset $\{\mathbf{x}_m\}_{m=1}^{M}$ of the full basis $\left\{\mathbf{x}_p\right\}_{p=1}^{P}$ in order to represent $\mathbf{y}$, where $M \leq P$. Although in this case there is no guarantee that the vector $\mathbf{y}$ lies completely in the span of the partial basis $\{\mathbf{x}_m\}_{m=1}^{M}$, we can still approximate $\mathbf{y}$ via a linear combination of its elements,

$$\sum_{m=1}^{M} \mathbf{x}_m w_m \approx \mathbf{y}. \tag{5.2}$$

[1] Throughout the remainder of this section we will be fairly loose in our discussion of function approximation, which is by nature a highly technical subject. The interested reader can see Section 5.7 for a short discussion and a list of more technical treatments of the subject.

Note that the approximation in (5.2) can be made to hold to any desired level of tolerance by making M larger.

The ideal set of weights $\{w_m\}_{m=1}^{M}$ to make the *partial basis approximation* in (5.2) hold as well as possible can then be determined by solving the related Least Squares problem:

$$\underset{w_1 \ldots w_M}{\text{minimize}} \left\| \sum_{m=1}^{M} \mathbf{x}_m w_m - \mathbf{y} \right\|_2^2, \tag{5.3}$$

which has a closed form solution as detailed in Section 3.1.3.

5.1.2 From vectors to continuous functions

Any vector $\mathbf{y}$ in $\mathbb{R}^P$ can be viewed as a "discrete function" on the unit interval [0, 1] after plotting its entries at equidistant points $\left\{x_p = p/P\right\}_{p=1}^{P}$ on the x-axis, as pairs $\left\{\left(x_p, y_p\right)\right\}_{p=1}^{P}$. We illustrate this idea in the top left panel of Fig. 5.2 using a $P = 4$ dimensional vector $\mathbf{y}$ defined entry-wise as $y_p = \sin\left(2\pi x_p\right)$, where $x_p = p/P$ and $p = 1 \ldots P$. Also shown in the top row of this figure is the vector $\mathbf{y}$, constructed in precisely the same manner, only this time with $P = 40$ (middle panel) and $P = 400$ (right panel). As can be seen in this figure, for larger values of P the collection of points $\left(x_p, y_p\right)$ closely resembles the continuous function $y(x) = \sin(2\pi x)$.

In other words, as $P \longrightarrow \infty$ the set of points $\left\{\left(x_p, y_p\right) = \left(p/P, \sin\left(2\pi x_p\right)\right)\right\}_{p=1}^{P}$ for all intents and purposes, precisely describes the continuous function $y(x) = \sin(2\pi x)$. Hence we can think of a continuous function (defined on the interval [0, 1]) as, *roughly*, an infinite dimensional vector.

This same intuition applies to functions $y(x)$ defined over an arbitrary interval $[a, b]$ as well, since we can make the same argument given above when $a = 0$ and $b = 1$ and approximate y as finely as desired using a discrete set of sampled points. Furthermore, we can employ a natural extension of this argument to say the same thing about general functions $y(\mathbf{x})$ where $\mathbf{x}$ is an N-dimensional vector defined over a hyperrectangle, that is where each entry of $\mathbf{x}$ lies in some interval $x_n \in [a_n, b_n]$. We do this by evaluating y over a finer and finer grid of evenly spaced points $\mathbf{x}_p$ covering the hyperrectangle of its input domain (illustrated with a particular example for two dimensional input in the bottom row of Fig. 5.2). Therefore in general we can roughly think about any continuous function $y(\mathbf{x})$, with bounded input $\mathbf{x}$ of length N, as an infinite length vector.

This perspective on continuous functions is especially helpful in framing the notion of function approximation, since the key concepts broadly follow the same shape as (finite length) vector approximation described in Section 5.1.1. As we discuss next, many of the defining ideas with vector approximation in $\mathbb{R}^P$, e.g., the notions of bases and Least Squares weight fitting, have direct analogs in the case of continuous function approximation.

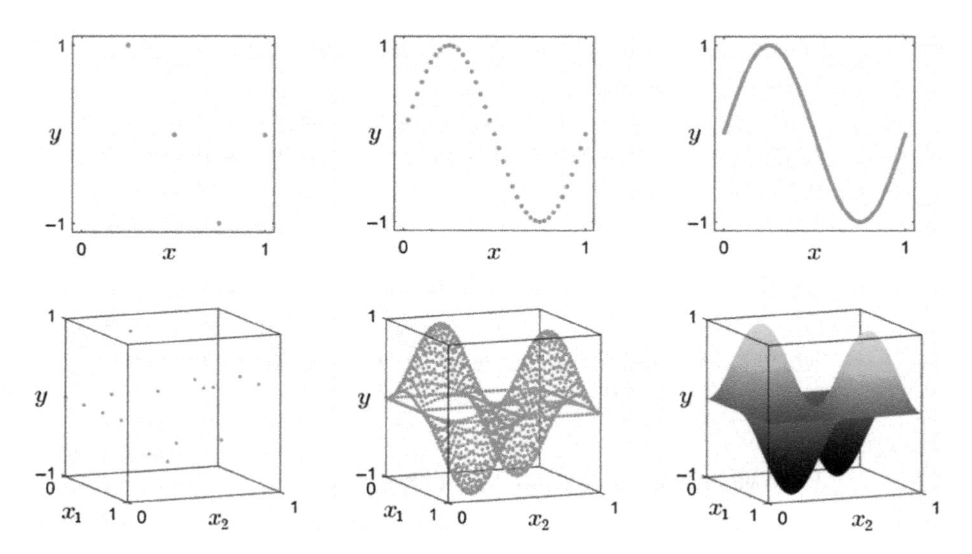

Fig. 5.2 (top row) The P-dimensional vector $\mathbf{y}$ with entries $y_p = \sin(2\pi x_p)$ where $x_p = p/P$ and $p = 1 \ldots P$, plotted as points (x_p, y_p) with (left) $P = 4$, (middle) $P = 40$, and (right) $P = 400$. The vector $\mathbf{y}$, as a discrete function, closely resembles the continuous function $y(x) = \sin(2\pi x)$, especially for larger values of P. (bottom row) An analogous example in three dimensions using the function $y(\mathbf{x}) = \sin(2\pi x_1)\sin(2\pi x_2)$ evaluated over a grid of (left) $P = 16$, (middle) $P = 1600$, and (right) $P = 160\,000$ evenly spaced points over the unit square. Note that the number of samples P required to maintain a certain resolution of the function grows exponentially with the input dimension. This unwanted phenomenon is often called "the curse of dimensionality."

5.1.3 Continuous function approximation

Like a discrete vector $\mathbf{y}$ in $\mathbb{R}^P$, any continuous function $y(\mathbf{x})$ with bounded N-dimensional input $\mathbf{x}$ can be completely decomposed over a variety of bases. For clarity and convenience we will suppose $\mathbf{x}$ lies in the N-dimensional unit hypercube $[0, 1]^N$ for the remainder of this chapter, that is each entry $x_n \in [0, 1]$ (however the discussion here holds for y with more general input as well). In any case, as with finite length vectors, we may write such a function $y(\mathbf{x})$ as a linear combination of basis elements which are themselves continuous functions. In terms of regression, the elements of a basis are basic features that we may combine in order to perfectly (or near-perfectly) approximate our continuous function input/output data $(\mathbf{x}, y(\mathbf{x}))$, for which we have for all $\mathbf{x} \in [0, 1]^N$.

As can be intuited by the rough description given previously of such a continuous function as an "infinite length vector," we must correspondingly use an infinite number of basis elements to represent any such desired function completely. Formally a basis in this instance is a set of basic feature transformations $\{f_m(\mathbf{x})\}_{m=1}^{\infty}$ such that at all points $\mathbf{x}$ in the unit hypercube we can express $y(\mathbf{x})$ perfectly as

$$\sum_{m=0}^{\infty} f_m(\mathbf{x})\, w_m = y(\mathbf{x}). \tag{5.4}$$

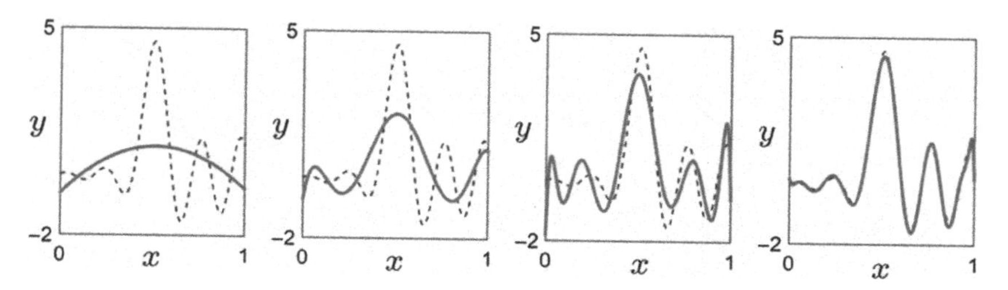

Fig. 5.3 Partial basis approximation (in blue) of an example function $y(x)$ (in dashed black), where (from left to right) $M = 2, 6, 10$, and 20, respectively. As the number of basis features is increased the approximation more closely resembles the underlying function.

Here for each m the weight w_m must be tuned properly for a given y so that the above equality will indeed hold. Take a moment to appreciate just how similar in both concept and shape this is to the analogous vector formula given in (5.1) which describes the decomposition of a (finite length) vector over a corresponding vector basis. Structurally, the set of vectors $\mathbb{R}^P$ and the set of continuous functions defined on the N-dimensional unit hypercube $[0, 1]^N$, which we shall denote by $\mathcal{C}^N$, have much in common.

As with vectors, for large enough M we can approximate y over its input domain as

$$\sum_{m=0}^{M} f_m(\mathbf{x})\, w_m \approx y(\mathbf{x}). \tag{5.5}$$

Again, this approximation can be made as finely as desired[2] by increasing the number of basis features M used and tuning the associated parameters appropriately (as detailed in Section 5.1.5). In other words, by increasing M we can design automatically (near) perfect features to represent $y(\mathbf{x})$. This is illustrated with a particular example in Fig. 5.3, where an increasing number of basis elements (in this instance polynomials) are used to approximate a given function.[3]

5.1.4 Common bases for continuous function approximation

Bases for continuous function approximation, sometimes referred to as *universal approximators*, can be distinguished by those whose elements are functions of the input $\mathbf{x}$ alone, and those whose elements are also functions of further internal parameters. The former variety, referred to as *fixed bases* due to their fixed shape and sole dependence on $\mathbf{x}$, include the polynomial and sinusoidal feature bases. For $N = 1$ elements of the polynomial basis consist of a constant term $f_0(x) = 1$ and the set of simple monomial features of the form

[2] Depending on both the function and the employed basis, the approximation in (5.5) may not improve at each and every point $\mathbf{x} \in [0, 1]^N$ as we increase the number of basis elements. However, this technicality does not concern us as the approximation can be shown to generally improve over the entire domain of the function by increasing M, which is sufficient for our purposes.

[3] The function approximated here is defined as $y(x) = e^{3x}\frac{\sin\left(3\pi^2(x-0.5)\right)}{3\pi^2(x-0.5)}$.

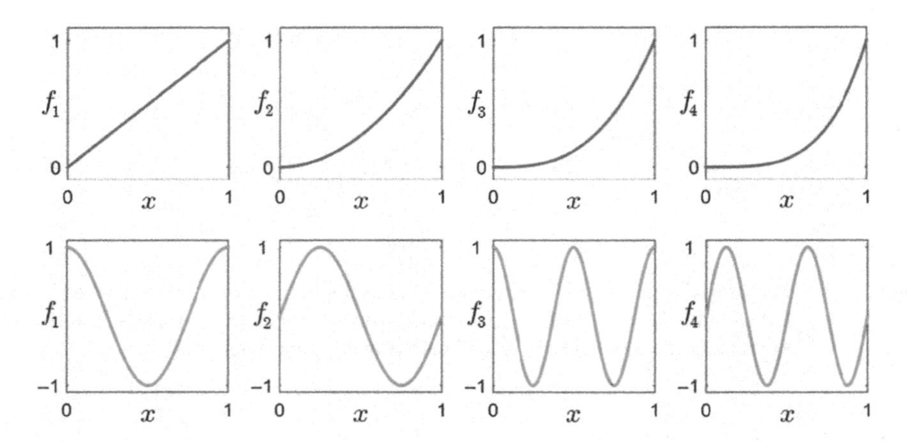

Fig. 5.4 (top row, from left to right) The first four non-constant elements of the polynomial basis. (bottom row, from left to right) The first four non-constant elements of the Fourier basis.

$$f_m(x) = x^m \quad \text{for all } m \geq 1. \tag{5.6}$$

One often encounters the polynomial basis when learning calculus, where it is employed in the form of a Taylor series approximation of a (many times) differentiable function $y(x)$.

Likewise for $N = 1$ the sinusoidal or *Fourier* basis, so named after its inventor Joseph Fourier who first used these functions in the early 1800s to study heat diffusion, consists of the constant term $f_0(x) = 1$ and the set of cosine and sine waves with ever increasing frequency of the form[4]

$$\begin{cases} f_{2m-1}(x) = \cos(2\pi mx) & \text{for all } m \geq 1 \\ f_{2m}(x) = \sin(2\pi mx) & \text{for all } m \geq 1. \end{cases} \tag{5.7}$$

The first four non-constant elements of both the polynomial and Fourier bases for $N = 1$ are shown in Fig. 5.4. With slightly more cumbersome notation both the polynomial and Fourier bases are likewise defined[5] for general N-dimensional input $\mathbf{x}$.

The second class of bases we refer to as *adjustable*, due to the fact that their elements have tunable internal parameters, and are more straightforward to define for general N-dimensional input. The simplest adjustable basis is what is commonly referred to as a *single hidden layer feed forward neural network*, coined by neuroscientists in the late 1940s who first created this sort of basis as a way to roughly model how the human brain processes information. With the exception of the constant term $f_0(\mathbf{x}) = 1$, this basis uses

[4] It is also common to write the Fourier basis approximation using classic *complex exponential* definitions of both cosine and sine, i.e., $\cos(\alpha) = \frac{1}{2}\left(e^{i\alpha} + e^{-i\alpha}\right)$ and $\sin(\alpha) = \frac{1}{2i}\left(e^{i\alpha} - e^{-i\alpha}\right)$. With these complex exponential definitions, one can show (see Exercise 5.5) that with a scalar input x, Fourier basis elements can be written in complex exponential form $f_m(x) = e^{2\pi imx}$.

[5] For a general N-dimensional input each polynomial feature takes the analogous form $f_m(\mathbf{x}) = x_1^{m_1} x_2^{m_2} \cdots x_N^{m_N}$. Likewise, using the complex exponential notation (see previous footnote) each multidimensional Fourier basis element takes the form $f_m(\mathbf{x}) = e^{2\pi im_1x_1}e^{2\pi im_2x_2} \ldots e^{2\pi im_Nx_N}$.

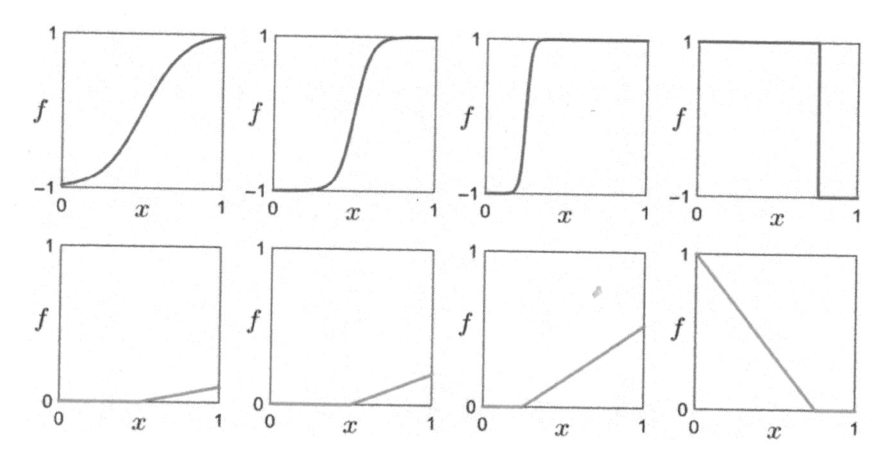

Fig. 5.5 Unlike a fixed basis, elements of an adjustable basis are free to change form by adjusting internal parameters. Here we show four instances of a single (top row) hyperbolic tangent and (bottom row) hinge basis function, with each instance corresponding to a different setting of its internal parameters.

a single type of parameterized function (often referred to as an *activation function*) for each basis feature. Common examples include the hyperbolic tangent function

$$f_m(\mathbf{x}) = \tanh\left(c_m + \mathbf{x}^T\mathbf{v}_m\right) \quad \text{for all } m \geq 1, \tag{5.8}$$

and the max or hinge function (also referred to as a "rectified linear unit" in this context)

$$f_m(\mathbf{x}) = \max\left(0,\ c_m + \mathbf{x}^T\mathbf{v}_m\right) \quad \text{for all } m \geq 1. \tag{5.9}$$

Note that the scalar parameter c_m as well as the N-dimensional vector parameter $\mathbf{v}_m$ are unique to each basis element f_m and are adjustable, giving each basis element a range of possible shapes to take depending on how the parameters are set.

For example, illustrated in Fig. 5.5 are four instances[6] of basis features in Equations (5.8) and (5.9) with scalar input (i.e., $N = 1$), where the $\tanh(\cdot)$ and $\max(0, \cdot)$ elements are shown in the top and bottom rows, respectively. For each type of basis the four panels show the form taken by a single basis element with four different settings for c_m and v_m.

Generally speaking the flexibility of each basis feature, gained by introduction of adjustable internal parameters, typically enables effective approximation using fewer neural network basis elements than a standard fixed basis. For example, in Fig. 5.6 we show the results of using a polynomial, Fourier, and single hidden layer neural network with $\tanh(\cdot)$ activation function respectively to approximate a particular function[7] $y(x)$ over $x \in [0, 1]$. In each row from left to right we use $M = 2$ and $M = 6$ basis elements of each type to approximate the function, and as expected the approximations improve for all basis types as we increase M. However, comparing the evolution of approximations

[6] Note that unlike the fixed basis functions shown in Fig. 5.4, which can be arranged and counted in a sequence of low to high "degree" elements, there is no such ordering within a set of adjustable basis functions.

[7] The function approximated here is $y(x) = e^x \cos(2\pi \sin(\pi x))$.

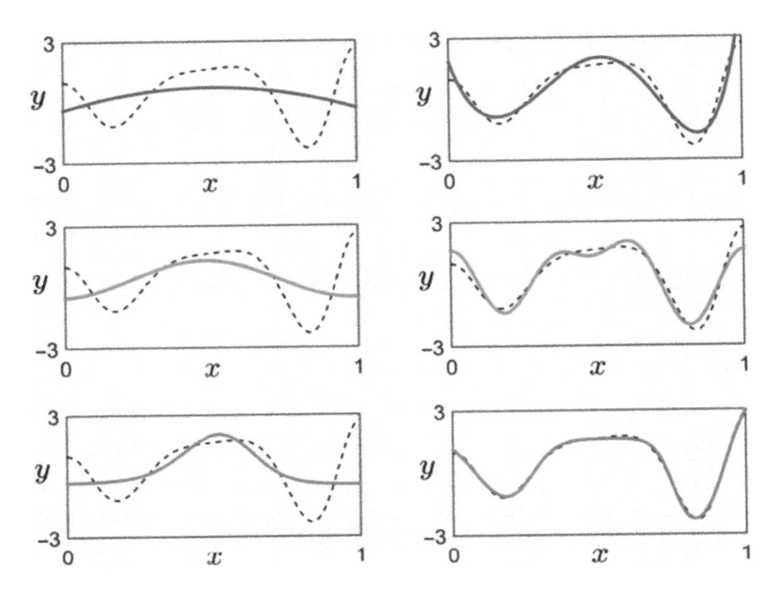

Fig. 5.6 From left to right, approximation of a continuous function (shown by the dashed black curve) over [0, 1], using $M = 2$ and $M = 6$ elements of (top row) polynomial, (middle row) Fourier, and (bottom row) single hidden layer neural network bases, respectively. While all three bases could approximate this function as finely as desired by increasing M, the neural network basis (with its adjustable internal parameters) approximates the underlying function more closely using the same number of basis elements compared to both fixed bases.

in each row one can see that the neural network basis (with the added flexibility of its internal parameters) better approximates y than either fixed bases using the same number of basis elements.

Even more flexible adjustable basis features are commonly constructed via *summation* and *composition* of activation functions.[8] For instance, to create a single basis element of a feed forward neural network with *two hidden layers* we take a weighted sum of single hidden layer basis features and pass the result through an activation function (of the same kind as used in the single layer basis). Doing this with the tanh basis in Equation (5.8) gives

$$f_m(\mathbf{x}) = \tanh\left(c_m^{(1)} + \sum_{m_2=1}^{M_2} \tanh\left(c_{m_2}^{(2)} + \mathbf{x}^T \mathbf{v}_{m_2}^{(2)}\right) v_{m_2,m}^{(1)}\right), \tag{5.10}$$

for all $m \geq 1$. Note that here for organizational purposes we have used superscripts on each internal parameter to indicate the layer it belongs to. Likewise, we can compose the hinge function with itself to create a two hidden layer basis function of the form

[8] Although one could think of designing more flexible basis elements than those of a single layer neural network in Equation (5.8) in a variety of ways, this is the most common approach (i.e., summation and composition using a single type of activation function).

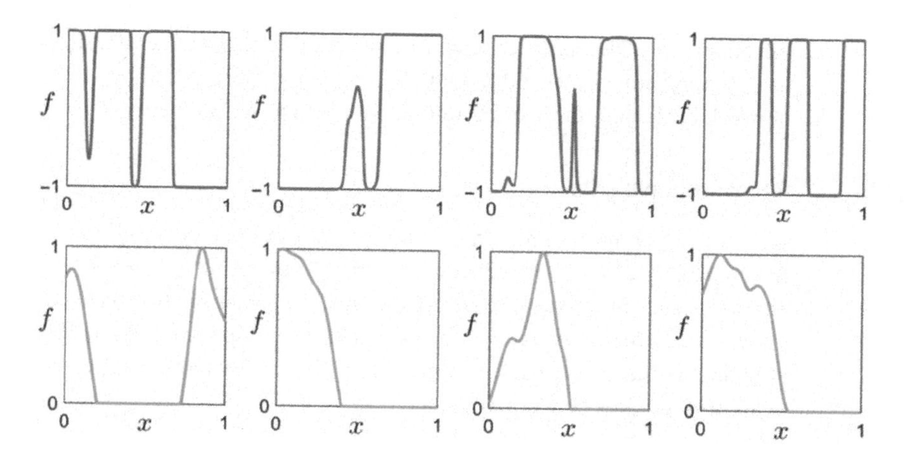

Fig. 5.7 Four instances of a two layer neural network basis function made by composing (top row) hyperbolic tangent and (bottom row) hinge functions. In each instance the internal parameters are set randomly. Note how the two layer basis elements are far more diverse in shape than those of a single layer basis shown in Fig. 5.5.

$$f_m(\mathbf{x}) = \max\left(0,\; c_m^{(1)} + \sum_{m_2=1}^{M_2} \max\left(0,\; c_{m_2}^{(2)} + \mathbf{x}^T \mathbf{v}_{m_2}^{(2)}\right) v_{m_2,m}^{(1)}\right), \tag{5.11}$$

for all $m \geq 1$. Note that with a two layer neural network basis, we have increased the number of internal parameters which are nonlinearly related layer by layer. Specifically, in addition to the first layer bias parameter $c_m^{(1)}$ and $M_2 \times 1$ weight vector $\mathbf{v}_m^{(1)}$, we also have in the second layer M_2 bias parameters and M_2 weight vectors each of length N.

In Fig. 5.7 we show four instances for each of the two basis function types[9] in Equations (5.10) and (5.11), where $N = 1$ and $M_2 = 1000$. Note that the two layer basis functions in Fig. 5.7 are clearly more diverse in shape than the single layer basis functions shown in Fig. 5.5.

To achieve even greater flexibility for individual basis features we can create a *neural network with three hidden layers* (or more) by simply repeating the procedure used to create the two layer basis from the single layer version. That is, we take a weighted sum of two hidden layer basis features and pass the result through an activation function (of the same kind as used in the two layer basis). For example, performing this procedure for the two layer hinge basis function in Equation (5.11) gives a three layer network basis function of the form

$$f_m(\mathbf{x}) = \max\left(0,\; c_m^{(1)} + \sum_{m_2=1}^{M_2} \max\left(0,\; c_{m_2}^{(2)} + \sum_{m_3=1}^{M_3} \max\left(0,\; c_{m_3}^{(3)} + \mathbf{x}^T \mathbf{v}_{m_3}^{(3)}\right) v_{m_3,m_2}^{(2)}\right) v_{m_2,m}^{(1)}\right), \tag{5.12}$$

[9] Although not a common choice, one can mix and match different activation functions for different layers, as in $f_m(\mathbf{x}) = \tanh\left(c_m^{(1)} + \sum_{m_2=1}^{M_2} \max\left(0,\; c_{m_2}^{(2)} + \mathbf{x}^T \mathbf{v}_{m_2}^{(2)}\right) v_{m_2,m}^{(1)}\right)$.

for all $m \geq 1$. This procedure can be repeated to produce a neural network basis with an arbitrary number of hidden layers. Currently the convention is to refer to a neural network basis with three or more hidden layers as a *deep network* [6–10].

5.1.5 Recovering weights

As with vector approximation, a common way to tune the weights $\{w_m\}_{m=0}^{M}$, as well as the possible internal parameters of the basis features themselves when neural networks are used, is to minimize a Least Squares cost function. In this instance we seek to minimize the difference between y and its partial basis approximation in Equation (5.5), over all points in the unit hypercube denoted by $[0, 1]^N$. Stated formally, the minimization of this Least Squares cost is written as

$$\underset{w_0,\ldots,w_M,\Theta}{\text{minimize}} \int_{\mathbf{x}\in[0,1]^N} \left(\sum_{m=0}^{M} f_m(\mathbf{x})\, w_m - y(\mathbf{x})\right)^2 d\mathbf{x}, \tag{5.13}$$

where the set Θ contains possible parameters of the basis elements themselves, which is empty if a fixed basis is used. Note that Equations (5.3) and (5.13) are entirely analogous Least Squares problems for learning the weights associated to a partial basis approximation, the former stated for vectors in $\mathbb{R}^P$ and the latter for continuous functions in C^N.

Unlike its vector counterpart, however, the Least Squares problem in (5.13) cannot typically[10] be solved in closed form due to intractability of the integrals involved. Instead, one typically solves an approximate form of the problem where each function is first discretized, as will be described in Section 5.2.1. This makes a tractable problem which, as we will soon see, naturally leads to a model for the general problem of regression.

5.1.6 Graphical representation of a neural network

It is common to represent the weighted sum of M neural network basis features,

$$r = b + \sum_{m=1}^{M} f_m(\mathbf{x})\, w_m, \tag{5.14}$$

graphically to visualize the compositional structure of each element. Here $f_m(\mathbf{x})$ can be a general multilayer neural network feature as described previously.

As a simple example, in Fig. 5.8 we represent graphically the mathematical expression $a(x_1v_1 + x_2v_2 + x_3v_3)$ where $a(\cdot)$ is any activation function. This graphical representation consists of: 1) weighted edges that represent the individual multiplications (i.e., of x_1 by v_1, x_2 by v_2, and x_3 by v_3); 2) a summation unit representing the sum $x_1v_1 + x_2v_2 + x_3v_3$ (shown as a small hollow circle); and finally 3) an activation unit representing the sum evaluated by the activation function (shown as a larger blue circle).

[10] While finding weights associated to a fixed basis is simple in theory (see chapter exercises), discretization is still usually required.

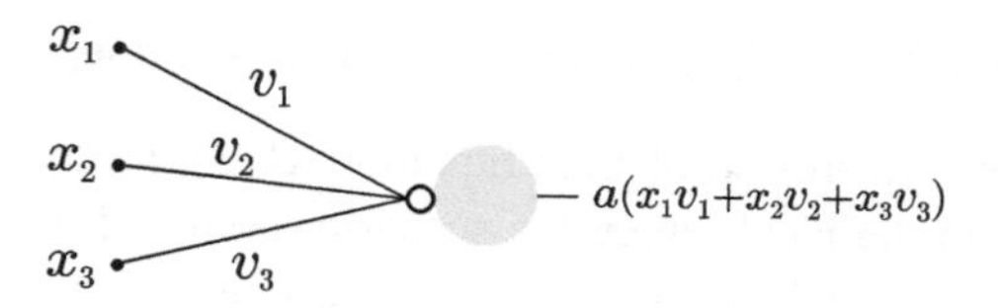

Fig. 5.8 Representation of a single activation function with $N = 3$-dimensional input. Each input is connected to the summation unit via a weighted edge. The summation unit (shown by a small hollow circle) takes in the weighted inputs, and outputs their sum to the activation unit (shown by a large blue circle). See text for further details.

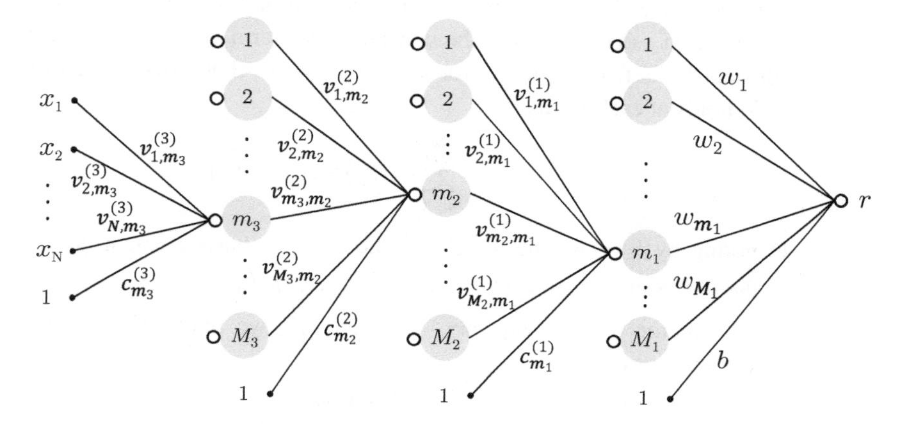

Fig. 5.9 A three hidden layer neural network. Note that for visualization purposes, we just show the edges connecting the units in one layer to only one of the units in the previous layer. See text for further details.

By cascading this manner of representing the input/output relationship of a single activation function we may similarly represent sums of multilayer neural network basis features. For example, in Fig. 5.9 we show a graphical representation of (5.14) with $M = M_1$ three hidden layer network basis features taking in general N dimensional input (like the one shown in Equation (5.12)). As in Fig. 5.8, the input $\mathbf{x}$ is shown entry-wise on the left and the output r on the right. In between are each of the three hidden layers, from right to left each consisting of M_1, M_2, and M_3 hidden units respectively. Some texts number these hidden layers in ascending order from left to right. While this slightly changes the notation we use in this book, all results and conclusions remain the same.

5.2 Automatic feature design for the real regression scenario

Here we describe how fixed and adjustable bases of features, introduced in the previous section, are applied to the automatic design of features in the real regression scenario. Although they lose their power as perfect feature design tools (which they had in the case of the ideal regression scenario), strong features can often be built by combining elements of bases for real instances of regression.

5.2.1 Approximation of discretized continuous functions

The Least Squares problem in Equation (5.13), for tuning the weights of a sum of basis features in the ideal regression scenario discussed in Section 5.1 is highly intractable. However, by finely discretizing all of the continuous functions involved we can employ standard optimization tools to solve a discrete version of the problem. Recall from Section 5.1.2 that given any function y defined over the unit hypercube, we may sample with a finely spaced grid over unit hypercube $[0,1]^N$ a potentially large (but finite) number of P points so that the collection of pairs $\left\{\left(\mathbf{x}_p, y\left(\mathbf{x}_p\right)\right)\right\}_{p=1}^{P}$ resembles the function $y(\mathbf{x})$ as well as desired. Using such a discretization scheme we can closely approximate the function $y(\mathbf{x})$, as well as the constant basis term $f_0(\mathbf{x}) = 1$ and any set of M non-constant basis features $f_m(\mathbf{x})$ for $m = 1 \ldots M$, so that a discretized form of Equation (5.5) holds at each $\mathbf{x}_p$, as

$$\sum_{m=0}^{M} f_m\left(\mathbf{x}_p\right) w_m \approx y\left(\mathbf{x}_p\right). \tag{5.15}$$

Denoting by $b = w_0$ the weight on the constant basis element, $y_p = y\left(\mathbf{x}_p\right)$, as well as the compact *feature vector* notation $\mathbf{f}_p = \left[\begin{array}{cccc} f_1\left(\mathbf{x}_p\right) & f_2\left(\mathbf{x}_p\right) & \cdots & f_M\left(\mathbf{x}_p\right)\end{array}\right]^T$ and weight vector $\mathbf{w} = \left[\begin{array}{cccc} w_1 & w_2 & \cdots & w_M\end{array}\right]^T$, we may write Equation (5.15) more conveniently as

$$b + \mathbf{f}_p^T \mathbf{w} \approx y_p. \tag{5.16}$$

In order for this approximation to hold we can then consider minimizing the squared difference between both sides over all P, giving the associated Least Squares problem

$$\underset{b,\mathbf{w},\Theta}{\text{minimize}} \sum_{p=1}^{P} \left(b + \mathbf{f}_p^T \mathbf{w} - y_p\right)^2. \tag{5.17}$$

Note that once again we denote by Θ the set of internal parameters of all basis features, which is empty in the case of fixed bases. This is precisely a discretized form of the continuous Least Squares problem shown originally in Equation (5.13). Also note that this is only a slight generalization of the Least Squares problem for regression discussed throughout Chapter 3.

We illustrate the idea of approximating a continuous function from a discretized version of it, using a particular example in Fig. 5.10, where the continuous function $y(x) = \sin(2\pi x)$ is shown in the left panel along with its discretized version in the middle panel. Employing a polynomial basis, and using weights provided by solving the Least Squares problem in Equation (5.17), we have an excellent approximation of the true function in the right panel.

5.2.2 The real regression scenario

The approximation of a finely discretized continuous function provides an almost ideal scenario for regression where the data, the sampled points, gives a clear (noiseless)

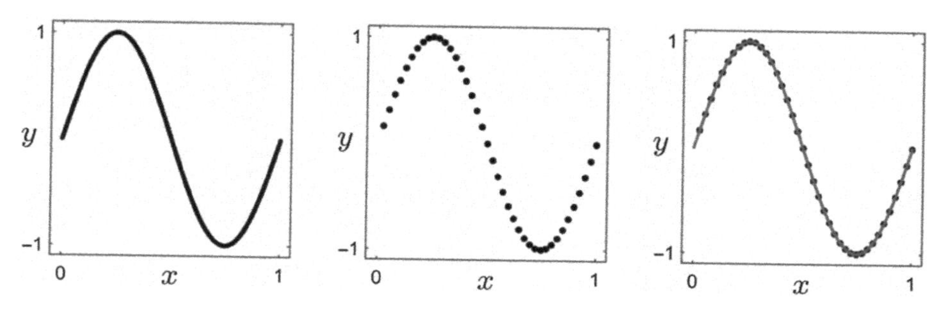

Fig. 5.10 (left panel) The continuous function $y(x) = \sin(2\pi x)$ defined on the unit interval and (middle panel) its discretized version made by evaluating $y(x)$ over a sequence of $P = 40$ evenly spaced points on the horizontal axis. (right panel) Fitting a degree $M = 12$ polynomial curve (in blue) to the discretized function via solving (5.17) provides an excellent continuous approximation to the original function.

picture of the underlying function over its entire domain. However, rarely in practice do we have access to such large quantities of noiseless data which span the entire input space of a phenomenon. Conversely, a dataset seen in practice may consist of only a small number of samples, these samples may not be distributed evenly in space, and they may be corrupted by measurement error or some other sort of "noise." Indeed most datasets for regression are akin to noisy samples of some unknown continuous function making the machine learning task of regression, in general, a function approximation problem based only on noisy samples of the underlying function.

> The general instance of regression is a function approximation problem based on noisy samples of the underlying function.

To illustrate this idea, in the right panel of Fig. 5.11 we show a simulated example of a realistic regression dataset consisting of $P = 21$ data points $\left\{\left(x_p, y_p\right)\right\}_{p=1}^{P}$. This dataset is made by taking samples of the function $y(x) = \sin(2\pi x)$, where each input x_p is chosen randomly on the interval $[0, 1]$ and evaluated by the function with the addition of noise ϵ_p as $y_p = y\left(x_p\right) + \epsilon_p$. Here this noise simulates common small errors made, for instance, in the collection of regression data. Also shown in this figure for comparison is the ideal dataset previously shown in the middle panel of Fig. 5.10.

With a realistic regression dataset, we use the same learning framework to find optimal parameters as we did with ideal discretized data, i.e., by solving the discrete Least Squares problem in (5.17). Note again that depending on the basis type (fixed or adjustable) used in (5.17), the design of the feature vector $\mathbf{f}_p$ changes.

In Fig. 5.12 we show various fits to the toy sinusoidal dataset in the right panel of Fig. 5.11, using a polynomial (where $M = 3$), Fourier (where $M = 1$), and single hidden layer neural network basis with $\tanh(\cdot)$ activation function (where $M = 4$). Details on

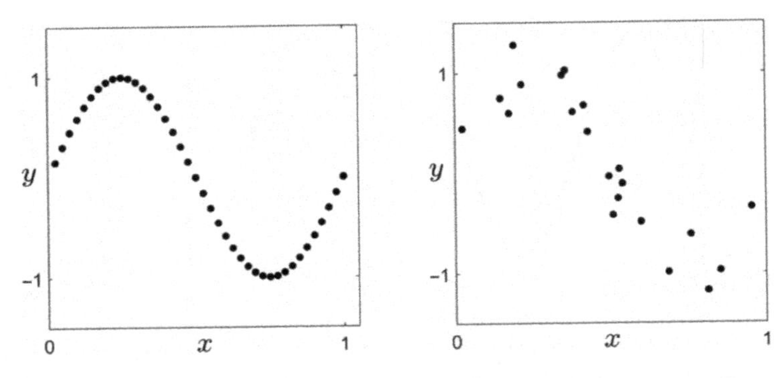

Fig. 5.11 (left panel) An ideal dataset for regression made by finely and evenly sampling the continuous function $y(x) = \sin(2\pi x)$ over the unit interval. (right panel) A more realistic simulated dataset for regression made by evaluating the same function at a smaller number of random input points and corrupting the result by adding noise to simulate e.g., errors made in data collection.

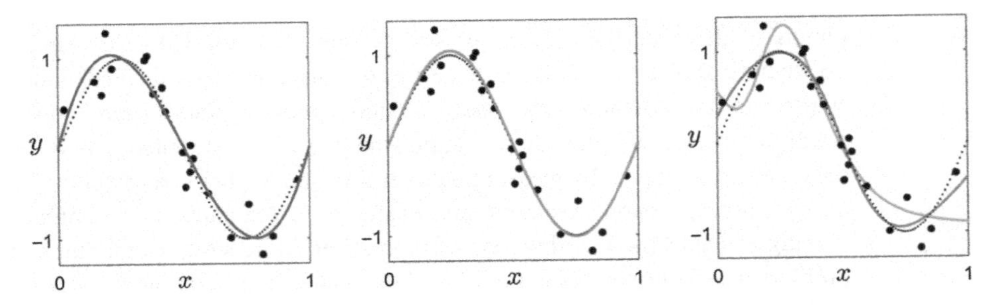

Fig. 5.12 A comparison of (left panel) polynomial, (middle panel) Fourier, and (right panel) single layer neural network fits to the regression dataset shown in the right panel of Fig. 5.11. The optimal weights in each case are found by minimizing the Least Squares cost function as described in Examples 5.1 (for polynomial and Fourier features) and 5.2 (for neural network features). Two solutions shown in the right panel correspond to the finding of poor (in green) and good (in magenta) stationary points when minimizing the non-convex Least Squares cost in the case of neural network features.

how these weights were learned in each instance are discussed in separate examples following the figure.

Example 5.1 Regression with fixed bases of features

To perform regression using a fixed basis of features (e.g., polynomials or Fourier) it is natural to choose a degree D and transform the input data using the associated basis functions. For example, employing a degree D polynomial or Fourier basis for a scalar input, we transform each input x_p to form an associated feature vector $\mathbf{f}_p = \begin{bmatrix} x_p & x_p^2 & \cdots & x_p^D \end{bmatrix}^T$ or $\mathbf{f}_p = \begin{bmatrix} \cos(2\pi x_p) & \sin(2\pi x_p) & \cdots & \cos(2\pi D x_p) \end{bmatrix}$ $\sin(2\pi D x_p) \big]^T$ respectively. For higher dimensions of input N fixed basis features can be similarly used; however, the sheer number of elements involved (the length of each $\mathbf{f}_p$)

explodes[11] for even moderate values of N and D (as we will see in Section 7.1, this issue can be ameliorated via the notion of a "kernal," however, this introduces a serious numerical optimization problem as the size of the data-set grows).

In any case, once feature vectors $\mathbf{f}_p$ have been constructed using the data we can then determine proper weights b and $\mathbf{w}$ by minimizing the Least Squares cost function as

$$\underset{b,\,\mathbf{w}}{\text{minimize}} \sum_{p=1}^{P} \left(b + \mathbf{f}_p^T \mathbf{w} - y_p\right)^2. \tag{5.18}$$

Using the compact notation $\tilde{\mathbf{w}} = \begin{bmatrix} b \\ \mathbf{w} \end{bmatrix}$ and $\tilde{\mathbf{f}}_p = \begin{bmatrix} 1 \\ \mathbf{f}_p \end{bmatrix}$ for each p we may rewrite the cost as $g\left(\tilde{\mathbf{w}}\right) = \sum_{p=1}^{P} \left(\tilde{\mathbf{f}}_p^T \tilde{\mathbf{w}} - y_p\right)^2$, and checking the first order condition then gives the linear system of equations

$$\left(\sum_{p=1}^{P} \tilde{\mathbf{f}}_p \tilde{\mathbf{f}}_p^T\right) \tilde{\mathbf{w}} = \sum_{p=1}^{P} \tilde{\mathbf{f}}_p y_p, \tag{5.19}$$

that when solved recovers an optimal set of parameters $\tilde{\mathbf{w}}$.

Example 5.2 Regression with a basis of single hidden layer neural network features

The feature vector of the input $\mathbf{x}_p$ made by using a basis of single hidden layer neural network features takes the form

$$\mathbf{f}_p = \begin{bmatrix} a\left(c_1 + \mathbf{x}_p^T \mathbf{v}_1\right) & a\left(c_2 + \mathbf{x}_p^T \mathbf{v}_2\right) & \cdots & a\left(c_M + \mathbf{x}_p^T \mathbf{v}_M\right) \end{bmatrix}^T, \tag{5.20}$$

where $a(\cdot)$ is any activation function as detailed in Section 5.1.4. However, unlike the case with fixed feature bases, the corresponding Least Squares problem

$$\underset{b,\,\mathbf{w},\,\Theta}{\text{minimize}} \sum_{p=1}^{P} \left(b + \mathbf{f}_p^T \mathbf{w} - y_p\right)^2, \tag{5.21}$$

cannot be solved in closed form due to the internal parameters (denoted all together in the set Θ) that are related in a nonlinear fashion with the basis weights b and $\mathbf{w}$. Moreover, this problem is almost always non-convex, and so several runs of gradient descent are typically made in order to ensure convergence to a good local minimum.

[11] For a general N-dimensional input $\mathbf{x}$, a degree D polynomial basis-feature transformation includes all monomials of the form $f_m(\mathbf{x}) = x_1^{m_1} x_2^{m_2} \cdots x_N^{m_N}$ where $0 \leq m_1 + m_2 + \cdots + m_N \leq D$. Similarly, a degree D Fourier expansion contains basis elements of the form $f_m(\mathbf{x}) = e^{2\pi i m_1 x_1} e^{2\pi i m_2 x_2} \cdots e^{2\pi i m_N x_N}$ where $-D \leq m_1, m_2, \cdots, m_N \leq D$. Containing all non-constant terms, one can easily show that the associated polynomial and Fourier feature vectors have length $M = \frac{(N+D)!}{N!D!} - 1$ and $M = (2D+1)^N - 1$, respectively. Note that in both cases the feature vector dimension grows extremely rapidly in N and D, which can lead to serious practical problems even with moderate amounts of N and D.

Calculating the full gradient of the cost g, we have a vector of length $Q = M(N+2)+1$ containing the derivatives of the cost with respect to each variable,

$$\nabla g = \left[\begin{array}{cccccccccc} \frac{\partial}{\partial b} g & \frac{\partial}{\partial w_1} g & \cdots & \frac{\partial}{\partial w_M} g & \frac{\partial}{\partial c_1} g \cdots & \frac{\partial}{\partial c_M} g & \nabla_{\mathbf{v}_1}^T g & \cdots & \nabla_{\mathbf{v}_M}^T g \end{array} \right]^T, \quad (5.22)$$

where the derivatives are easily calculated using the chain rule (see Exercise 5.9).

Example 5.3 Regression with a basis of multiple hidden layer neural network features

As discussed in Section 5.1.4, by increasing the number of layers in a neural network each basis function gains more flexibility. In their use with machine learning, this added flexibility comes at the practical expense of making the corresponding cost function more challenging to minimize. In terms of numerical optimization the primary challenge with deep net features is that the associated cost function can become highly non-convex (i.e., having many local minima and/or saddle points). Proper choice of activation function can help ameliorate this problem, e.g., the hinge or rectified linear function $a(x) = \max(0, x)$ has been shown to work well for deep nets (see e.g., [36]). Furthermore, as discussed in Section 3.3.2, regularization can also be used to improve this problem.

A practical implementation issue with using deep net features is that computing the gradient, for use with gradient descent (often referred to in the machine learning community as *the backpropagation algorithm*) of associated cost functions becomes more cumbersome as additional layers are added, requiring several applications of the chain rule and careful book-keeping to ensure no errors are made. For the interested reader we provide an organized derivation of the gradient for a cost function that employs deep net features in Section 7.2. To avoid potential errors in computing the derivatives of a deep net cost function by hand, computational techniques like automatic differentiation [59] are often utilized when using deep nets in practice.

5.3 Cross-validation for regression

In the ideal instance of regression, where we look to approximate a continuous function using a fixed or adjustable (neural network) basis of features, we saw in Section 5.1 that using more elements of a basis results in a better approximation (see e.g., Fig. 5.3). In short, in the context of continuous function approximation more (basis elements) is always better. Does the same principle apply in the real instance of regression, i.e., in the case of a noisily sampled function approximation problem? Unfortunately, *no*.

Take for example the semi-ideal and realistic sinusoidal datasets shown in Fig. 5.13, along with polynomial fits of degrees three (in blue) and ten (in purple). In the left panel of this figure, which shows the discrete sinusoidal dataset with evenly spaced points, by increasing the number of basis features M from 3 to 10 the corresponding polynomial model fits the data *and* the underlying function better. Conversely, in the right panel while the model fits the data better as we increase the number of polynomial features

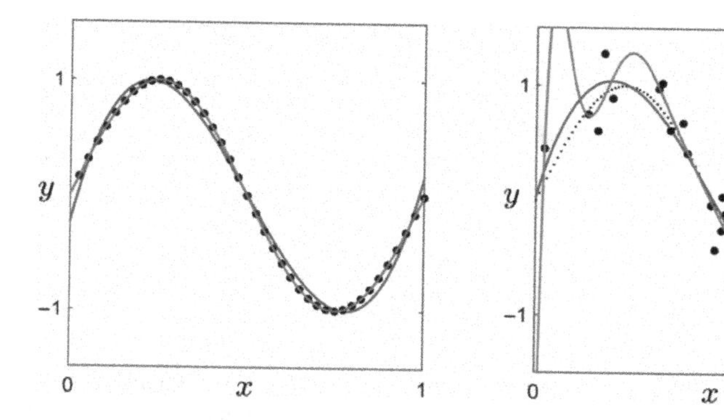

Fig. 5.13 Plots of (left panel) discretized and (right panel) noisy samples of the data-generating function $y(x) = \sin(2\pi x)$, along with its degree three and degree ten polynomial approximations in blue and purple, respectively. While the higher degree polynomial does a better job at modeling both the discretized data and underlying function, it only fits the noisy sample data better, providing a worse approximation of the underlying data-generating function than the lower degree polynomial. Using cross-validation we can determine the more appropriate model, in this case the degree three polynomial, for such a dataset.

from $M = 3$ to 10, the representation of the underlying data-generating function actually gets *worse*. Since the underlying function is the object we truly wish to understand, this is a problem.

The phenomenon illustrated through this simple example is in fact true more generally: by increasing the number M of any type of basis features (fixed or neural network) we can indeed produce better fitting models of a dataset, but at the potential cost of creating poorer representations of the data-generating function we care foremost about. Stated formally, given any dataset we can drive the value of the Least Squares cost to zero via solving the minimization problem

$$\underset{b,\,\mathbf{w},\Theta}{\text{minimize}} \sum_{p=1}^{P} \left(b + \mathbf{f}_p^T \mathbf{w} - y_p\right)^2, \tag{5.23}$$

by increasing M where $\mathbf{f}_p = \begin{bmatrix} f_1(\mathbf{x}_p) & f_2(\mathbf{x}_p) & \cdots & f_M(\mathbf{x}_p) \end{bmatrix}^T$. Therefore, choosing M correctly is extremely important. Note in the language of machine learning, a model corresponding to too large a choice of M is said to *overfit* the data. Likewise when choosing M too small[12] the model is said to *underfit* the data.

In this section we describe *cross-validation*, an effective framework for choosing the proper value for M automatically and intelligently so as to prevent the problem of underfitting/overfitting. For example, in the case of the data shown in the right panel of Fig. 5.13 cross-validation will determine $M = 3$ the better model, as opposed to $M = 12$. This discussion will culminate in the description of a specific procedure known as *k-fold cross-validation* which is commonly used in practice.

[12] For instance, using a degree $M = 1$ polynomial feature we can only find the best linear fit to the data in Fig. 5.13, which would be not only a poor fit to the observed data but also a poor representation of the underlying sinusoidal pattern.

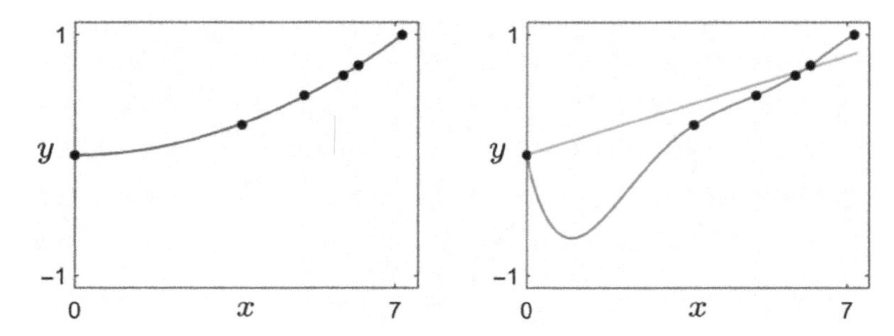

Fig. 5.14 Data from Galileo's simple ramp experiment from Example 3.3, exploring the relationship between time and the distance an object falls due to gravity. (left panel) Galileo fit a simple quadratic to the data. (right panel) A linear model (shown in green) is not flexible enough and as a result, underfits the data. A degree twelve polynomial (shown in magenta) overfits the data, being too complicated and unnatural (between the start and 0.25 of the way down the ramp the ball travels a negative distance!) to be a model of a simple natural phenomenon.

Example 5.4 Overfitting and underfitting Galileo's ramp data

In the left panel of Fig. 5.14 we show the data from Galileo's classic ramp experiment, initially described in Example 1.7, performed in order to understand the relationship between time and the acceleration of an object due to (the force we today know as) gravity. Also shown in this figure is (left panel) the kind of quadratic fit Galileo used to describe the underlying relationship traced out by the data, along with two other possible model choices (right panel): a linear fit in green, as well as a degree 12 polynomial fit in magenta. Of course the linear model is inappropriate, as with this data any line would have large squared error (see e.g., Fig. 3.3) and would thus be a poor representation of the data. On the other hand, while the degree 12 polynomial fits the data-set perfectly, with corresponding squared error value of zero, the model itself just "looks wrong."

Examining the right panel of this figure why, for example, when traveling between the beginning and a quarter of the way down the ramp, does the distance the ball travels become negative! This kind of behavior does not at all match our intuition or expectation about how gravity should operate on an object. This is why Galileo chose a quadratic, rather than a higher order degree polynomial, to fit such a data-set: because he *expected* that the rules which govern our universe are explanatory yet simple.

This principle, that the rules we use to describe our universe should be flexible yet simple, is often called *Occam's Razor* and lies at the heart of essentially all scientific inquiry past and present. Since machine learning can be thought of as a set of tools for making sense of arbitrary kinds of data, i.e., not only data relating to a physical system or law, we want the relationship learned in solving a regression (or classification) problem to also satisfy this basic Occam's Razor principle. In the context of machine learning, Occam's Razor manifests itself geometrically, i.e., we expect the model (or function) underlying our data to be simple yet flexible enough to explain the data we have. The linear model in Fig. 5.14, being too rigid and inflexible to establish the relationship between time and the distance an object falls due to gravity, fits very poorly. As previously mentioned, in machine learning such a model is said to underfit the data we have. On the other hand, the degree 12 polynomial model is needlessly complicated, resulting in a very close fit to the data we have, but is far too oscillatory to be representative of the underlying phenomenon and is said to overfit the data.

5.3.1 Diagnosing the problem of overfitting/underfitting

A reasonable diagnosis of the overfitting/underfitting problems is that both fail at representing *new* data, generated via the same process by which the current data was made, that we can potentially receive in the future. For example, the overfitting degree ten polynomial shown in the right panel of Fig. 5.13 would poorly model any future data generated by the same process since it poorly represents the underlying data-generating function (a sinusoid). This data-centric perspective provokes a practical criterion for determining an ideal choice of M for a given dataset: the number M of basis features used should be such that the corresponding model fits well to both the current dataset as well as to new data we will receive in the future.

5.3.2 Hold out cross-validation

While we of course do not have access to any "new data we will receive in the future," we can *simulate* such a scenario by splitting our data into two subsets: a larger *training set* of data we already have, and a smaller *testing set* of data that we "will receive in the future." Then, we can try a range of values for M by fitting each to the training set of known data, and pick the one that performs the best on our testing set of unknown data. By keeping a larger portion of the original data as the training set we can safely assume that the learned model which best represents the testing data will also fit the training set fairly well. In short, by employing this sort of procedure for comparing a set of models, referred to as *hold out cross-validation*, we can determine a candidate that approximately satisfies our criterion for an ideal well-fitting model.

What portion of our dataset should we save for testing? *There is no hard rule*, and in practice typically between $1/10$ to $1/3$ of the data is assigned to the testing set. One general rule of thumb is that the larger the dataset (given that it is relatively clean and well distributed) the bigger the portion of the original data may be assigned to the testing set (e.g., $1/3$ may be placed in the testing set) since the data is plentiful enough for the training data to still accurately represent the underlying phenomenon. Conversely, in general with smaller or less rich (i.e., more noisy or poorly distributed) datasets we should assign a smaller portion to the testing set (e.g., $1/10$ may be placed in the testing set) so that the relatively larger training set retains what little information of the underlying phenomenon was captured by the original data.

In general the larger/smaller the original dataset the larger/smaller the portion of the original data that should be assigned to the testing set.

Fig. 5.15 Hold out cross-validation. The original data (left panel) shown here as the entire circular mass is split randomly (middle panel) into k non-overlapping sets (here $k = 3$). (right panel) One piece, or $1/k$ of the original dataset, is then taken randomly as the testing set with the remaining pieces, or $k-1/k$ of the original data, taken as the training set.

As illustrated in Fig. 5.15, to form the training and testing sets we split the original data randomly into k non-overlapping parts and assign 1 portion for testing ($1/k$ of the original data) and $k - 1$ portions to the training set ($k-1/k$ of the original data).

Regardless of the value we choose for k, we train our model on the training set using a range of different values of M. We then evaluate how well each model (or in other words, each value of M) fits to both the training and testing sets, via measuring the model's *training error* and *testing error*, respectively. The best-fitting model is chosen as the one providing the lowest testing error or the best fit to the "unseen" testing data. Finally, in order to leverage the full power of our data we use the optimal number of basis features M to train our model, this time using the entire data (both training and testing sets).

Example 5.5 Hold out for regression using Fourier features

To solidify these details, in Fig. 5.16 we show an example of applying hold out cross-validation using a dataset of $P = 30$ points generated via the function $y(x)$ shown in Fig. 5.3. To perform hold out cross-validation on this dataset we randomly partition it into $k = 3$ equal-sized (ten points each) non-overlapping subsets, using two partitions together as the training set and the final part as testing set, as illustrated in the left panel of Fig. 5.16. The points in this panel are colored blue and yellow indicating that they belong to the training and testing sets respectively. We then train our model on the training set (blue points) by solving several instances of the Least Squares problem in (5.18). In particular we use a range of even values for M Fourier features $M = 2, 4, 6, \ldots, 16$ (since Fourier elements naturally come in pairs of two as shown in Equation (5.7)) which corresponds to the range of degrees $D = 1, 2, 3, \ldots, 8$ (note that for clarity panels in the figure are indexed by D).

Based on the models learned for each value of M (see the middle set of eight panels of the figure) we plot training and testing errors (in the panel second from the right), measuring how well each model fits the training and testing data respectively, over the entire range of values. Note that unlike the testing error, the training error always decreases as we increase M (which occurs more generally regardless of the dataset/feature basis used). The model that provides the smallest testing error ($M^{\star} = 10$ or equivalently

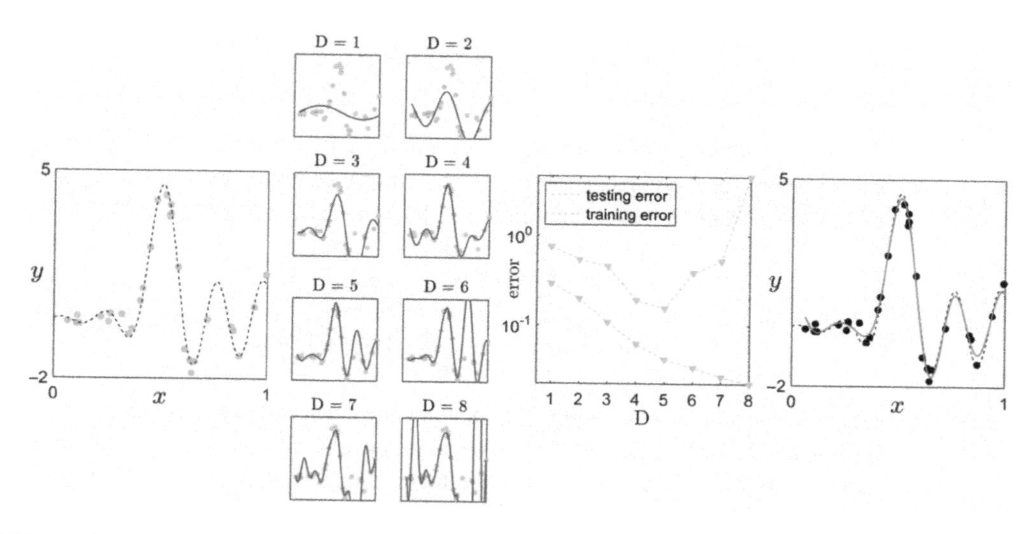

Fig. 5.16 An example of hold out cross-validation applied to a simple dataset using Fourier features. (left panel) The original data split into training and testing sets, with the points belonging to each set colored blue and yellow respectively. (middle eight panels) The fit resulting from each set of degree D Fourier features in the range $D = 1, 2, \ldots, 8$ is shown in blue in each panel. Note how the lower degree fits underfit the data, while the higher degree fits overfit the data. (second from right panel) The training and testing errors, in blue and yellow respectively, of each fit over the range of degrees tested. From this we see that $D^{\star} = 5$ (or $M^{\star} = 10$) provides the best fit. Also note how the training error always decreases as we increase the degree/number of basis elements, which will always occur regardless of the dataset/feature basis type used. (right panel) The final model using $M^{\star} = 10$ trained on the entire dataset (shown in red) fits the data well and closely matches the underlying data generating function (shown in dashed black).

$D^{\star} = 5$) is then trained again on the entire dataset, giving the final regression model shown in red in the rightmost panel of Fig. 5.16.

5.3.3 Hold out calculations

Here we give a complete set of hold out cross-validation calculations in a general setting. We denote the collection of points belonging to the training and testing sets respectively by their indices as

$$\begin{aligned}\Omega_{\text{train}} &= \left\{p \mid \left(\mathbf{x}_p,\, y_p\right) \text{ belongs to the training set}\right\}\\ \Omega_{\text{test}} &= \left\{p \mid \left(\mathbf{x}_p,\, y_p\right) \text{ belongs to the testing set}\right\}.\end{aligned} \tag{5.24}$$

We then choose a basis type (e.g., polynomial, Fourier, neural network) and choose a range for the number of basis features over which we search for an ideal value for M. To determine the training and testing error of each value of M tested we first form the corresponding feature vector $\mathbf{f}_p = \left[\, f_1\left(\mathbf{x}_p\right) \;\; f_2\left(\mathbf{x}_p\right) \;\; \cdots \;\; f_M\left(\mathbf{x}_p\right) \,\right]^T$ and fit a corresponding model to the training set by solving the corresponding[13] Least Squares problem

[13] Once again, for a fixed basis this problem may be solved in closed form since Θ is empty, while for neural networks it must be solved via gradient descent (see e.g., Example 5.2).

$$\underset{b,\mathbf{w},\Theta}{\text{minimize}} \sum_{p\in\Omega_{\text{train}}} \left(b+\mathbf{f}_p^T\mathbf{w}-y_p\right)^2. \tag{5.25}$$

Denoting a solution to the problem above as $\left(b_M^\star,\ \mathbf{w}_M^\star,\ \Theta_M^\star\right)$ we find the training and testing errors for the current value of M by simply computing the mean squared error using these parameters over the training and testing sets, respectively

$$\begin{aligned}\text{Training error} &= \frac{1}{|\Omega_{\text{train}}|}\sum_{p\in\Omega_{\text{train}}}\left(b_M^\star+\mathbf{f}_p^T\mathbf{w}_M^\star-y_p\right)^2\\ \text{Testing error} &= \frac{1}{|\Omega_{\text{test}}|}\sum_{p\in\Omega_{\text{test}}}\left(b_M^\star+\mathbf{f}_p^T\mathbf{w}_M^\star-y_p\right)^2,\end{aligned} \tag{5.26}$$

where the notation $|\Omega_{\text{train}}|$ and $|\Omega_{\text{test}}|$ denotes the cardinality or number of points in the training and testing sets, respectively. Once we have performed these calculations for all values of M we wish to test, we choose the one that provides the lowest testing error, denoted by $M^\star$.

Finally we form the feature vector $\mathbf{f}_p = \left[\ f_1\left(\mathbf{x}_p\right)\ \ f_2\left(\mathbf{x}_p\right)\ \ \cdots\ \ f_{M^\star}\left(\mathbf{x}_p\right)\ \right]^T$ for all the points in the entire dataset, and solve the Least Squares problem over the entire dataset to form the final model

$$\underset{b,\mathbf{w},\Theta}{\text{minimize}} \sum_{p=1}^{P}\left(b+\mathbf{f}_p^T\mathbf{w}-y_p\right)^2. \tag{5.27}$$

5.3.4 *k*-fold cross-validation

While the hold out method previously described is an intuitive approach to determining proper fitting models, it suffers from an obvious flaw: having been chosen at random, the points assigned to the training set may not adequately describe the original data. However, we can easily extend and robustify the hold out method as we now describe.

As illustrated in Fig. 5.17 for $k = 3$, with *k-fold cross-validation* we once again randomly split our data into k non-overlapping parts. By combining $k-1$ parts we can, as with the hold out method, create a large training set and use the remaining single fold as a test set. With k-fold cross-validation we will repeat this procedure k times (each instance being referred to as a *fold*), in each instance using a different single portion of the split as testing set and the remaining $k-1$ parts as the corresponding training set, and computing the training and testing errors of all values of M as described in the previous section. We then choose the value of M that has the lowest *average testing error*, a more robust choice than the hold out method provides, that can average out a scenario where one particular choice of training set inadequately describes the original data.

Note, however, that this advantage comes at a cost: k-fold cross-validation is (approximately) k times more computationally costly than its hold out counterpart. In fact performing k-fold cross-validation is often the most computationally expensive process performed to solve a regression problem.

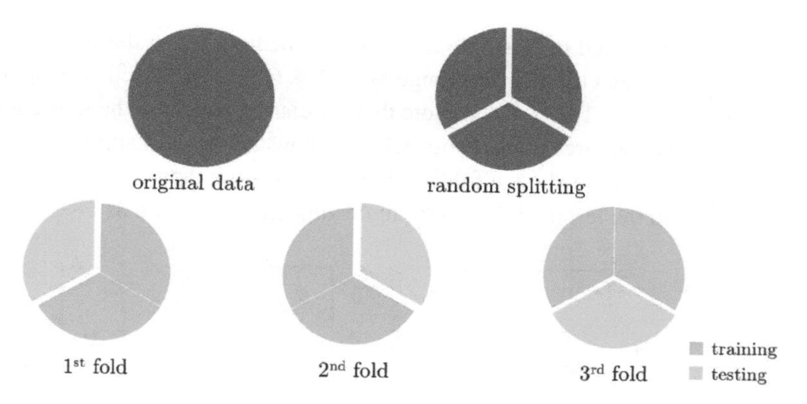

Fig. 5.17 k-fold cross-validation for $k = 3$. The original data shown here as the entire circular mass (top left) is split into k non-overlapping sets (top right) just as with the hold out method. However with k-fold cross-validation we repeat the hold out calculations k times (bottom), once per "fold," in each instance, keeping a different portion of the split data as the testing set while merging the remaining $k - 1$ pieces as the training set.

> Performing k-fold cross-validation is often the most computationally expensive component in solving a general regression problem.

There is again no universal rule for the number k of non-overlapping partitions (or the number of folds) to break the original data into. However, the same intuition previously described for choosing k with the hold out method also applies here, as well as the same convention with popular values of k ranging from $k = 3 \ldots 10$ in practice.

For convenience we provide a pseudo-code for applying k-fold cross-validation in Algorithm 5.1.

Algorithm 5.1 k-fold cross-validation pseudo-code

Input: Data-set $\left\{\left(\mathbf{x}_p, y_p\right)\right\}_{p=1}^{P}$, k (number of folds), a range of values for M to try, and a type of basis feature
Split the data into k equal (as possible) sized folds
for $s = 1 \ldots k$
  for each M (in the range of values to try)
    1) Train a model with M basis features on sth fold's training set
    2) Compute corresponding testing error on this fold
Return: value $M^{\star}$ with lowest *average* testing error over all k folds

---

Example 5.6 k-fold cross-validation for regression using Fourier features

In Fig. 5.18 we illustrate the result of applying k-fold cross-validation to choose the ideal number M of Fourier features for the dataset shown in Example 5.5, where it was originally used to illustrate the hold out method. As in the previous example, here we set $k = 3$ and try M in the range $M = 2, 4, 6, \ldots, 16$, which corresponds to the range of degrees $D = 1, 2, 3, \ldots, 8$ (note that for clarity, panels in the figure are indexed by D).

In the top three rows of Fig. 5.18 we show the result of applying hold out on each fold. In each row we show a fold's training and testing data colored blue and yellow respectively in the left panel, the training/testing errors for each M on the fold (as computed in

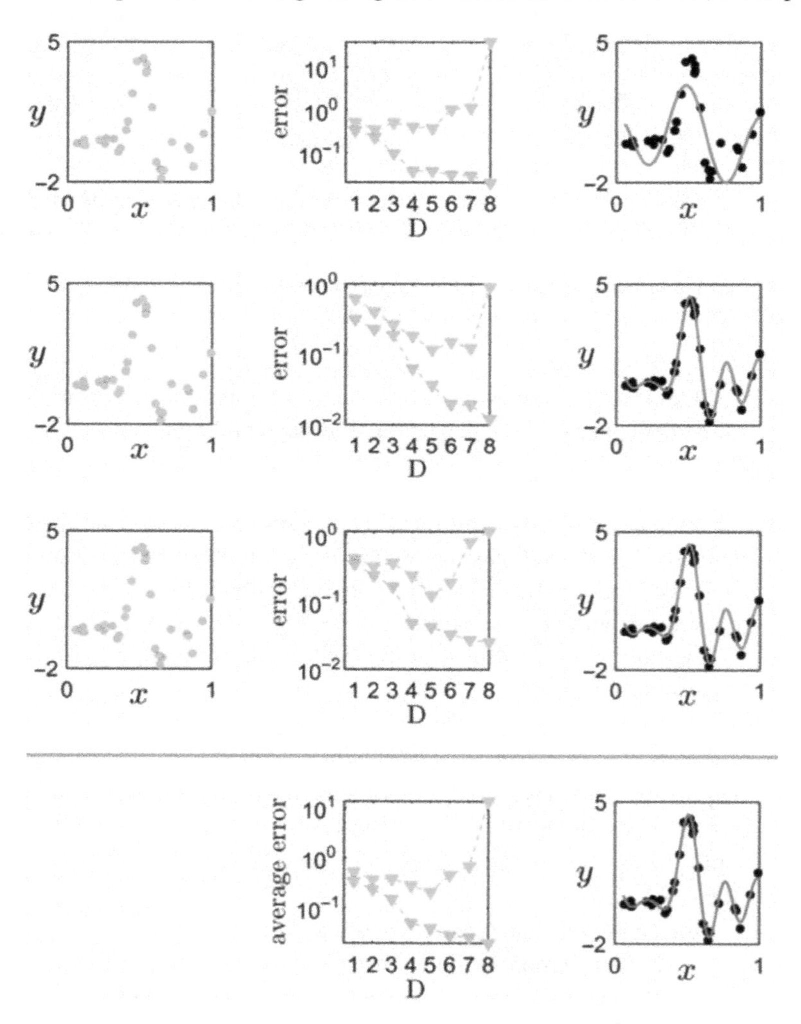

Fig. 5.18 Result of performing k-fold cross-validation with $k = 3$ (see text for further details). The top three rows display the result of performing the hold out method on each fold. The left, middle, and right columns show each fold's training/testing sets (colored blue and yellow respectively) training and testing errors over the range of M tried, and the final model (fit to the entire dataset) chosen by picking the value of M providing the lowest testing error. Due to the split of the data, performing hold out on the first fold (top row) results in a poor underfitting model for the data. However, as illustrated in the final row, by averaging the testing errors (bottom middle panel) and choosing the model with minimum associated average test error, we average out this problem (finding that $D^{\star} = 5$ or $M^{\star} = 10$) and determine an excellent model for the phenomenon (as shown in the bottom right panel).

Equation (5.26)) in the middle panel, and the final model (learned to the entire dataset) provided by the choice of M with lowest testing error. As can be seen in the top row, the particular split of the first fold leads to too low a value of M being chosen, and thus an underfitting model. In the middle panel of the final row we show the result of averaging the training/testing errors over all $k = 3$ folds, and in the right panel the result of choosing the overall best $M^\star = 10$ (or equivalently $D^\star = 5$) providing the lowest average testing error. By taking this value we average out the poor choice determined on the first fold, and end up with a model that fits both the data and underlying function quite well.

Example 5.7 Leave-one-out cross-validation for Galileo's ramp data

In Fig. 5.19 we show how using $k = P$ fold cross-validation (since we have only $P = 6$ data points, intuition suggests, see Section 5.3.2, that we use a large value for k), sometimes referred to as *leave-one-out cross-validation*, allows us to recover precisely the quadratic fit Galileo made by eye. Note that by choosing $k = P$ this means that every data point will take a turn being the testing set. Here we search over the polynomial basis features of degree $M = 1 \ldots 6$. While not all of the hold out models over the six folds fit the data well, the average k-fold result is indeed the $M^\star = 2$ quadratic polynomial fit originally proposed by Galileo!

5.4 Which basis works best?

While some guidance can be given in certain situations regarding the best basis to employ, no general rule exists for which basis one should use in all instances

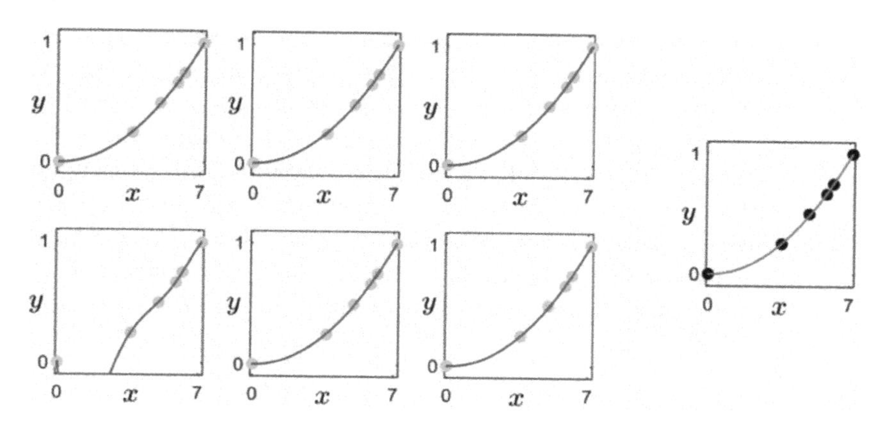

Fig. 5.19 (six panels on the left) Each fold of training/testing sets shown in blue/yellow respectively of a k-fold run on the Galileo's ramp data, along with their individual hold out model (shown in blue). Only the model learned on the fourth fold overfits the data. By choosing the model with minimum average testing error over the $k = 6$ folds we recover the desired quadratic $M^\star = 2$ fit originally proposed by Galileo (shown in magenta in the right panel).

of regression. Indeed for an arbitrary dataset it may very well be the case that no basis/feature map is especially better than any other. However, in some instances understanding of the phenomenon underlying the data and practical considerations can lead one to a particular choice of basis or at least eliminate potential candidates.

5.4.1 Understanding of the phenomenon underlying the data

Gathering some understanding of a phenomenon, while not always easy, is generally the most effective way of deducing the particular effectiveness of a specific basis. For example, the gravitational phenomenon underlying Galileo's ramp dataset, shown in e.g., Fig. 5.19, is extremely well understood as quadratic in nature, implying the appropriateness of a polynomial basis [58]. Fourier basis/feature map are intuitively appropriate if dealing with data generated from a known periodic phenomenon. Often referred to as "time series" data (due to the input variable being time), periodic behavior arises in a variety of disciplines including speech processing and financial modeling (see e.g., [37, 40, 67]). Fourier and neural network bases/features are often employed with image and audio data, in the latter case due to a belief in the correspondence of neural network bases and the way such data is processed by the human brain or the compositional structure of certain problems (see e.g., [17, 19, 36] and references therein). Importantly, note that the information used to favor a particular basis need not come from a regression (or more broadly a machine learning) problem itself but rather from scientific understanding of a phenomenon more broadly.

5.4.2 Practical considerations

Practical considerations can also guide the choice of basis/feature type. For example if, given the nature of a dataset's input variable, it does not make sense to normalize its values to lie in the range [0, 1], then polynomials can be a very poor choice of basis/features. This is due to the fact that polynomials grow rapidly for values outside this range, e.g., $x^{20} = 3\,486\,784\,401$ when $x = 3$, making the corresponding (closed form) calculations for solving the Least Squares problem difficult if not impossible to perform on a modern computer. The amount of engineering involved in effectively employing a kernelized form of a fixed basis (see Chapter 7 for further details) or a deep neural network (e.g., the number of layers as well as the number of activation functions within each layer, type of activation function, etc.) can also guide the choice of basis in practice.

5.4.3 When the choice of basis is arbitrary

In the (rare) scenario where data is plentiful, relatively noise free, and nicely distributed throughout the input space, the choice of basis/feature map is fairly arbitrary, since in such a case the data carves out the entire underlying function fairly well (and so we

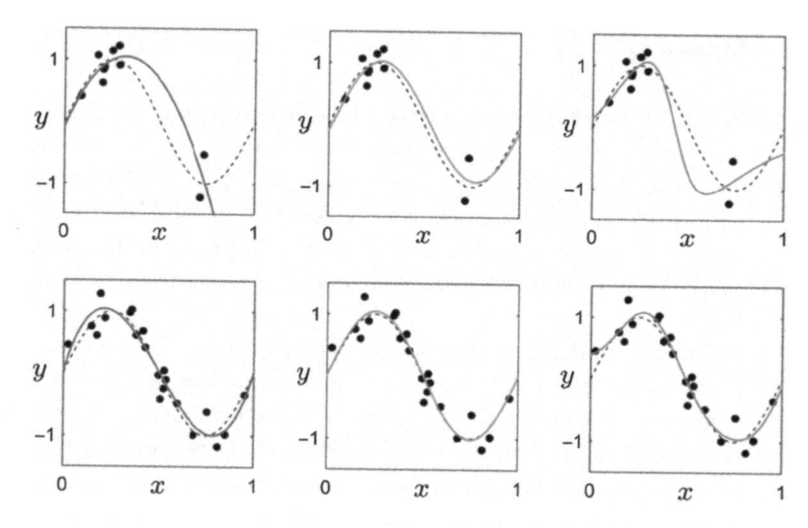

Fig. 5.20 A comparison of the k-fold found best polynomial (left column), Fourier (middle column), and single hidden layer neural network (right column) fit to two datasets (see text for further details). Since the data generating function belongs to the Fourier basis itself, the Fourier cross-validated model fits the underlying function better on the smaller dataset (top row). On the larger dataset the choice of basis is less crucial to finding a good cross-validated model, as the relatively large amount of data carves out the entire underlying function fairly well.

essentially "revert" to the problem of continuous function approximation described in Section 5.1 where all bases are equally effective).

Example 5.8 A simulated dataset where one basis is more appropriate than others

In Fig. 5.20 we illustrate (using simulated datasets) the scenario where one basis is more effective than others, as well as when the data is large/clean/well distributed enough to make the choice of basis a moot point. In this figure we show two datasets of $P = 10$ (top row) and $P = 21$ (bottom row) points generated using the underlying function $y(x) = \sin(2\pi x)$, and the result of k-fold cross validated polynomial (left column), Fourier (middle column), and single hidden layer neural network with $\tanh(\cdot)$ activation (right column) bases.[14] Because the data generating function itself belongs to the Fourier basis, the Fourier cross-validated model fits the underlying model better than the other two bases on the smaller dataset. However, on the larger dataset the choice is less important, with all three cross-validated models performing well.

[14] For each dataset the same training/testing sets were used in performing k-fold cross-validation for each basis feature type. For the smaller dataset we have used $k = 5$, while for the larger dataset we set $k = 3$. For all three basis types the range of degrees/number of hidden units was in the range $M = 1, 2, 3, 4, 5$.

5.5 Summary

In this chapter we have described how features may be designed automatically for the general problem of regression, by viewing it as a noisily sampled function approximation problem. Beginning in Section 5.1.3 with the perfect but unrealistic scenario where we have the data generating function itself, we introduced both fixed and adjustable feed forward neural network bases of fundamental features whose elements can be combined in order to automatically generate features to approximate any such function.

The corresponding Least Squares problem, given in Equation (5.13), for learning proper weights of these bases being intractable, we then saw in Section 5.2 how discretization of both the bases and desired function leads to a corresponding discretized Least Squares problem that closely approximates the original and can be solved using numerical optimization. In this same section we next saw how the realistic case of nonlinear regression (where the data is assumed to be noisy samples of some underlying function) is described mathematically using the same discretized framework, and thus how the same discretized Least Squares problem can be used for regression.

However, we saw in the third section that while increasing the number of basis features creates a better fitting model for the data we currently have, this will overfit the data giving a model that poorly represents data we might receive in the future/the underlying data generating function. This motivated the technique of cross-validation, culminating in the highly useful but computationally costly k-fold cross-validation method, for choosing the proper number of basis features in order to prevent this.

In the final section we discussed the proper choice of basis/feature map. While generally speaking we can say little about which basis will work best in all circumstances, understanding of the phenomenon underlying the data as well as practical considerations can be used to choose a proper basis (or at least narrow down potential candidates) in many important instances.

5.6 Exercises

Section 5.1 exercises

Exercises 5.1 The convenience of orthogonality

A basis $\{\mathbf{x}_m\}_{m=1}^{M}$ is often chosen to be *orthogonal* (typically to simplify calculations) meaning that

$$\mathbf{x}_k^T \mathbf{x}_j = \begin{cases} S & \text{if } k = j \\ 0 & \text{else,} \end{cases} \tag{5.28}$$

for some $S > 0$. For example, the "standard basis" defined for each $m = 1 \ldots M$ as

$$e_{m,j} = \begin{cases} 1 & \text{if } m = j \\ 0 & \text{else.} \end{cases} \tag{5.29}$$

is clearly orthogonal. Another example is the Discrete Cosine Transform basis described in Example 9.3. An orthogonal basis provides much more easily calculable weights for representing a given vector $\mathbf{y}$ than otherwise, as you will show in this exercise.

Show in this instance that the ideal weights are given simply as $w_j = \frac{1}{S}\mathbf{x}_j^T\mathbf{y}$ for all j by solving the Least Squares problem

$$\underset{w_1 \ldots w_M}{\text{minimize}} \quad \left\| \sum_{m=1}^{M} \mathbf{x}_m w_m - \mathbf{y} \right\|_2^2 \tag{5.30}$$

for the jth weight w_j. Briefly describe how this compares to the Least Squares solution.

Exercises 5.2 Orthogonal basis functions

In analogy to orthogonal bases in the case of N-dimensional vectors, a set of basis functions $\{f_m\}_{m=0}^{\infty}$, used in approximating a function $y(x)$ over the interval $[0, 1]$ is *orthogonal* if

$$\langle f_m, f_j \rangle = \int_0^1 f_m(x) f_j(x)\, dx = \begin{cases} S & \text{if } m = j \\ 0 & \text{else} \end{cases} \tag{5.31}$$

for some $S > 0$. The integral quantity above defines the continuous inner product between two functions, and is a generalization of the vector inner product.

Show, as in the finite dimensional case, that orthogonality provides an easily expressible set of weights of the form $w_j = \frac{1}{S}\int_0^1 f_j(x)\, y(x)\, dx$ as solutions to the corresponding Least Squares problem:

$$\underset{w_0, w_1, \ldots}{\text{minimize}} \quad \int_0^1 \left(\sum_{m=0}^{\infty} f_m(x)\, w_m - y(x) \right)^2 dx, \tag{5.32}$$

for the optimal jth weight w_j. *Hint: you may pass each derivative* $\frac{\partial}{\partial w_j}$ *through both the integral and sum.*

Exercises 5.3 The Fourier basis is orthogonal

Using the fact that basis functions $\sin(2\pi kx)$ and $\cos(2\pi jx)$ are orthogonal functions, i.e.,

$$\begin{aligned} &\int_0^1 \sin(2\pi kx)\cos(2\pi jx)\, dx = 0 \text{ for all } k, j \\ &\int_0^1 \sin(2\pi kx)\sin(2\pi jx)\, dx = \begin{cases} 1/2 & \text{if } k = j \\ 0 & \text{else} \end{cases} \\ &\int_0^1 \cos(2\pi kx)\cos(2\pi jx)\, dx = \begin{cases} 1/2 & \text{if } k = j \\ 0 & \text{else} \end{cases}, \end{aligned} \tag{5.33}$$

find the ideal Least Squares coefficients, that is the solution to

$$\underset{w_0,w_1,\ldots,w_{2K}}{\text{minimize}} \int_0^1 \left(w_0 + \sum_{k=1}^{K} \left(w_{2k-1}\sin\left(2\pi kx\right) + w_{2k}\cos\left(2\pi kx\right)\right) - y\left(x\right)\right)^2 dx. \quad (5.34)$$

These weights will be expressed as simple integrals.

Exercises 5.4 Least Squares weights for polynomial approximation

The Least Squares weights for a degree D polynomial basis of a scalar input function $y(x)$ over $[0, 1]$ (note there are $M = D + 1$ terms in this case) are determined by solving the Least Squares problem

$$\underset{w_0,w_1,\ldots,w_D}{\text{minimize}} \int_0^1 \left(\sum_{m=0}^{D} x^m w_m - y\left(x\right)\right)^2 dx. \quad (5.35)$$

Show that solving the above by setting the derivatives of the cost function in each w_j equal to zero results in a linear system of the form

$$\mathbf{P}\mathbf{w} = \mathbf{d}. \quad (5.36)$$

In particular show that $\mathbf{P}$ (often referred to as a Hilbert matrix) takes the explicit form

$$\mathbf{P} = \begin{bmatrix} 1 & 1/2 & 1/3 & \cdots & 1/D+1 \\ 1/2 & 1/3 & 1/4 & \cdots & 1/D+2 \\ 1/3 & 1/4 & 1/5 & \cdots & 1/D+3 \\ \vdots & \vdots & \vdots & \ddots & \vdots \\ 1/D+1 & 1/D+2 & 1/D+3 & \cdots & 1/2D+1 \end{bmatrix}. \quad (5.37)$$

Hint: $\int_0^1 x^j dx = \frac{1}{j+1}$.

Exercises 5.5 Complex Fourier representation

Verify that using complex exponential definitions of cosine and sine functions, i.e., $\cos\left(\alpha\right) = \frac{1}{2}\left(e^{i\alpha} + e^{-i\alpha}\right)$ and $\sin\left(\alpha\right) = \frac{1}{2i}\left(e^{i\alpha} - e^{-i\alpha}\right)$, we can write the partial Fourier expansion

$$w_0 + \sum_{m=1}^{M} \cos\left(2\pi mx\right) w_{2m-1} + \sin\left(2\pi mx\right) w_{2m} \quad (5.38)$$

equivalently as

$$\sum_{m=-M}^{M} e^{2\pi imx} w'_m, \quad (5.39)$$

where the complex weights $\left\{w'_m\right\}_{m=-M}^{M}$ are given in terms of the real weights $\left\{w_m\right\}_{m=0}^{2M}$ as

$$w'_m = \begin{cases} \frac{1}{2}\left(w_{2m-1} - iw_{2m}\right) & \text{if } m > 0 \\ w_0 & \text{if } m = 0 \\ \frac{1}{2}\left(w_{1-2m} + iw_{-2m}\right) & \text{if } m < 0. \end{cases} \quad (5.40)$$

Exercises 5.6 Graphical representation of a neural network

a) Use Fig. 5.9 or Equation (5.12) to count the total number of parameters Q (including both internal parameters and feature weights) in a three hidden layer neural network basis approximation. Can you generalize this to find a formula for Q in a neural network with L hidden layers? *Hint: you may find it convenient to define* $M_0 = 1$, $M_1 = M$, *and* $M_{L+1} = N$.

b) Based on your answer in part a), how well does a neural network basis scale to large datasets? More specifically, how does the input dimension N contribute to the number of parameters Q (a.k.a. the dimension of the optimization problem)? How does the number of parameters change with the number of data points P?

Section 5.2 exercises

Exercises 5.7 Polynomial basis feature regression for scalar valued input

In this exercise you will explore how various degree D polynomial basis features fit the sinusoidal dataset shown in Fig. 5.12. You will need the wrapper *poly_regression_hw* and the data located in *noisy_sin_samples.csv*.

a) Use the description of polynomial basis features given in Example 5.1 to transform the input using a general degree D polynomial. Write this feature transformation in the module

$$\mathbf{F} = \text{poly\_features}\,(\mathbf{x}, D) \tag{5.41}$$

located in the wrapper. Here $\mathbf{x}$ is the input data, D the degree of the polynomial features, and $\mathbf{F}$ the corresponding degree D feature transformation of the input (note your code should be able to transform the input to any degree D desired).

b) With your module complete you may run the wrapper. Two figures will be generated: the first shows the data along with various degree D polynomial fits, and the second shows the mean squared error (MSE) of each fit to the dataset. Discuss the results shown in these two figures. In particular describe the relationship between a model's MSE and how well it seems to represent the phenomenon generating the data as D increases over the range shown.

Exercises 5.8 Fourier basis feature regression for scalar valued input

In this exercise you will explore how various degree D Fourier basis features fit the sinusoidal dataset shown in Fig. 5.12. For this exercise you will need the wrapper *fourier_regression_hw* and the data located in *noisy_sin_samples.csv*.

a) Use the description of Fourier basis features given in Example 5.1 to transform the input using a general degree D Fourier basis (remember that there are $M = 2D$ basis elements in this case, a $\cos(2\pi mx)$ and $\sin(2\pi mx)$ pair for each $m = 1, \ldots, D$). Write this feature transformation in the module

$$\mathbf{F} = \text{fourier\_features}\left(\mathbf{x},\, D\right) \tag{5.42}$$

located in the wrapper. Here $\mathbf{x}$ is the input data, D the degree of the Fourier features, and $\mathbf{F}$ the corresponding degree D feature transformation of the input (note your code should be able to transform the input to any degree D desired).

b) With your module complete you may run the wrapper. Two figures will be generated: the first shows the data along with various degree D Fourier fits, and the second shows the MSE of each fit to the data-set. Discuss the results shown in these two figures. In particular describe the relationship between a model's MSE and how well it seems to represent the phenomenon generating the data as D increases over the range shown.

Exercises 5.9 Single hidden layer network regression with scalar valued input

In this exercise you will explore how various initializations affect the result of an $M = 4$ neural network basis features fit to the sinusoidal dataset shown in Fig. 5.12. For this exercise you will need the wrapper *tanh_regression_hw* and the data located in *noisy_sin_samples.csv*.

a) Using the chain rule, verify that the gradient of the Least Squares problem shown in Equation (5.21) with general activation function $a(\cdot)$ is given as

$$\begin{aligned}
\frac{\partial}{\partial b} g &= 2\sum_{p=1}^{P}\left(b + \sum_{m=1}^{M} a\left(c_m + \mathbf{x}_p^T \mathbf{v}_m\right) w_m - y_p\right) \\
\frac{\partial}{\partial w_n} g &= 2\sum_{p=1}^{P}\left(b + \sum_{m=1}^{M} a\left(c_m + \mathbf{x}_p^T \mathbf{v}_m\right) w_m - y_p\right) a\left(c_n + \mathbf{x}_p^T \mathbf{v}_n\right) \\
\frac{\partial}{\partial c_n} g &= 2\sum_{p=1}^{P}\left(b + \sum_{m=1}^{M} a\left(c_m + \mathbf{x}_p^T \mathbf{v}_m\right) w_m - y_p\right) a'\left(c_n + \mathbf{x}_p^T \mathbf{v}_n\right) w_n \\
\nabla_{\mathbf{v}_n} g &= 2\sum_{p=1}^{P}\left(b + \sum_{m=1}^{M} a\left(c_m + \mathbf{x}_p^T \mathbf{v}_m\right) w_m - y_p\right) a'\left(c_n + \mathbf{x}_p^T \mathbf{v}_n\right) w_n \mathbf{x}_p,
\end{aligned} \tag{5.43}$$

where $a'(\cdot)$ is the derivative of the activation with respect to its input.

b) This gradient can be written more efficiently for programming languages like Python and MATLAB/OCTAVE that have especially good implementations of matrix/vector operations by writing it more compactly. Supposing that $a = \tanh(\cdot)$ is the activation function (meaning $a' = \text{sech}^2(\cdot)$ is the hyperbolic secant function squared) verify that the derivatives from part a) can be written more compactly for a scalar input as

$$\frac{\partial}{\partial b}g = 2 \cdot \mathbf{1}_{P\times 1}^{T}\mathbf{q}$$
$$\frac{\partial}{\partial w_n}g = 2 \cdot \mathbf{1}_{P\times 1}^{T}\left(\mathbf{q}\odot\mathbf{t}_n\right)$$
$$\frac{\partial}{\partial c_n}g = 2 \cdot \mathbf{1}_{P\times 1}^{T}\left(\mathbf{q}\odot\mathbf{s}_n\right)w_n \tag{5.44}$$
$$\frac{\partial}{\partial v_n}g = 2 \cdot \mathbf{1}_{P\times 1}^{T}\left(\mathbf{q}\odot\mathbf{x}\odot\mathbf{s}_n\right)w_n,$$

where $q_p = \left(b + \sum_{m=1}^{M} w_m \tanh\left(c_m + x_p v_m\right) - y_p\right)$, $t_{np} = \tanh\left(c_n + x_p v_n\right)$, and $s_{np} = \operatorname{sech}^2\left(c_n + x_p v_n\right)$, and $\mathbf{q}$, $\mathbf{t}_n$, and $\mathbf{s}_n$ are the P length vectors containing these entries. Note that $\mathbf{a}\odot\mathbf{b}$ denotes the entry-wise product of vectors $\mathbf{a}$ and $\mathbf{b}$.

c) Plug in the form of the gradient from part b) into the gradient descent module called

$$\left[b, \mathbf{w}, \mathbf{c}, \mathbf{v}, \text{obj\_val}\right] = \text{tanh\_grad\_descent}\left(\mathbf{x}, \mathbf{y}, i\right) \tag{5.45}$$

located in the wrapper. Here $\mathbf{x}$ and $\mathbf{y}$ are the input and output data respectively, i is a counter that will load an initialization for all variables, and b, $\mathbf{w}$, $\mathbf{c}$, and $\mathbf{v}$ are the optimal variables learned via gradient descent. Use the maximum iteration stopping condition with 15 000 iterations, and a fixed step length $\alpha = 10^{-3}$. Initializations are already given in the wrapper.

With this module completed the wrapper will execute gradient descent three times using three different initializations, displaying for each run the corresponding fit to the data achieved at the last step of gradient descent, as well as the objective value calculated at each iteration. Briefly discuss the results shown in this figure.

Section 5.3 exercises

Exercises 5.10 Four guys and four error plots

Eric, Stanley, Kyle, and Kenneth used hold out cross-validation to find the best degree polynomial fit to their respective datasets. Based on the error plots they have made (Fig. 5.21), what advice would you give to each of them as to what their next step should be.

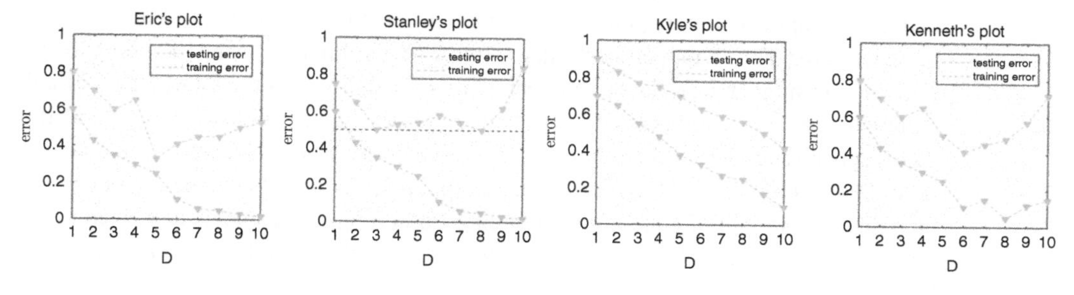

Fig. 5.21 Hold out cross-validation error plots for four different datasets.

Exercises 5.11 Practice the hold out method

In this exercise you will perform hold out cross-validation on the dataset shown in Fig. 5.16 as described in Example 5.5 of the text. Start by randomly splitting the dataset, located in *wavy_data.csv*, into $k = 3$ equal sized folds (keeping 2 folds as training, and 1 fold as testing data). Use the Fourier basis features M in the range $M = 2, 4, 6, \ldots, 16$ (or likewise D in the range $D = 1, 2, \ldots, 8$) and produce a graph showing the training and testing error for each D like the one shown in Fig. 5.16, as well as the best (i.e., the lowest test error) model fit to the data.

Note: your results may appear slightly different than those of the figure given that you will likely use a different random partition of the data. Note: you may find it very useful here to re-use code from previous exercises e.g., functions that compute Fourier features, plot curves, etc.

Exercises 5.12 Code up k-fold cross-validation

In this exercise you will perform k-fold cross-validation on the dataset shown in Fig. 5.19 as described in Example 5.7 of the text. Start by randomly splitting the dataset of $P = 6$ points, located in *galileo_ramp_data.csv*, into $k = 6$ equal sized folds (keeping 5 folds as training, and 1 fold as testing data during each round of cross-validation). The value of $k = P$ has been chosen in this instance due to the small size of the dataset (this is sometimes called "leave one out" cross-validation since each training set consists of all but one point from the original dataset).

Use the polynomial basis features and M in the range $M = 1, 2, \ldots, 6$ and produce a graph showing the average training and testing error for each M, as well as the best (i.e., the lowest average test error) model fit to the data.

Note: you may find it very useful here to re-use code from previous exercises, e.g., functions that split a dataset into k random parts, that compute training/testing errors, polynomial features, plot curves, etc.

Section 5.4 exercises

Exercises 5.13 Comparing all bases

In this exercise you will reproduce the k-fold cross-validation result shown in Fig. 5.12 using the wrapper *compare_maps_regression_hw* and the corresponding datasets shown in the figure. This wrapper performs k-fold cross-validation using the polynomial, Fourier, and single hidden layer tanh feature maps and produces the figure. It is virtually complete, i.e., the code necessary to generate the associated plot is already provided in the wrapper, save four modules you will need to include. Insert the data splitting module described in Exercise 5.11, as well as the polynomial, Fourier, and single hidden layer tanh modules for solving their corresponding Least Squares problems described in Exercises 5.7, 5.8, and 5.9 respectively. With these modules installed you should be able to run the wrapper.

Note that some of your results should appear different than those of the figure given that you will use a different random partition of the data in each instance.

5.7 Notes on continuous function approximation

Polynomials were the first provable universal approximators, this having been shown in 1885 via the so-called (Stone–) Weierstrass approximation theorem (see e.g., [71]). The Fourier basis (and its discrete derivatives) is an extremely popular function approximation tool, used particularly in physics, signal processing, and engineering fields. The convergence behavior of Fourier series has been studied for centuries, and much can be said about its convergence on larger classes of functions beyond $\mathcal{C}^N$ (see e.g., [61, 74] for a sample of results). The universal approximation properties of popular adjustable bases like the single-layer and multilayer neural networks were shown in the late 1980s and early 1990s [28, 38, 63]. Interestingly, an evolutionary step between fixed and adjustable bases, a random fixed basis where internal parameters of a given adjustable basis type are randomized, leaving only the external linear weights to be learned, has been shown to be a universal approximator more recently than deep architectures (see e.g., [69]).

6 Automatic feature design for classification

In Chapter 6 we mirror closely the exposition given in the previous chapter on regression, beginning with the approximation of the underlying data generating function itself by bases of features, and going on to finally describing cross-validation in the context of classification. In short we will see that all of the tools from the previous chapter can be applied to the automatic design of features for the problem of classification as well.

6.1 Automatic feature design for the ideal classification scenario

In Fig. 6.1 we illustrate a prototypical dataset on which we perform the general task of two class classification, where the two classes can be effectively separated using a nonlinear boundary. In contrast to those examples given in Section 4.5, where visualization or scientific knowledge guided the fashioning of a feature transformation to capture this nonlinearity, in this chapter we suppose that this cannot be done due to the complexity and/or high dimensionality of the data. At the heart of the two class classification framework is the tacit assumption that the data we receive are in fact noisy samples of some underlying indicator function, a nonlinear generalization of the step function briefly discussed in Section 4.5, like the one shown in the right panel of Fig. 6.1. Akin to regression, our goal with classification is then to approximate this data-generating indicator function as well as we can using the data at our disposal.

In this section we will assume the impossible: that we have clean and complete access to every data point in the space of a two class classification environment, whose labels take on values in $\{-1, 1\}$, and hence access to its associated indicator function $y(\mathbf{x})$. Although an indicator function is *not* continuous, the same bases of continuous features discussed in the previous chapter can be used to represent it (near) perfectly.

6.1.1 Approximation of piecewise continuous functions

In Section 5.1 we saw how fixed and adjustable neural network bases of features can be used to approximate *continuous* functions. These bases can also be used to effectively approximate the broader class of *piecewise continuous* functions, composed of fragments of continuous functions with gaps or jumps between the various pieces. Shown

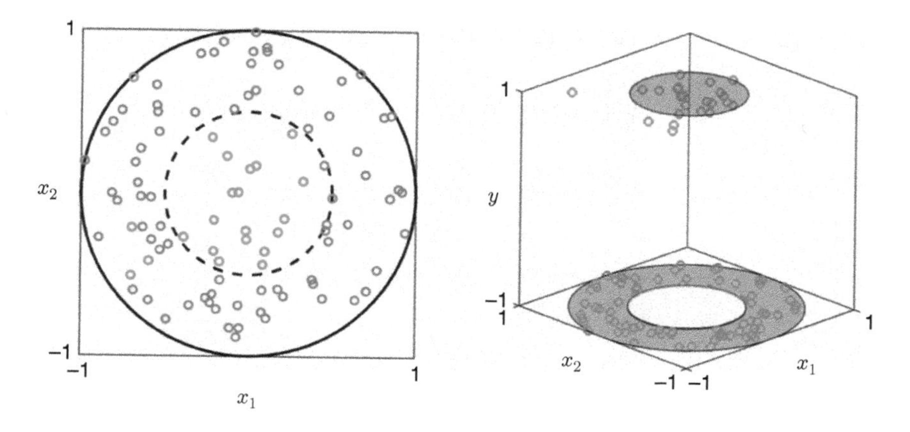

Fig. 6.1 (left panel) A realistic dataset for two class classification shown in the original space from above and made by taking noisy samples from the data generating indicator function $y(\mathbf{x}) = \begin{cases} +1 & \text{if} \quad 0 \leq \|\mathbf{x}\|_2^2 \leq 0.5 \\ -1 & \text{if} \;\; 0.5 < \|\mathbf{x}\|_2^2 \leq 1, \end{cases}$ over the unit circle. (right panel) View of the original space from the side with the data-generating indicator function shown in gray.

in Fig. 6.2 are two example piecewise continuous functions[1] (in black) along with their polynomial, Fourier, and single hidden layer neural network basis approximations. For each instance in the figure we have used as many of the respective basis elements as needed to give a visually close approximation.

As with continuous functions, adding more basis elements generally produces a finer approximation of any piecewise function defined over a bounded space (for convenience we will take the domain of y to be the unit hypercube, again denoted as $[0, 1]^N$, but any bounded domain would also suffice, e.g., a hypersphere with finite radius). That is, by increasing the number of basis elements M we generally have that the approximation

$$\sum_{m=0}^{M} f_m(\mathbf{x})\, w_m \approx y(\mathbf{x}) \tag{6.2}$$

improves overall.[2] Note here that once again $f_0(\mathbf{x}) = 1$ is the constant basis element, with the remaining $f_m(\mathbf{x})$ basis elements of any type desired.

[1] The piecewise continuous functions in the top and bottom panels are defined over the unit interval, respectively as

$$y(x) = \begin{cases} +1 & 0.33 \leq x \leq 0.67 \\ -1 & \text{else,} \end{cases} \qquad y(x) = \begin{cases} 0.25\left(\left(1 - 10x^2\right)\cos(10\pi x) + 1\right) & 0 \leq x < 0.33 \\ 0.8 & 0.33 \leq x \leq 0.67 \\ 4\left(x^2 - 0.75\right) + 0.4 & 0.67 < x \leq 1. \end{cases} \tag{6.1}$$

[2] As with continuous function approximation, the details of this statement are quite technical, and we do not dwell on them here. Our goal is to provide an intuitive high level understanding of this sort of function approximation. See Section 5.7 for further information and reading.

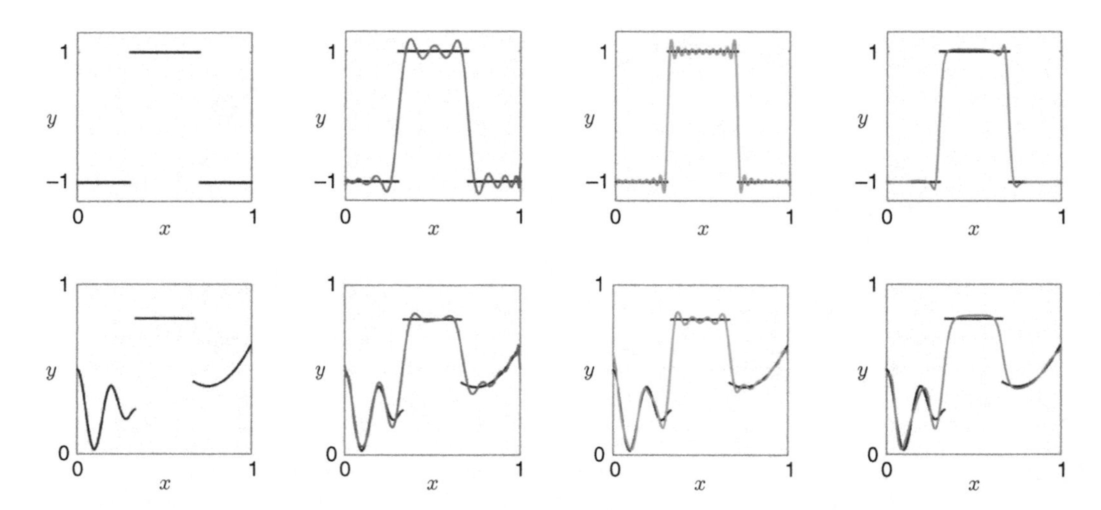

Fig. 6.2 Two piecewise continuous functions (in black). The first function is constant in each piece (top panels) while the second is a more complicated function consisting of several different types of continuous pieces (bottom panels). Also shown are polynomial (in blue), Fourier (in red), and single hidden layer neural network (in purple) basis approximations of each function. Using more basis features in each case will increase the quality of the approximation, just as with continuous function approximation.

6.1.2 The formal definition of an indicator function

Formally speaking an indicator function, like the one shown in Fig. 6.1, is a simple tool for identifying points $\mathbf{x}$ in a set $\mathcal{S}$ by assigning them some unique constant value while assigning all other points that are not in $\mathcal{S}$ a different constant value.[3] Using $+1$ to indicate that a point is in a set $\mathcal{S}$ and -1 otherwise, we can define an indicator function on $\mathcal{S}$ as

$$y(\mathbf{x}) = \begin{cases} +1 & \text{if } \mathbf{x} \in \mathcal{S} \\ -1 & \text{if } \mathbf{x} \notin \mathcal{S}. \end{cases} \tag{6.3}$$

For instance, the one-dimensional function plotted in the top panels of Fig. 6.2 is such an indicator function on the interval $\mathcal{S} = [0.33, 0.67]$. The two-dimensional function shown at the beginning of the section in Fig. 6.1 is an indicator function on the set $\mathcal{S} = \{\mathbf{x} \mid 0 \leq \|\mathbf{x}\|_2^2 \leq 0.5\}$. Moreover the general step function discussed in Section 3.3.1 is another example of such an indicator function on the set $\mathcal{S} = \{\mathbf{x} \mid b + \mathbf{x}^T\mathbf{w} > 0\}$, taking the form

$$y(\mathbf{x}) = \text{sign}\left(b + \mathbf{x}^T\mathbf{w}\right) = \begin{cases} +1 & \text{if } b + \mathbf{x}^T\mathbf{w} > 0 \\ -1 & \text{if } b + \mathbf{x}^T\mathbf{w} < 0. \end{cases} \tag{6.4}$$

[3] Although $\mathcal{S}$ can take any arbitrary shape in general, in the context of classification we are only interested in sets that, loosely speaking, have an *interior* to them. This excludes degenerate cases such as the union of isolated points.

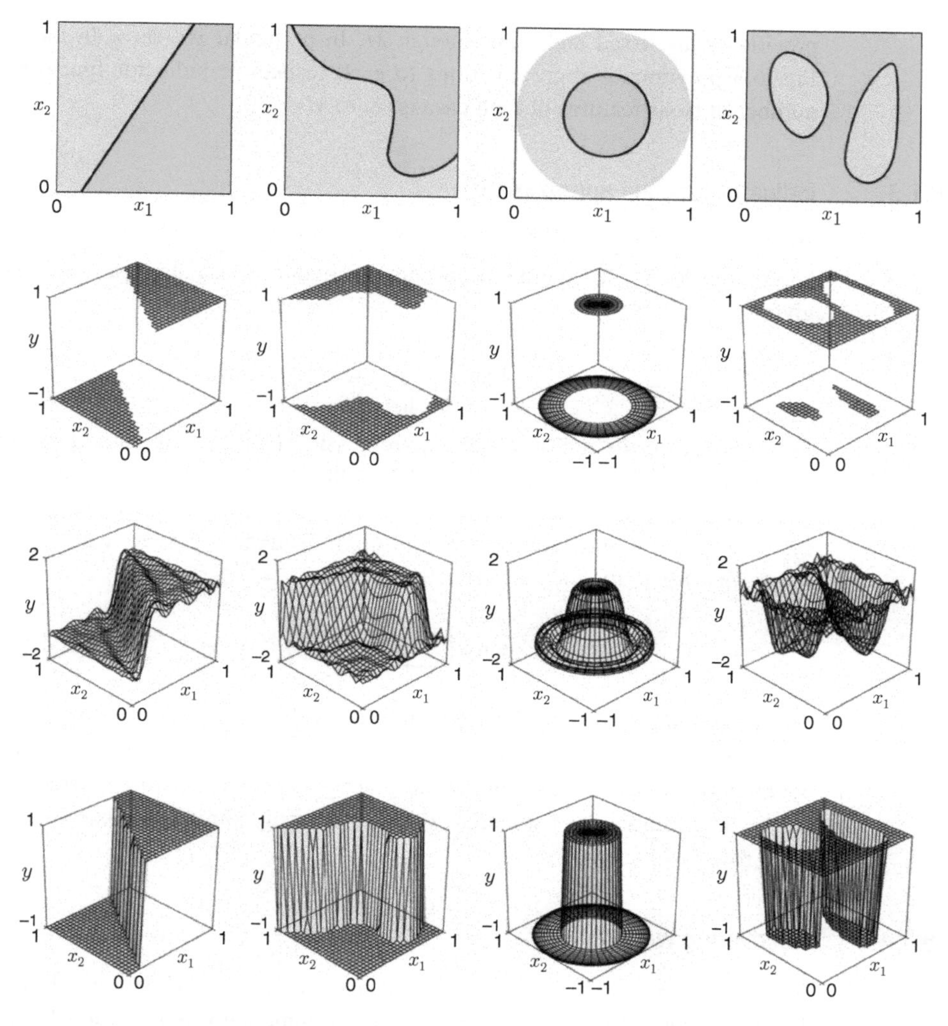

Fig. 6.3 (top row) Four instances of the type of sets $\mathcal{S}$ that concern us in machine learning. Shown from left to right a half space, a general region with nonlinear boundary, a circular region, and the unit square with two ovoid shapes removed from it. For visualization purposes the points that are in the set $\mathcal{S}$ are colored red in each instance, while the rest are colored blue. (second row) The corresponding indicator functions shown in the data space. (third row) Respective polynomial approximations to each indicator function where $M = 30$ in each instance. (bottom row) The associated logistic approximations match the original indicator functions very closely.

We show four instances of such sets in the top row of Fig. 6.3 where the points that are in the set $\mathcal{S}$ are colored red while the rest are colored blue in each instance. From a classification perspective, the red and blue regions indicate the domain of class $+1$ and class -1 in the input space, respectively. Plotted in the second row of this figure are the corresponding indicator functions.

Since indicator functions are piecewise constant, a special subclass of piecewise continuous functions, they can be approximated effectively using any of the familiar bases previously discussed and large enough M. In particular we show in the third row of Fig. 6.3 polynomial approximations to each respective indicator function, where the number of basis features in each case is set to $M = 30$.

6.1.3 Indicator function approximation

Given the fact that an indicator function $y(\mathbf{x})$ takes on values ± 1 we can refine our approximation by passing the basis sum in Equation (6.2) through the sign $(\cdot)$ function giving

$$\text{sign}\left(\sum_{m=0}^{M} f_m(\mathbf{x})\, w_m\right) \approx y(\mathbf{x}). \tag{6.5}$$

Fundamental to the notion of logistic regression (and to two class classification more broadly), as discussed in Section 4.2.2, is that the smooth logistic function $\tanh(\alpha t)$ can be made to approximate $\text{sign}(t)$ as finely as desired by increasing α. By absorbing a large constant α into each weight w_m in Equation (6.5) we can write the *logistic approximation* to the indicator function $y(\mathbf{x})$ as

$$\tanh\left(\sum_{m=0}^{M} f_m(\mathbf{x})\, w_m\right) \approx y(\mathbf{x}). \tag{6.6}$$

In the bottom row of Fig. 6.3 we show the result of learning weights so that Equation (6.6) holds, again using a polynomial basis with $M = 30$ for each of the four indicator functions. As can be seen, the logistic approximation provides a better resemblance to the actual indicator compared to the direct polynomial approximation seen in the third row.

6.1.4 Recovering weights

Since $y(\mathbf{x})$ takes on values in $\{\pm 1\}$ at each $\mathbf{x}$ and the approximation in Equation (6.6) is linear in the weights w_m, using precisely the argument given in Section 4.2.2 (in deriving the softmax cost function in the context of logistic regression) we may rewrite Equation (6.6) equivalently as

$$\log\left(1 + e^{-y(\mathbf{x}) \sum\limits_{m=0}^{M} f_m(\mathbf{x}) w_m}\right) \approx 0. \tag{6.7}$$

Therefore in order to properly tune the weights $\{w_m\}_{m=0}^{M}$, as well as any internal parameters Θ when employing a neural network basis, we can formally minimize the logistic approximation in Equation (6.7) over all $\mathbf{x}$ in the unit hypercube as

$$\underset{w_0 \ldots w_M, \Theta}{\text{minimize}} \int\limits_{\mathbf{x} \in [0,1]^N} \log\left(1 + e^{-y(\mathbf{x}) \sum\limits_{m=0}^{M} f_m(\mathbf{x}) w_m}\right) d\mathbf{x}. \tag{6.8}$$

As with the Least Squares problem with continuous function approximation previously discussed in Section 5.1.5, this is typically not solvable in closed form due to the intractability of the integrals involved. Instead, once again by discretizing the functions involved we will see how this problem reduces to a general problem for two class classification, and how this framework is directly applicable to real classification datasets.

6.2 Automatic feature design for the real classification scenario

In this section we discuss how fixed and neural network feature bases are applied to the automatic design of features for real two-class classification datasets. Analogous to their incorporation into the framework of regression discussed in Section 5.2, here we will see that the concept of feature bases transfers quite easily from the ideal scenario discussed in the previous section.

6.2.1 Approximation of discretized indicator functions

As with the discussion in Section 5.1.2 for continuous functions, *discretizing* an indicator function $y(\mathbf{x})$ finely over its domain gives a close facsimile of the true indicator. For example, shown in Fig. 6.4 is the circular indicator previously shown in Fig. 6.1 and 6.3, along with a closely matching discretized version made by sampling the function over a fine grid of evenly spaced points in its input domain.

Formally, by taking a fine grid of P evenly spaced points $\left\{\left(\mathbf{x}_p, y\left(\mathbf{x}_p\right)\right)\right\}_{p=1}^{P}$ over the input domain of an indicator function $y(\mathbf{x})$ we can then say for a given number M of basis features that the condition in (6.7) essentially holds for each p as

$$\log\left(1+e^{-y(\mathbf{x}_p)\sum\limits_{m=0}^{M} f_m(\mathbf{x}_p)w_m}\right)\approx 0. \tag{6.9}$$

By denoting $y_p = y\left(\mathbf{x}_p\right)$, using the feature vector notation $\mathbf{f}_p = \left[\begin{array}{cccc} f_1\left(\mathbf{x}_p\right) & f_2\left(\mathbf{x}_p\right) & \cdots & f_M\left(\mathbf{x}_p\right)\end{array}\right]^T$, and reintroducing the bias notation $b = w_0$ we may write Equation (6.9) more conveniently as

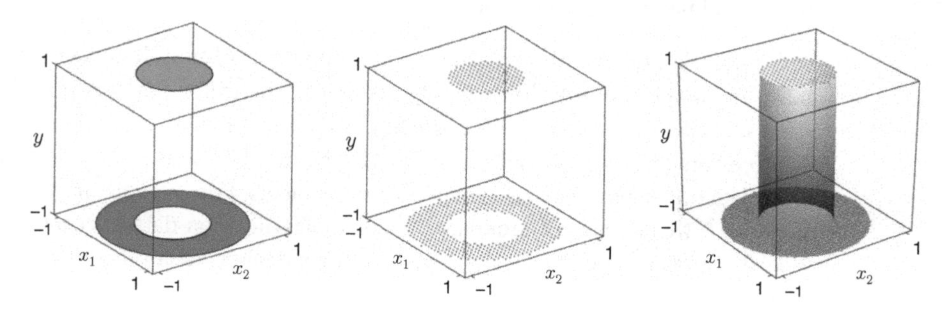

Fig. 6.4 (left panel) An indicator function $y(\mathbf{x})$ defined over the unit circle, and (middle panel) a discretized facsimile made by evaluating y over a fine grid of evenly spaced points. (right panel) We can fit a smooth approximation to the discretized indicator via minimizing the softmax cost function.

$$\log\left(1+e^{-y_p\left(b+\mathbf{f}_p^T\mathbf{w}\right)}\right)\approx 0. \tag{6.10}$$

Similarly to the discussion of logistic regression in Section 4.2.2, by minimizing the sum of these terms,

$$g\left(b,\mathbf{w},\Theta\right)=\sum_{p=1}^{P}\log\left(1+e^{-y_p\left(b+\mathbf{f}_p^T\mathbf{w}\right)}\right), \tag{6.11}$$

we can learn parameters $(b,\mathbf{w},\Theta)$ to make Equation (6.10) hold as well as possible for all p. Note how this is precisely the logistic regression problem discussed in Section 4.2.2, only here each original data point $\mathbf{x}_p$ has been replaced with its feature transformed version $\mathbf{f}_p$. Stating this minimization formally,

$$\underset{b,\,\mathbf{w},\,\Theta}{\text{minimize}}\ \sum_{p=1}^{P}\log\left(1+e^{-y_p\left(b+\mathbf{f}_p^T\mathbf{w}\right)}\right), \tag{6.12}$$

we can see that it is in fact a discrete form of the original learning problem in Equation (6.8) for the ideal classification scenario. In addition, recall from Section 4.1 that there are many highly related costs to the softmax function that recover similar weights when properly minimized. Therefore we can use e.g., the squared margin cost in place of the softmax, and instead solve

$$\underset{b,\,\mathbf{w},\,\Theta}{\text{minimize}}\ \sum_{p=1}^{P}\max^2\left(0,1-y_p\left(b+\mathbf{f}_p^T\mathbf{w}\right)\right) \tag{6.13}$$

to determine optimal parameters. Regardless of the cost function and feature basis used, the corresponding problem can be minimized via numerical techniques like gradient descent (for details see the next section). Also note that regardless of the cost function or feature basis used in producing a *nonlinear* boundary in the original feature space, we are simultaneously determining a *linear* boundary in the transformed feature space in the bias b and weight vector $\mathbf{w}$, as we showed visually with the elliptical dataset in Example 4.7.

6.2.2 The real classification scenario

Very rarely in practice can we acquire large quantities of noiseless data which span the entire input space evenly like a finely discretized indicator function. On the contrary, often we have access to a limited amount of data that is not so evenly distributed and, due to errors in its acquisition, is noisy. In the case of classification, noisy means that some data points have been assigned the wrong labels. For example, in the right column of Fig. 6.5 we show three simulated realistic classification datasets each composed of P noisy samples[4] $\left\{\left(\mathbf{x}_p,y\left(\mathbf{x}_p\right)\right)\right\}_{p=1}^{P}$ of the three indicator functions shown discretized in the figure's left column.

[4] In each instance we take $P=99$ points randomly from the respective input domain, and evaluate each input $\mathbf{x}_p$ in its indicator function giving the associated label $y\left(\mathbf{x}_p\right)$. We then add noise to the first two datasets by randomly switching the output (or label) $y\left(\mathbf{x}_p\right)$ of 4 and 6 points respectively, that is, for these points we replace $y\left(\mathbf{x}_p\right)$ with $-y\left(\mathbf{x}_p\right)$, and then refer to the (potentially noisy) label of each $\mathbf{x}_p$ as simply y_p.

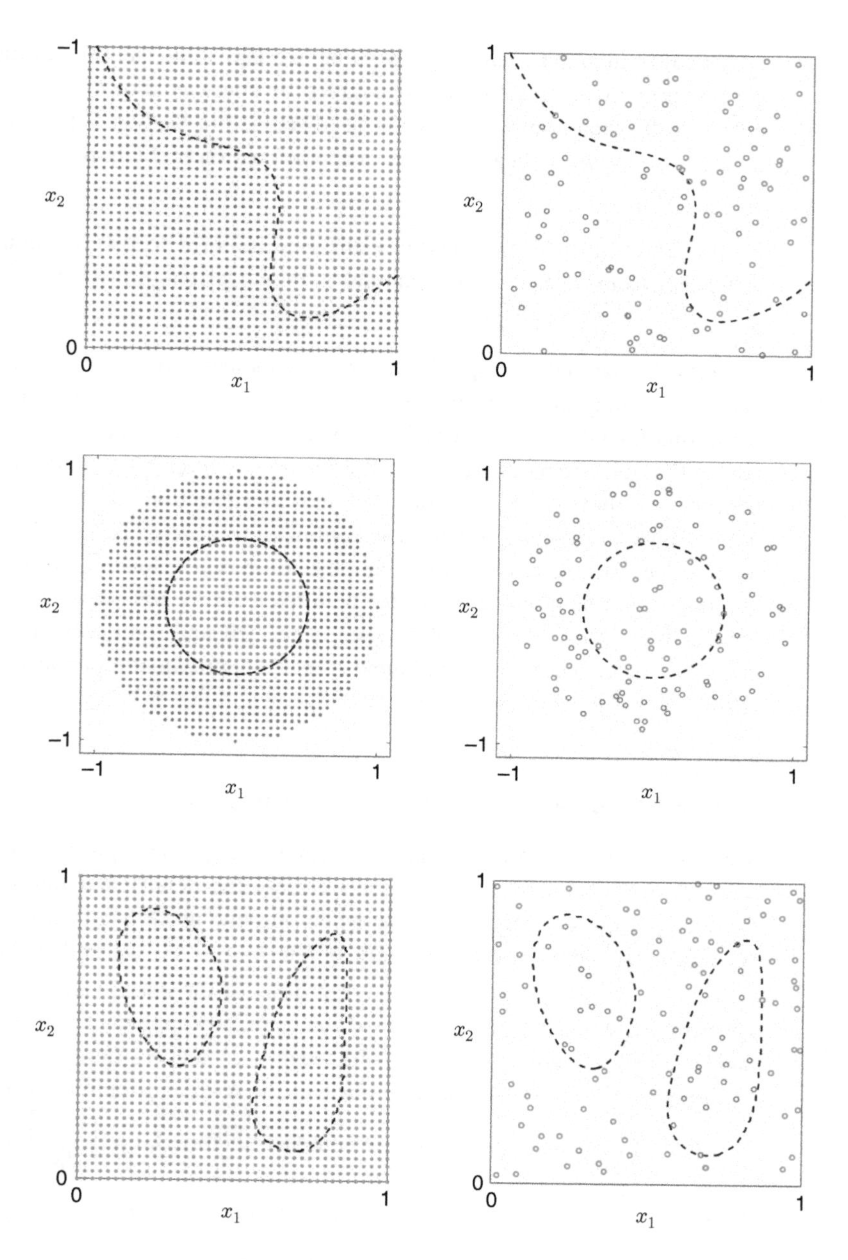

Fig. 6.5 (left column) Three ideal datasets for general two class classification shown in the original feature space (bird's-eye view). (right column) Examples of realistic datasets for two class classification, as noisy samples of each respective indicator.

With this in mind, generally speaking we can think about classification datasets encountered in practice as noisy samples of some unknown indicator function. In other words, analogous to regression (as discussed in Section 5.2.2), general two class classification is an (indicator) function approximation problem based on noisy samples.

The general case of two class classification is an (indicator) function approximation problem based on noisy samples of the underlying function.

As with the data consisting of a discretized indicator function, for real classification datasets we may also approximate the underlying indicator function/determine class separating boundaries by leveraging fixed and neural network feature bases. For example, in Fig. 6.6 we show the result of employing degree[5] 3, 2, and 5 polynomial basis features to approximate the underlying indicator functions of the noisy datasets originally shown in the top, middle, and bottom right panels of Fig. 6.5. In each case we learn proper parameters by minimizing the softmax cost function as in Equation (6.12), the details of which we describe (for both fixed and neural network feature bases) following this discussion. As can be seen in the left and right columns of Fig. 6.6, this procedure determines nonlinear boundaries and approximating indicator functions[6] that closely mimic those of the true indicators shown in Fig. 6.3.

Example 6.1 Classification with fixed bases of features

The minimization of the softmax or squared margin perceptron cost functions using a fixed feature transformation follows closely the details first outlined in Examples 4.1 and 4.2 respectively. Foremost when employing M fixed basis features, the associated cost, being a function only of the bias $b = w_0$ and weight vector $\mathbf{w} = \begin{bmatrix} w_1 & w_2 & \ldots & w_M \end{bmatrix}^T$, is convex regardless of the cost function used. Hence both gradient descent and Newton's method can be readily applied. The form of the gradients and Hessians are entirely the same, with the only cosmetic difference being the use of the M-dimensional feature vector $\mathbf{f}_p$ in place of the N-dimensional input $\mathbf{x}_p$ (see Exercises 6.1 and 6.2). For example, using the compact notation $\tilde{\mathbf{f}}_p = \begin{bmatrix} 1 \\ \mathbf{f}_p \end{bmatrix}$ and $\tilde{\mathbf{w}} = \begin{bmatrix} b \\ \mathbf{w} \end{bmatrix}$, the softmax cost in Equation (6.11) can be written as $g\left(\tilde{\mathbf{w}}\right) = \sum_{p=1}^{P} \log\left(1 + e^{-y_p \tilde{\mathbf{f}}_p^T \tilde{\mathbf{w}}}\right)$, whose gradient is given by

$$\nabla_{\tilde{\mathbf{w}}} g\left(\tilde{\mathbf{w}}\right) = -\sum_{p=1}^{P} \sigma\left(-y_p \tilde{\mathbf{f}}_p^T \tilde{\mathbf{w}}\right) y_p \tilde{\mathbf{f}}_p. \tag{6.14}$$

[5] Note that a set of degree D polynomial features for input of dimension $N > 1$ consists of all monomials of the form $f_m(\mathbf{x}) = x_1^{m_1} x_2^{m_2} \cdots x_N^{m_N}$ (see footnote 5 of Chapter 5 where this was first introduced) where $0 \leq m_1 + m_2 + \cdots + m_N \leq D$. There are a total of $M = \frac{(N+D)!}{N!D!} - 1$ such terms excluding the constant feature and hence this is the length of the corresponding feature vector.

[6] The general equations defining both the learned boundaries and corresponding indicator functions shown in the figures of this section are explicitly defined in Section 6.2.3.

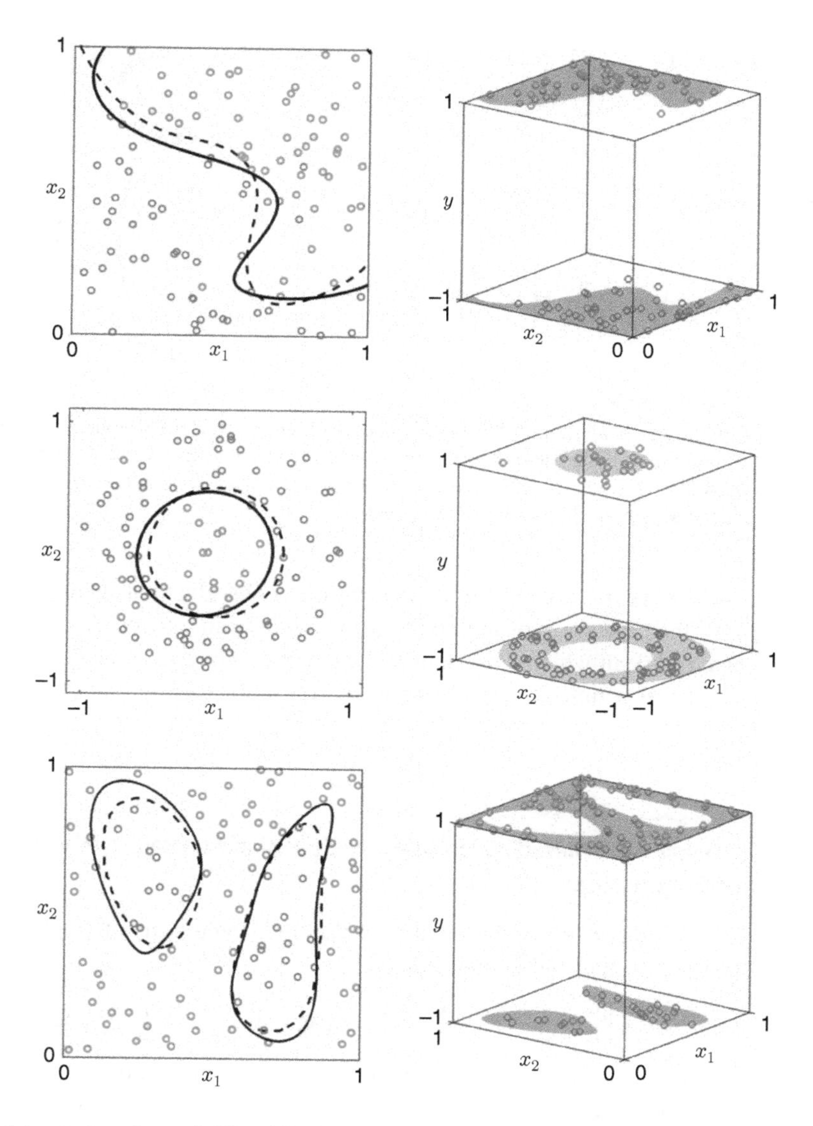

Fig. 6.6 Three datasets first shown in Fig. 6.5. Each dataset is shown in the original feature space from above (left panels) where both the original boundaries (dashed black) and learned boundaries (black) using polynomial basis features are shown, and from the side (right panels) where the indicator functions corresponding to each learned boundary are shown (in gray). The general equations defining both the learned boundaries and corresponding indicator functions are defined in Section 6.2.3.

In Fig. 6.7 we show two additional datasets along with nonlinear separators formed using fixed feature bases, in particular degree 2 polynomial and Fourier features[7] respectively, which in each case provides perfect classification. As was the case with regression

[7] Note that a degree D Fourier set of basis features for input $\mathbf{x}_p$ of dimension $N > 1$ consists of all monomial features of the form $f(\mathbf{x}) = e^{2\pi i m_1 x_1} e^{2\pi i m_2 x_2} \cdots e^{2\pi i m_N x_N}$ (see footnote 5 of Chapter 5) where $-D \leq m_1, m_2, \cdots, m_N \leq D$. There are a total of $M = (2D+1)^N - 1$ such terms excluding the constant feature, and hence this is the length of the corresponding feature vector.

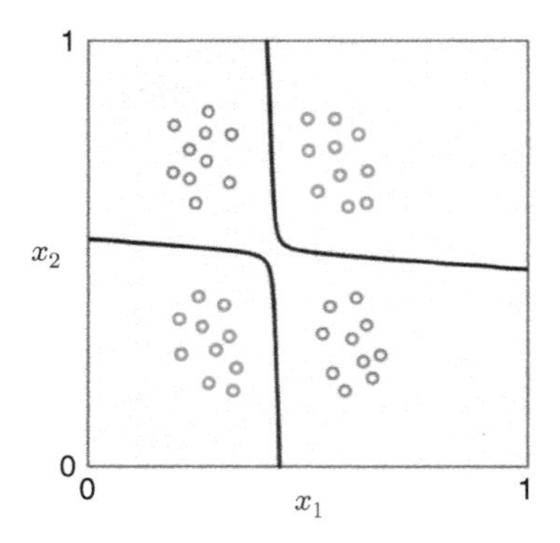

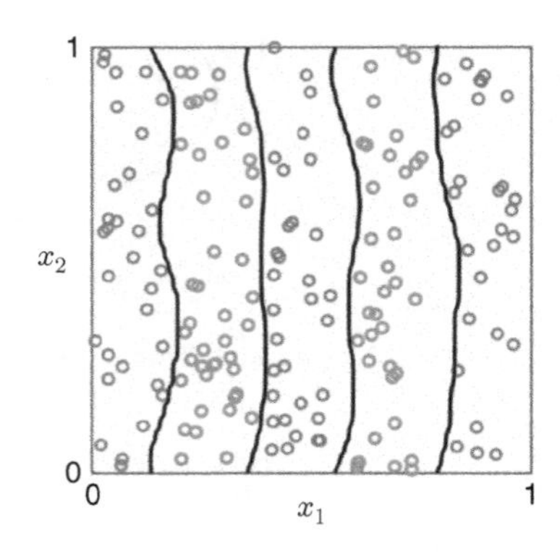

Fig. 6.7 (left panel) A version of the so-called XOR dataset with learned nonlinear boundary using the squared margin cost and degree 2 polynomial features. (right panel) Degree 2 Fourier features used for classification of a dataset consisting of consecutive bands of differing class points. In each case the data is perfectly separated.

in Example 5.1, the number of fixed bases for classification again grows combinatorially with the dimension of the input. As we will see in Section 7.1, this problem can once again be dealt with via the notion of a "kernel," however, this again introduces a serious numerical optimization problem as the size of the data-set grows.

Example 6.2 Classification with a basis of single hidden layer neural network features

The feature vector of the input $\mathbf{x}_p$ made by using a basis of single hidden layer neural network features takes the form

$$\mathbf{f}_p = \left[\; a\left(c_1 + \mathbf{x}_p^T\mathbf{v}_1\right) \quad a\left(c_2 + \mathbf{x}_p^T\mathbf{v}_2\right) \quad \cdots \quad a\left(c_M + \mathbf{x}_p^T\mathbf{v}_M\right) \; \right]^T, \tag{6.15}$$

where $a\left(\cdot\right)$ is any activation function as detailed in Section 5.1.4. However, unlike the case with fixed basis features, when using neural networks a cost like the softmax is non-convex, and thus effectively solving e.g.,

$$\underset{b,\,\mathbf{w},\,\Theta}{\text{minimize}} \sum_{p=1}^{P} \log\left(1 + e^{-y_p\left(b + \mathbf{f}_p^T\mathbf{w}\right)}\right) \tag{6.16}$$

requires running gradient descent several times (using a different initialization in each instance) in order to find a good local minimum (see Exercise 6.4).

This issue is illustrated in Fig. 6.8, where we have used a single hidden layer basis with the tanh activation to classify the datasets originally shown in Fig. 6.7. In Fig. 6.8 we show the result of running gradient descent, with (top) $M = 2$ and (bottom) $M = 4$ basis features respectively, three times with three random initializations. Note that in each case one of these initializations leads gradient descent to a bad stationary point

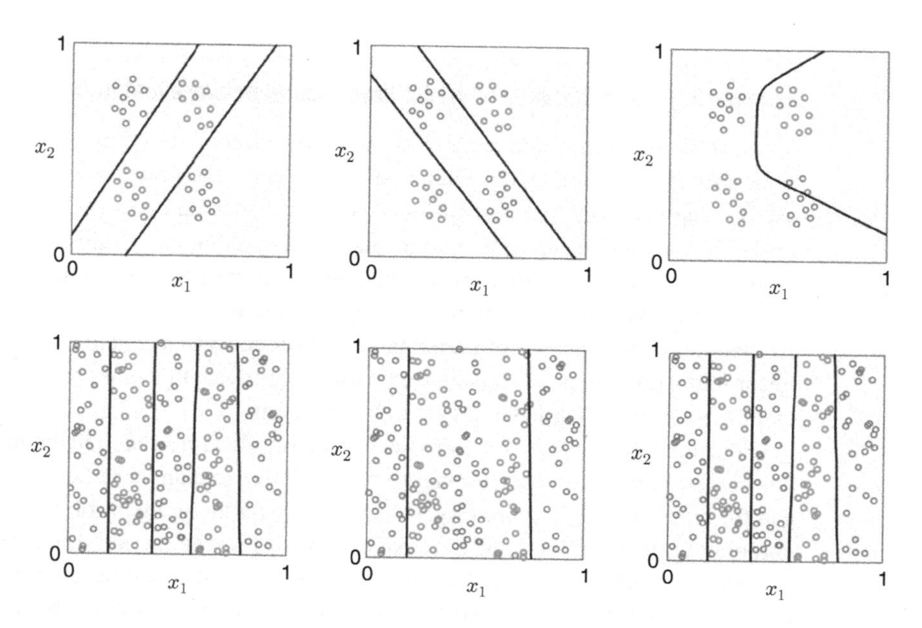

Fig. 6.8 The single hidden layer feature basis with tanh activation applied to classifying "XOR" and "stripe" datasets first shown in Fig. 6.7. For each dataset gradient descent is run three times with a different random initialization in each instance. (top panels) The first two resulting learned boundaries for the XOR dataset classify the data perfectly, while the third fails. (bottom panels) The first and last run of gradient descent provide a perfect separating boundary for the dataset consisting of consecutive class stripes, while the second does not.

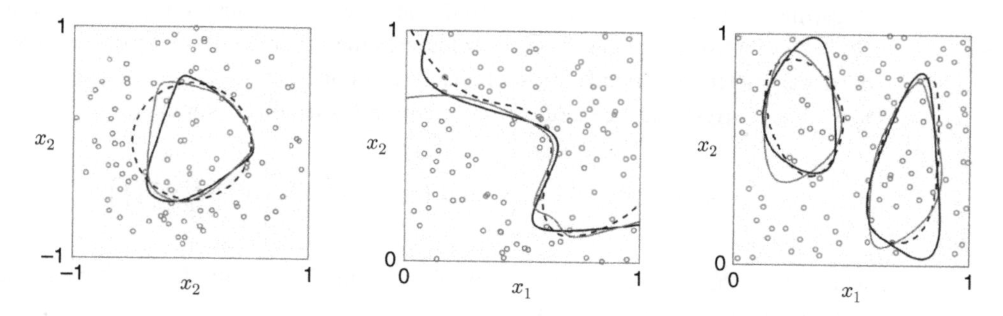

Fig. 6.9 The datasets first shown in Fig. 6.6 classified using a single hidden layer basis with (left panel) $M = 3$, (middle panel) $M = 4$, and (right panel) $M = 6$ basis features. For each case the boundaries of two successful runs of gradient descent, with random initializations, are shown in black and gray. Also plotted in dashed black are the true boundaries.

resulting in a poor fit to the respective dataset. In Fig. 6.9 we show the result of applying the same basis type with $M = 3$ (left panel), $M = 4$ (middle panel), and $M = 6$ (right panel) basis features respectively to the three datasets shown in Fig. 6.6.

Example 6.3 Classification with multilayer neural network features

To construct multilayer neural network bases we can sum and compose simple functions as described in Section 5.1.4. Doing so creates the same practical tradeoff as with the use of deep net basis features for regression, as discussed in Example 5.3. This tradeoff being that adding more hidden layers, while making each adjustable basis feature more flexible, comes at the cost of making the corresponding minimization problem more non-convex and thus more challenging to solve. However, the same ideas, i.e., regularization and the use of particularly helpful activation functions like the rectified linear unit $a(t) = \max(0, t)$, can also be used to mitigate this non-convexity issue when employing deep net basis features for classification.

Unfortunately, implementing gradient descent (often referred to in the machine learning community as *the backpropagation algorithm*) remains a tedious task due to the careful book-keeping required to correctly compute the gradient of a cost function incorporating deep net features. Because of this we provide the interested reader with an organized presentation of gradient computation for cost functions employing deep net basis features in Section 7.2. To avoid potential errors in computing the derivatives of a deep net cost function by hand, computational techniques like automatic differentiation [59] are often utilized when using deep nets in practice.

6.2.3 Classifier accuracy and boundary definition

Regardless of the cost function used, once we have learned proper parameters $(b^\star, \mathbf{w}^\star, \Theta^\star)$ using a given set of basis features the accuracy of a learned classifier with these parameters is defined almost precisely, as in Section 4.1.5, by replacing each input $\mathbf{x}_p$ with its corresponding feature representation $\mathbf{f}_p$. We first compute the counting cost

$$g_0\left(b^\star, \mathbf{w}^\star, \Theta^\star\right) = \sum_{p=1}^{P} \max\left(0,\ \text{sign}\left(-y_p\left(b^\star + \mathbf{f}_p^T\mathbf{w}^\star\right)\right)\right), \tag{6.17}$$

which gives the number of misclassifications of the learned model, and this defines the final accuracy of the classifier as

$$\text{accuracy} = 1 - \frac{g_0}{P}. \tag{6.18}$$

The learned nonlinear boundary (like those shown in each figure of this section) is then defined by the set of $\mathbf{x}$ where

$$b^\star + \sum_{m=1}^{M} f_m(\mathbf{x})\, w_m^\star = 0. \tag{6.19}$$

Likewise the final approximation of the true underlying indicator function (like those shown in the right column of Fig. 6.6) is given as

$$y(\mathbf{x}) = \text{sign}\left(b^{\star} + \sum_{m=1}^{M} f_m(\mathbf{x})\, w_m^{\star}\right), \tag{6.20}$$

which is itself an indicator function.

6.3 Multiclass classification

In this section we briefly describe how feature bases may be used to automate the design of features for both popular methods of multiclass classification first introduced in Section 4.4. Note that throughout this section we will suppose that we have a dataset $\left\{\left(\mathbf{x}_p, y_p\right)\right\}_{p=1}^{P}$ consisting of C distinct classes, where $y_p \in \{1, 2, \ldots, C\}$.

6.3.1 One-versus-all multiclass classification

Generalizing the one-versus-all (OvA) approach (introduced in Section 4.4.1) to multiclass classification is quite straightforward. Recall that with OvA we decompose the multiclass problem into C individual two class subproblems, and having just discussed how to incorporate feature bases into such problems in the previous section we can immediately employ them with each OvA subproblem.

To learn the classifier distinguishing class c from all other classes we assign temporary labels to all the points: points in classes c and "not-c" are assigned temporary labels $+1$ and -1, respectively. Choosing a type of feature basis we transform each $\mathbf{x}_p$ into an M_c length feature (note this length can be chosen independently for each subproblem) vector $\mathbf{f}_p^{(c)}$ and solve the associated two class problem, as described in the previous section, giving parameters $(b_c, \mathbf{w}_c, \Theta_c)$.

To determine the class of a point $\mathbf{x}$ we combine the resulting C individual classifiers (via the fusion rule in (4.47) properly generalized) as

$$y = \underset{j=1...C}{\text{argmax}}\left(b_j + \left(\mathbf{f}_p^{(j)}\right)^T \mathbf{w}_j\right). \tag{6.21}$$

Example 6.4 One-versus-all classification using fixed feature bases

In Fig. 6.10 we show the result of applying OvA to two multiclass datasets each containing $C = 3$ classes. For the first dataset (shown in the top left panel) we use a degree 4 polynomial for each subproblem, and likewise for the second dataset (shown in the bottom left panel) we use a degree 2 for all subproblems. The following three panels in each example show the resulting fit on each individual subproblem, with the final panel displaying the final combined boundary using Equation (6.21). Note in the second example especially that one of the subproblem classifiers is quite poor, but nonetheless the combined classifier perfectly separates all classes.

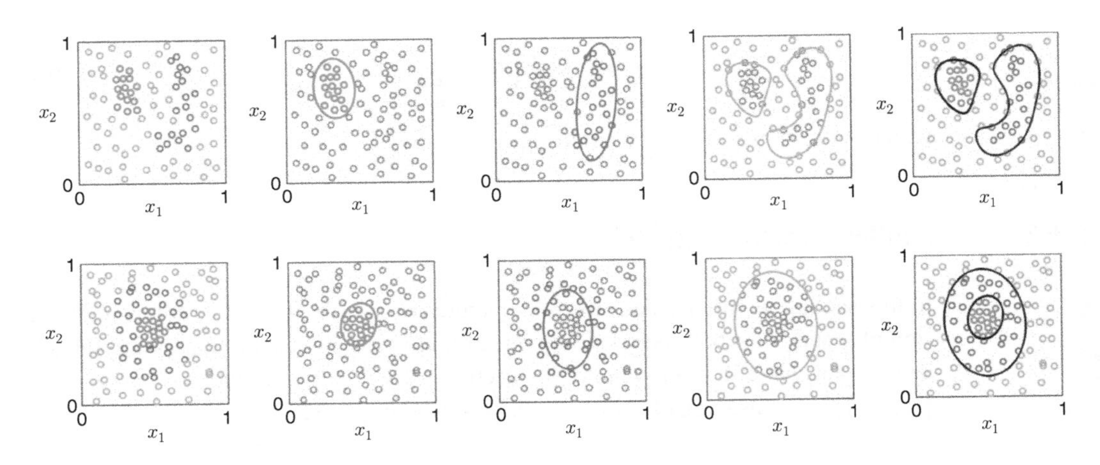

Fig. 6.10 Result of applying the OvA framework to two $C = 3$ class problems (shown in the top and bottom left panels respectively). In each case the following three panels show the result for the red class versus all, blue class versus all, and green class versus all subproblems. A degree 4 (top) and 2 (bottom) polynomial was used respectively for each subproblem. The panels on the extreme right show the combined boundary determined by Equation (6.21), which in both cases perfectly separates the three classes (even though the blue versus all classifier in the second example performs quite poorly).

6.3.2 Multiclass softmax classification

To apply the multiclass softmax framework discussed in Section 4.4 we transform all input data via a single fixed or neural network feature basis, and denote by $\mathbf{f}_p$ the resulting M length feature map of the point $\mathbf{x}_p$. We then minimize the corresponding version of the multiclass softmax cost function, first shown in Equation (4.53), on the transformed data as

$$g\left(b_1, \ldots, b_C, \mathbf{w}_1, \ldots, \mathbf{w}_C, \Theta\right) = -\sum_{c=1}^{C}\sum_{p\in\Omega_c}\left[\left(b_c + \mathbf{f}_p^T\mathbf{w}_c\right) - \log\left(\sum_{j=1}^{C} e^{b_j + \mathbf{f}_p^T\mathbf{w}_j}\right)\right]. \tag{6.22}$$

Note here that each $\mathbf{w}_c$ now has length M, and that the parameter set Θ as always contains internal parameters of a neural network basis feature if it is used (and is otherwise empty if using a fixed feature map). When employing a fixed feature basis this can then be minimized precisely as described for the original in Example 4.6, i.e., the gradient is given precisely as in Equation (4.57) replacing each $\mathbf{x}_p$ with $\mathbf{f}_p$. Computing the gradient is more complicated when employing a neural network feature basis (requiring careful bookkeeping and many uses of the chain rule) and additionally the corresponding cost function above becomes non-convex (while it remains convex when using any fixed feature basis).

6.4 Cross-validation for classification

In the previous chapter we saw how using more basis elements generally results in a better approximation of a continuous function. However, as we saw with regression in

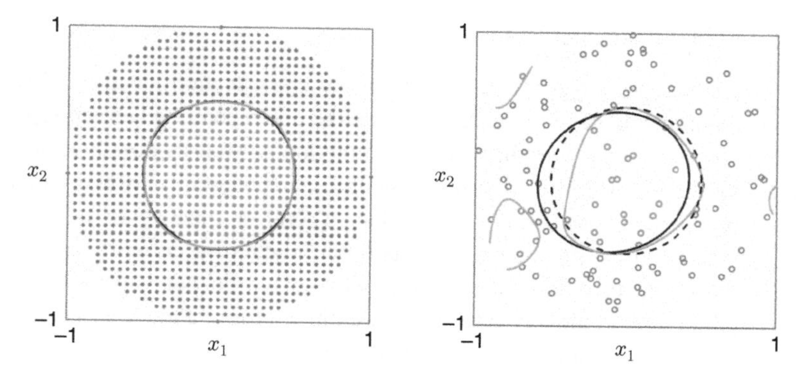

Fig. 6.11 Discretized (left panel) and noisy samples (right) from a generating function with class boundary shown (the circle of radius one-half) in dashed black, along with fits of degree 2 (in black) and degree 5 (in green) polynomial features. Like regression, while increasing the number of basis elements produces a better fit in the discretized case, for more realistic cases like the dataset on the right this can lead to overfitting. In the right panel the lower degree polynomial produces a classifier that matches the true boundary fairly well, while the higher degree polynomial leads to a classifier that overfits the data encapsulating falsely labeled red points outside the true boundary, thus leading to a poorer representation of the underlying generating function (see text for further details).

Section 5.3, while it is true that our approximation of a dataset itself improves as we add more basis features, this can substantially decrease our estimation of the underlying data generating function (a phenomenon referred to as *overfitting*). Unfortunately, the same overfitting problem presents itself in the case of classification as well. Similarly to regression, in the ideal classification scenario discussed in Section 6.1 using more basis elements generally improves our approximation. However, in general instances of classification, analogous to what we saw with regression, adding more basis features (increasing M) can result in fitting closely to the data we have while poorly to the underlying function (a phenomenon once again referred to as overfitting).

We illustrate the overfitting issue with classification using a particular dataset in Fig. 6.11, where we show the discretized indicator (left panel) along with the related noisy dataset (right panel) originally shown together in Fig. 6.5. For each dataset we show the resulting fit provided by both a degree 2 and a degree 5 polynomial (shown in black and green respectively). While the degree 2 features produce a classifier in each case that closely matches the true boundary, the higher degree 5 polynomial creates an overfitting classifier which encapsulates mislabeled points outside of the half circle boundary of the true function, leading to a poorer representation.

In this section we outline the use of cross-validation, culminating once again with the *k-fold cross-validation* method, for the intelligent automatic choice of M for both two class and multiclass classification problems. As with regression, here once again the k-fold method[8] provides a way of determining a proper value of M, however, once again this comes at significant computational cost.

[8] Readers particularly interested in using fixed bases with high dimensional input, deep network features, as well as multiclass softmax classification using feature bases should also see Section 7.3, where a variation of k-fold cross-validation is introduced that is more appropriate for these instances.

6.4.1 Hold out cross-validation

Analogous to regression, in the case of classification ideally we would like to choose a number M of basis features so that the corresponding learned representation matches the true data generating indicator function as well as possible. Of course because we only have access to this true function via (potentially) noisy samples, this goal must be pursued based solely on the data. Therefore once again we aim to choose M such that the corresponding model fits both the data we currently have, as well as the data we might receive in the future. Because we do not have access to any future data points this intuitively directs us to employ cross-validation, where we tune M, so that the corresponding model fits well to an unseen portion of our original data (i.e., the testing set).

Thus we can do precisely what was described for regression in Section 5.3 and perform k-fold cross-validation to determine M. In other words, we can simulate this desire by splitting our original data into k evenly sized pieces and merge $k - 1$ of them into a training set and use the remaining piece as a testing set. Furthermore the same intuition for choosing k introduced for regression also holds here, with common values in practice being in the range $k = 3$–10.

Example 6.5 Hold out for classification using polynomial features

For clarity we first show an example of the hold out method, followed by explicit computations, which are then simply repeated on each fold (averaging the results) in performing the k-fold method. In Fig. 6.12 we show the result of applying hold out cross-validation to the dataset first shown in the bottom panels of Fig. 6.5. Here we use $k = 3$, use the softmax cost, and M in the range $M = 2, 5, 9, 14, 20, 27, 35, 44$ which corresponds (see footnote 5) to polynomial degrees $D = 1, 2, \ldots, 8$ (note that for clarity panels in the figure are indexed by D).

Based on the models learned for each value of M (see the middle set of eight panels of the figure) we plot training and testing errors (in the panel second to the right), measuring how well each model fits the training and testing data respectively, over the entire range of values. Note that unlike the testing error, the training error always decreases as we increase M (which occurs more generally regardless of the dataset/feature basis used). The model that provides the smallest testing error ($M^\star = 14$ or equivalently $D^\star = 4$) is then trained again on the entire dataset, giving the final classification model shown in black in the rightmost panel of the figure.

6.4.2 Hold out calculations

Here we give a complete set of hold out cross-validation calculations in a general setting, which closely mirrors the version given for regression in Section 5.3.3. We denote the collection of points belonging to the training and testing sets respectively by their indices as

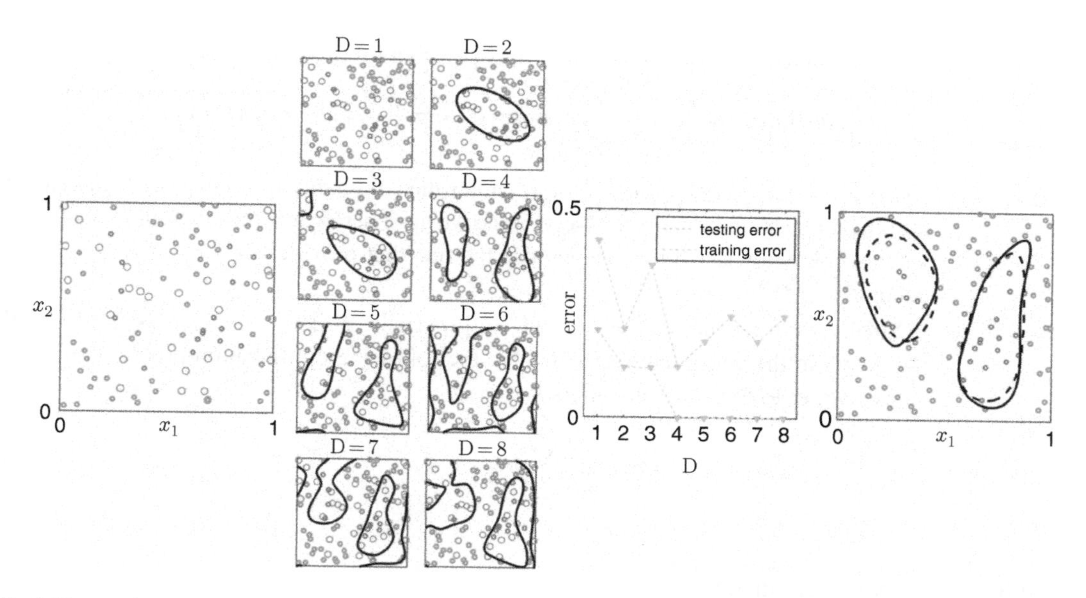

Fig. 6.12 An example of hold out cross-validation applied using polynomial features. (left panel) The original data split into training and testing sets, with the points belonging to each set drawn as smaller thick and larger thin points respectively. (middle eight panels) The fit resulting from each set of degree D polynomial features in the range $D = 1, 2, \ldots, 8$ shown in black in each panel. Note how the lower degree fits underfit the data, while the higher degree fits overfit the data. (second from right panel) The training and testing errors, in blue and yellow respectively, of each fit over the range of degrees tested. From this we see that $D^{\star} = 4$ (or $M^{\star} = 14$) provides the best fit. Also note how the training error always decreases as we increase the degree/number of basis elements, which will always occur regardless of the dataset/feature basis type used. (right panel) The final model using $M^{\star} = 14$, trained on the entire dataset (shown in black), fits the data well and closely matches the boundary of the underlying data generating function (shown in dashed black).

$$\begin{aligned} \Omega_{\text{train}} &= \left\{ p \mid \left(\mathbf{x}_p,\, y_p\right) \text{ belongs to the training set} \right\} \\ \Omega_{\text{test}} &= \left\{ p \mid \left(\mathbf{x}_p,\, y_p\right) \text{ belongs to the testing set} \right\} \end{aligned} . \tag{6.23}$$

We then choose a basis type (e.g., polynomial, Fourier, neural network) and choose a range for the number of basis features over which we search for an ideal value for M. To determine the training and testing error of each value of M tested we first form the corresponding feature vector $\mathbf{f}_p = \left[\begin{array}{cccc} f_1\left(\mathbf{x}_p\right) & f_2\left(\mathbf{x}_p\right) & \cdots & f_M\left(\mathbf{x}_p\right) \end{array} \right]^T$ and fit a corresponding model to the training set by minimizing e.g., the softmax or squared margin cost. For example employing the softmax we solve

$$\underset{b,\, \mathbf{w}, \Theta}{\text{minimize}} \sum_{p \in \Omega_{\text{train}}} \log\left(1 + e^{-y_p\left(b + \mathbf{f}_p^T \mathbf{w}\right)}\right). \tag{6.24}$$

Denoting a solution to the problem above as $\left(b_M^{\star},\, \mathbf{w}_M^{\star},\, \Theta_M^{\star}\right)$ we find the training and testing errors for the current value of M using these parameters over the training and testing sets using the counting cost (see Section 6.2.3), respectively:

$$\text{Training error} = \frac{1}{|\Omega_{\text{train}}|} \sum_{p \in \Omega_{\text{train}}} \max\left(0,\ \text{sign}\left(-y_p\left(b^\star + \mathbf{f}_p^T \mathbf{w}^\star\right)\right)\right)$$
$$\text{Testing error} = \frac{1}{|\Omega_{\text{test}}|} \sum_{p \in \Omega_{\text{test}}} \max\left(0,\ \text{sign}\left(-y_p\left(b^\star + \mathbf{f}_p^T \mathbf{w}^\star\right)\right)\right), \tag{6.25}$$

where the notation $|\Omega_{\text{train}}|$ and $|\Omega_{\text{test}}|$ denotes the cardinality or number of points in the training and testing sets, respectively. Once we have performed these calculations for all values of M we wish to test, we choose the one that provides the lowest testing error, denoted by $M^\star$.

Finally we form the feature vector $\mathbf{f}_p = \begin{bmatrix} f_1(\mathbf{x}_p) & f_2(\mathbf{x}_p) & \cdots & f_{M^\star}(\mathbf{x}_p) \end{bmatrix}^T$ for all the points in the entire dataset, and solve the following optimization problem over the entire dataset to form the final model

$$\underset{b,\,\mathbf{w},\,\Theta}{\text{minimize}} \sum_{p=1}^{P} \log\left(1 + e^{-y_p\left(b + \mathbf{f}_p^T \mathbf{w}\right)}\right). \tag{6.26}$$

6.4.3 *k*-fold cross-validation

As introduced in Section 5.3.4, *k-fold cross-validation* is a robust extension of the hold out method whereby the procedure is repeated k times where in each instance (or *fold*) we treat a different portion of the split as a testing set and the remaining $k-1$ portions as the training set. The hold out calculations are then made, as detailed previously, on each fold and the value of M with the lowest *average* testing error is chosen. This produces a more robust choice of M, because potentially poor hold out choices on individual folds can be averaged out, producing a stronger model.

Example 6.6 *k*-fold cross-validation for classification using polynomial features

In Fig. 6.13 we illustrate the result of applying k-fold cross-validation to choose the ideal number M of polynomial features for the dataset shown in Example 6.5, where it was originally used to illustrate the hold out method. As in the previous example, here we set $k = 3$, use the softmax cost, and try M in the range $M = 2, 5, 9, 14, 20, 27, 35, 44$ which corresponds (see footnote 5) to polynomial degrees $D = 1, 2, \ldots, 8$ (note that for clarity panels in the figure are indexed by D).

In the top three rows of Fig. 6.13 we show the result of applying hold out on each fold. In each row we show a fold's training and testing data in the left panel, the training/testing errors for each M on the fold (as computed in Equation (6.25)) in the middle panel, and the final model (learned to the entire dataset) provided by the choice of M with lowest testing error. As can be seen, the particular split leads to an overfitting result on the first two folds and an underfitting result on the third fold. In the middle panel of the final row we show the result of averaging the training/testing errors over all $k = 3$ folds, and in the right panel the result of choosing the overall best $M^\star = 14$ (or equivalently $D^\star = 4$) providing the lowest average

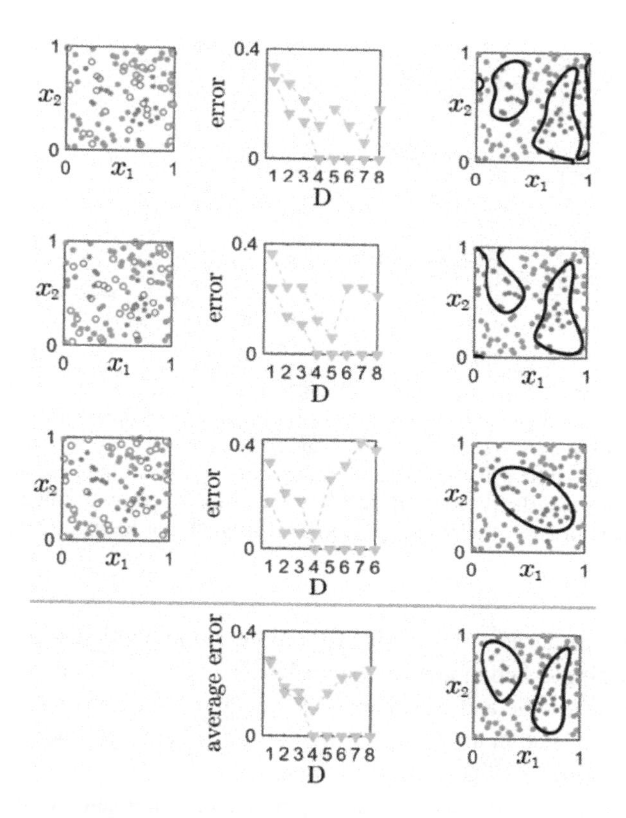

Fig. 6.13 Result of performing k-fold cross-validation with $k = 3$ (see text for further details). The top three rows display the result of performing the hold out method on each fold. The left, middle, and right columns show each fold's training/testing sets (drawn as thick and thin points respectively), training and testing errors over the range of M tried, and the final model (fit to the entire dataset) chosen by picking the value of M providing the lowest testing error. Due to the split of the data, performing hold out on each fold results in a poor overfitting (first two folds) or underfitting (final fold) model for the data. However, as illustrated in the final row, by averaging the testing errors (bottom middle panel) and choosing the model with minimum associated average test error we average out these problems (finding that $D^\star = 4$ or $M^\star = 14$) and determine an excellent model for the phenomenon (as shown in the bottom right panel).

testing error. By taking this value we average out the poor choices determined on each fold, and end up with a model that fits both the data and underlying function quite well.

Example 6.7 Warning examples

When a k-fold determined set of features performs poorly this is almost always indicative of a poorly structured dataset (i.e., there is little relationship between the input/output data), like the one shown in the left panel of Fig. 6.14. However, there are

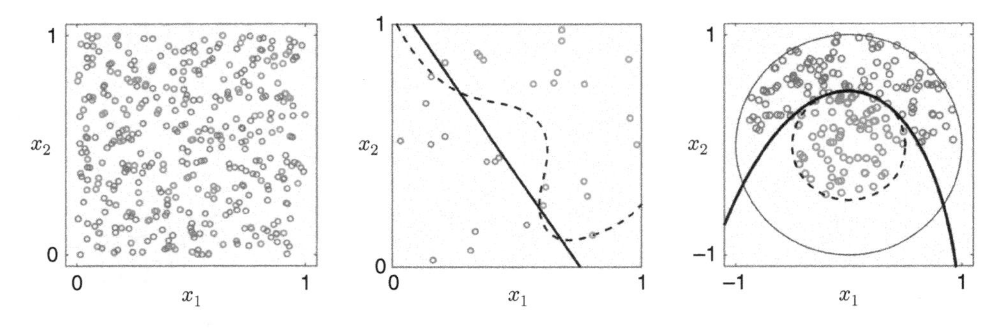

Fig. 6.14 (left panel) A low accuracy k-fold fit to a dataset indicates that it has little structure (i.e., that there is little to no relationship between the input and output). It is possible that a high accuracy k-fold fit fails to capture the true nature of an underlying function, as when (middle panel) we have too little data (the k-fold linear separator is shown in black, and the true nonlinear separator is shown dashed) and (right panel) when we have poorly distributed data (again the k-fold separator is shown in black, the original separator dashed). See text for further details.

also instances, when we have too little or too poorly distributed data, when a *high* performing k-fold model can be misleading as to how well we understand a phenomenon. In the middle and right panels of the figure we show two such instances that the reader should keep in mind when using k-folds, where we either have too little (middle panel) or poorly distributed data (right panel).

In the first instance we have generated a small sample of points based on the second indicator function shown in Fig. 6.3, which has a nonlinear boundary in the original feature space. However, the sample of data is so small that it is perfectly linearly separable, and thus applying e.g., k-fold cross-validation with polynomial basis features will properly (due to the small selection of data) recover a line to distinguish between the two classes. However, clearly data generated via the same underlying process in the future will violate this linear boundary, and thus our model will perform poorly. This sort of problem arises in applications such as automatic medical diagnosis (see Example 1.6) where access to data is limited. Unless we can gather additional data to fill out the space (making the nonlinear boundary more visible) this problem is unavoidable.

In the second instance shown in the right panel of the figure, we have plenty of data (generated using the indicator function originally shown in Fig. 6.4) but it is poorly distributed. In particular, we have no samples from the blue class in the lower half of the space. In this case the k-fold method (again here using polynomial features) properly determines a separating boundary that perfectly distinguishes the two classes. However, many of the blue class points we would receive in the future in the lower half of the space will be misclassified given the learned k-fold model. This sort of issue can arise in practice, e.g., when performing face detection (see Example 1.4), if we do not collect a thorough dataset of blue (e.g., "non-face") examples. Again, unless we can gather further data to fill out the space this problem is unavoidable.

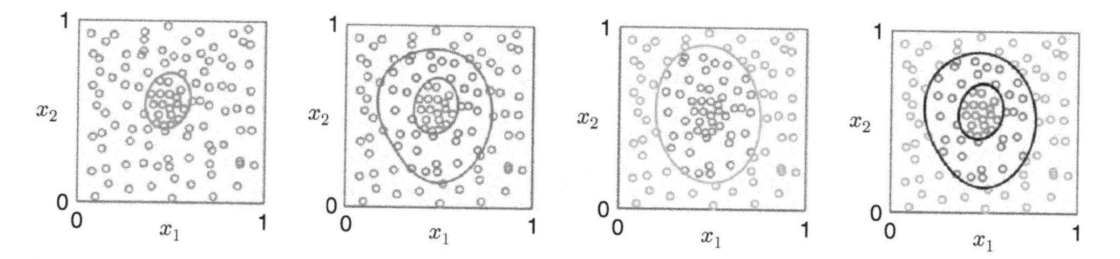

Fig. 6.15 Result of performing $k = 3$ fold cross-validation on the $C = 3$ class dataset first shown in Fig. 6.10 using OvA (see text for further details). The left three panels show the result for the red class versus all, blue class versus all, and green class versus all subproblems. For the red/green versus all problems the optimal degree found was $D^{\star} = 2$, while for the blue versus all $D^{\star} = 4$ (note how this produces a better fit than the $D = 2$ fit shown originally in Fig. 6.10). The right panel shows the combined boundary determined by Equation (6.21), which perfectly separates the three classes.

6.4.4 k-fold cross-validation for one-versus-all multiclass classification

Employing the one-versus-all (OvA) framework for multiclass classification, we can immediately apply the k-fold method described previously. For a C class problem we simply apply the k-fold method to each of the C two class classification problems, and combine the resulting classifiers as shown in Equation (6.21). We show the result of applying $k = 3$ fold cross-validation with OvA on two datasets with $C = 3$ and $C = 5$ classes respectively in Fig. 6.15 and 6.16, where we have used polynomial features with $M = 2, 5, 9, 14, 20, 27, 35, 44$ or equivalently of degree $D = 1, 2, \ldots, 8$ for each two class subproblem. Displayed in each figure are the nonlinear boundaries determined for each fold, as well as the combined result in the right panel of each figure. In both instances the combined boundaries separate the different classes of data very well.

6.5 Which basis works best?

For an arbitrary classification dataset we cannot say whether a particular feature basis will provide better results than others. However, as with the case of regression discussed in Section 5.4, we can say something about the choice of bases in particular instances. For example, in the instance where data is plentiful and well distributed throughout the input space we can expect comparable performance among different feature bases (this was illustrated for the case of regression in Fig. 5.20). Practical considerations can again guide the choice of basis as well.

Due to the nature of classification problems it is less common (than with regression) that domain knowledge leads to a particular choice of basis. Rather, in practice it is more common to employ knowledge in the design of a feature transformation (like those discussed for text, image, or audio data in Section 4.6), and then determine possible nonlinear boundaries in this transformed data using feature bases as described in this chapter. For certain data types such as image data one can incorporate a parameterized transformation that outlines the sort of edge detection/histogramming

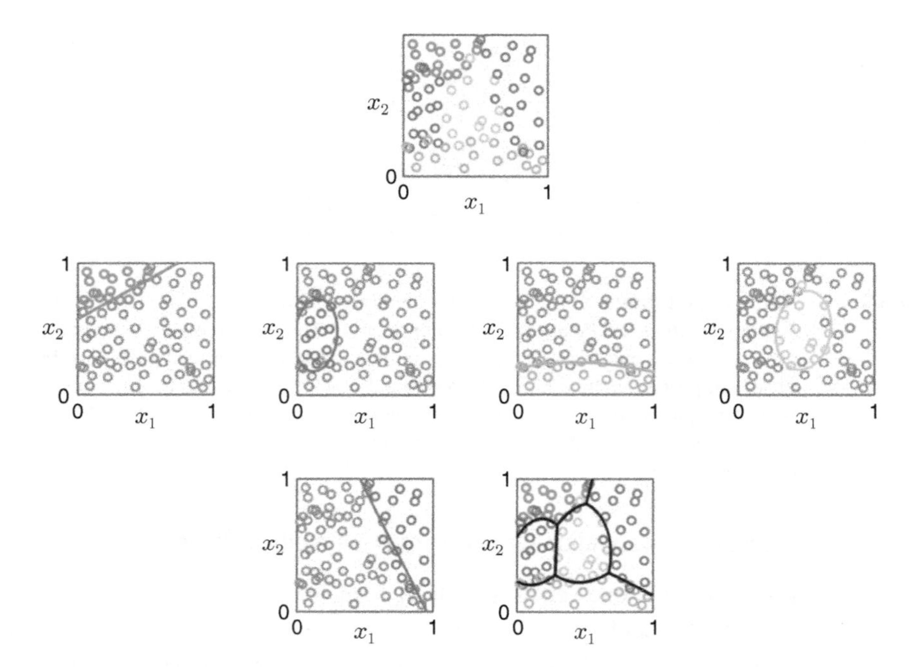

Fig. 6.16 Result of performing $k = 3$ fold cross-validation on an overlapping $C = 5$ class classification dataset (top panel) using OvA. The middle four panels show the result for the red, blue, green, and yellow class versus all subproblems respectively. The bottom two panels show the (left) purple class versus all and (right) the final combined boundary. For the red/purple versus all problems the k-fold found degree was $D^{\star} = 1$, while for the remaining subproblems $D^{\star} = 2$.

operations outlined in Section 4.6.2 directly into basis elements themselves. Parameters of this transformation are then learned simultaneously with those of the weighted basis sum itself. A popular example of this sort of approach is the *convolutional network* (see e.g., [42–44] and references therein), which incorporates such (parameterized) knowledge-driven features into a standard feed forward neural network basis.

Regardless of how knowledge is integrated, having some understanding of a phenomenon can significantly lessen the amount of data required to produce a k-fold representation that properly traces out a data generating function. On the other hand, broadly speaking if we have no understanding of a phenomenon we will typically require a significant amount of data in order to ensure that the features we have designed through the k-fold process are truly representative. What constitutes a "significant amount"? There is no precise formula in general, but due to the curse of dimensionality (see Fig. 5.2) we can say that the higher the dimension of the input the exponentially more data we will need to properly understand the underlying function.

This data–knowledge tradeoff is illustrated symbolically in Fig. 6.17.

6.6 Summary

We have seen that, analogous to regression, the general problem of classification is one of function approximation based on noisy samples. In the instance of classification,

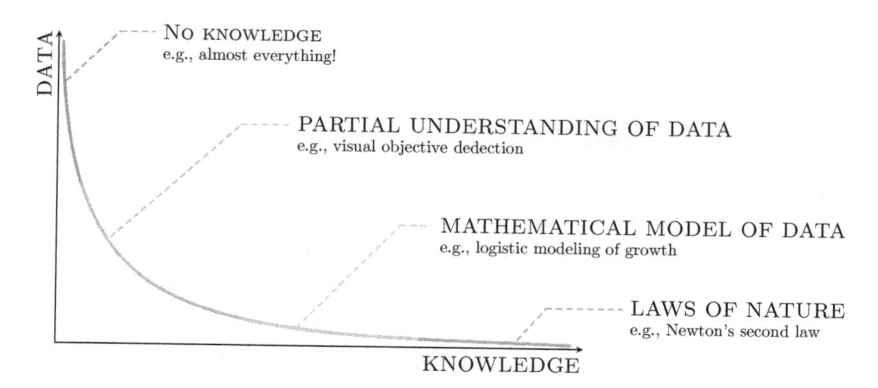

Fig. 6.17 A symbolic representation of the data–knowledge spectrum. The more knowledge we can incorporate into the design of features the less data is required to determine a strong k-fold cross-validated set of features. At the other end of the spectrum, if we know nothing regarding the underlying phenomenon we are modeling we will need a significant amount of data in order to forge strong cross-validated features.

however, the underlying data generating function is a piecewise continuous indicator function. As in the previous chapter, we began in the first section by investigating how to approximate such a data generating function itself, leading to both the familiar fixed and adjustable neural network bases we have seen previously.

In the second and third sections we described how real instances of classification datasets can be thought of as noisy samples from a true underlying indicator. We then saw how we can use a tractable minimization problem to learn the parameters of a weighted sum of basis features to fit general classification datasets. This idea was also shown to be easily integrated into both one-versus-all and multiclass softmax classification schemes, leading to natural nonlinear extensions of both.

In Section 6.4 we saw (as with regression) how overfitting is a problem when using too many features for classification. Cross-validation, culminating with the k-fold method, was then reintroduced in the context of classification as a way of preventing overfitting. Again, as with regression, it is (k-fold) cross-validation that often uses the bulk of computational resources in practice when solving general classification problems.

6.7 Exercises

Section 6.1 exercises

Exercises 6.1 Softmax cost gradient/Hessian calculations with fixed basis features

a) Assuming a fixed feature basis verify, using the compact notation $\tilde{\mathbf{f}}_p = \begin{bmatrix} 1 \\ \mathbf{f}_p \end{bmatrix}$ and $\tilde{\mathbf{w}} = \begin{bmatrix} b \\ \mathbf{w} \end{bmatrix}$, that the gradient of the softmax cost given in Equation (6.14) is correct. Furthermore, verify that the Hessian of the softmax in this case is given by

$$\nabla^2_{\tilde{\mathbf{w}}} g\left(\tilde{\mathbf{w}}\right) = \sum_{p=1}^{P} \sigma\left(-y_p \tilde{\mathbf{f}}_p^T \tilde{\mathbf{w}}\right)\left(1 - \sigma\left(-y_p \tilde{\mathbf{f}}_p^T \tilde{\mathbf{w}}\right)\right) \tilde{\mathbf{f}}_p \tilde{\mathbf{f}}_p^T. \tag{6.27}$$

Both gradient and Hessian here are entirely similar to the originals given in Example 4.1, replacing each $\mathbf{x}_p$ with its corresponding feature vector $\mathbf{f}_p$.

b) Show that the softmax cost using M elements of any fixed feature basis is still convex by verifying that it satisfies the second order condition for convexity. *Hint: the Hessian is a weighted sum of outer product matrices like the one described in Exercise 2.10.*

Exercises 6.2 Squared margin cost gradient/Hessian calculations with fixed feature basis

a) Assuming a fixed feature basis verify, using the compact notation $\tilde{\mathbf{f}}_p = \begin{bmatrix} 1 \\ \mathbf{f}_p \end{bmatrix}$ and $\tilde{\mathbf{w}} = \begin{bmatrix} b \\ \mathbf{w} \end{bmatrix}$, that the gradient and Hessian of the squared margin cost

$$g\left(\tilde{\mathbf{w}}\right) = \sum_{p=1}^{P} \max^2\left(0, 1 - y_p \tilde{\mathbf{f}}_p^T \tilde{\mathbf{w}}\right), \tag{6.28}$$

are given as

$$\begin{aligned} \nabla_{\tilde{\mathbf{w}}} g\left(\tilde{\mathbf{w}}\right) &= -2 \sum_{p=1}^{P} y_p \tilde{\mathbf{f}}_p \max\left(0,\ 1 - y_p \tilde{\mathbf{f}}_p^T \tilde{\mathbf{w}}\right) \\ \nabla^2_{\tilde{\mathbf{w}}} g\left(\tilde{\mathbf{w}}\right) &= 2 \sum_{p \in \Omega_{\tilde{\mathbf{w}}}} \tilde{\mathbf{f}}_p \tilde{\mathbf{f}}_p^T, \end{aligned} \tag{6.29}$$

where $\Omega_{\tilde{\mathbf{w}}}$ is the index set $\Omega_{\tilde{\mathbf{w}}} = \left\{p \mid 1 - y_p \tilde{\mathbf{f}}_p^T \tilde{\mathbf{w}} > 0\right\}$. These are entirely similar to the calculations given in Example 4.2 except for using the feature map $\mathbf{f}_p$ in place of the input $\mathbf{x}_p$.

b) Show that the squared margin cost using M elements of any fixed feature basis is convex. *Hint: see Exercise 4.6.*

Exercises 6.3 Polynomial basis features and the softmax cost

In this exercise you will explore how various degree D polynomial basis features fit using the softmax cost and the dataset shown in the bottom panel of Fig. 6.6. For this exercise you will need the wrapper *poly_classification_hw* and the data located in *2eggs_data.csv*.

a) Use the description of the two-dimensional polynomial basis features given in footnote 5 to transform the input using a general degree D polynomial. Write this feature transformation in the module

$$\mathbf{F} = \text{poly\_features}\left(\mathbf{X}, D\right) \tag{6.30}$$

located in the wrapper. Here $\mathbf{X}$ is the input data, D the degree of the polynomial features, and $\mathbf{F}$ the corresponding degree D feature transformation of the input (note your code should be able to transform the input to any degree D desired).

b) With your module complete you may run the wrapper. Two figures will be generated: the first shows the data along with various degree D polynomial fits, and the second shows the average number of misclassifications of each fit to the data-set. Discuss the results shown in these two figures. In particular, describe the relationship between a model's average number of misclassifications and how well it seems to represent the phenomenon generating the data as D increases over the range shown.

Exercises 6.4 Calculate the gradient using a single hidden layer basis

When employing M single hidden layer basis features (using any activation $a(\cdot)$) the full gradient of a cost g (e.g., the softmax) is a vector of length $Q = M(N+2)+1$ containing the derivatives of the cost with respect to each variable,

$$\nabla g = \begin{bmatrix} \frac{\partial}{\partial b} g & \frac{\partial}{\partial w_1} g & \cdots & \frac{\partial}{\partial w_M} g & \frac{\partial}{\partial c_1} g \cdots & \frac{\partial}{\partial c_M} g & \nabla^T_{\mathbf{v}_1} g & \cdots & \nabla^T_{\mathbf{v}_M} g \end{bmatrix}^T, \quad (6.31)$$

where the derivatives are easily calculated using the chain rule.

a) Using the chain rule verify that the derivatives of this gradient (using the softmax cost) are given by

$$\begin{aligned}
\frac{\partial}{\partial b} g &= -\sum_{p=1}^{P} \sigma\left(-y_p\left(b + \sum_{m=1}^{M} w_m a\left(c_m + \mathbf{x}_p^T \mathbf{v}_m\right)\right)\right) y_p \\
\frac{\partial}{\partial w_n} g &= -\sum_{p=1}^{P} \sigma\left(-y_p\left(b + \sum_{m=1}^{M} w_m a\left(c_m + \mathbf{x}_p^T \mathbf{v}_m\right)\right)\right) a\left(c_n + \mathbf{x}_p^T \mathbf{v}_n\right) y_p \\
\frac{\partial}{\partial c_n} g &= -\sum_{p=1}^{P} \sigma\left(-y_p\left(b + \sum_{m=1}^{M} w_m a\left(c_m + \mathbf{x}_p^T \mathbf{v}_m\right)\right)\right) a'\left(c_n + \mathbf{x}_p^T \mathbf{v}_n\right) w_n y_p \\
\nabla_{\mathbf{v}_n} g &= -\sum_{p=1}^{P} \sigma\left(-y_p\left(b + \sum_{m=1}^{M} w_m a\left(c_m + \mathbf{x}_p^T \mathbf{v}_m\right)\right)\right) a'\left(c_n + \mathbf{x}_p^T \mathbf{v}_n\right) \mathbf{x}_p w_n y_p.
\end{aligned} \quad (6.32)$$

b) This gradient can be written more efficiently for programming languages like Python and MATLAB/OCTAVE that have especially good implementations of matrix/vector operations by writing it more compactly. Supposing that $a = \tanh(\cdot)$ is the activation function (meaning $a' = \text{sech}^2(\cdot)$ is the hyperbolic secant function squared), verify that the derivatives from part a) may be written more compactly as

$$\begin{aligned}
\frac{\partial}{\partial b} g &= -\mathbf{1}_{P\times 1}^T \mathbf{q} \odot \mathbf{y} \\
\frac{\partial}{\partial w_n} g &= -\mathbf{1}_{P\times 1}^T \left(\mathbf{q} \odot \mathbf{t}_n \odot \mathbf{y}\right) \\
\frac{\partial}{\partial c_n} g &= -1_{P\times 1}^T \left(\mathbf{q} \odot \mathbf{s}_n \odot \mathbf{y}\right) w_n \\
\nabla_{\mathbf{v}_n} g &= -\mathbf{X} \cdot \mathbf{q} \odot \mathbf{s}_n \odot \mathbf{y} w_n,
\end{aligned} \quad (6.33)$$

where $\odot$ denotes the component-wise product and denoting $q_p = \sigma\left(-y_p\left(b + \sum_{m=1}^{M} w_m \tanh\left(c_m + \mathbf{x}_p^T \mathbf{v}_m\right)\right)\right)$, $t_{np} = \tanh\left(c_n + \mathbf{x}_p^T \mathbf{v}_n\right)$, $s_{np} = \text{sech}^2\left(c_n + \mathbf{x}_p^T \mathbf{v}_n\right)$, and $\mathbf{q}$, $\mathbf{t}_n$, and $\mathbf{s}_n$ the P length vectors containing these entries.

Exercises 6.5 Code up gradient descent using single hidden layer bases

In this exercise you will reproduce the classification result using a single hidden layer feature basis with tanh activation shown in the middle panel of Fig. 6.9.

a) Plug the gradient from Exercise 6.4 into the gradient descent function

$$\mathbf{T} = \text{tanh\_softmax}\left(\mathbf{X}, \mathbf{y}, M\right) \tag{6.34}$$

located in the wrapper *single_layer_classification_hw* and the dataset *genreg_data.csv*, both of which may be downloaded from the book website. Here $\mathbf{T}$ is the set of optimal weights learned via gradient descent, $\mathbf{X}$ is the input data matrix, $\mathbf{y}$ contains the associated labels, and M is the number of basis features to employ.

Almost all of this function has already been constructed for you, e.g., various initializations, step length, etc., and you need only enter the gradient of the associated cost function. All of the additional code necessary to generate the associated plot is already provided in the wrapper. Due to the non-convexity of the associated cost function when using neural network features, the wrapper will run gradient descent several times and plot the result of each run.

b) Try adjusting the number of basis features M in the wrapper and run it several times. Is there a value of M other than $M = 4$ that seems to produce a good fit to the underlying function?

Exercises 6.6 Code up the k-nearest neighbors (k-NN) classifier

The *k-nearest neighbors* (k-NN) is a local classification scheme that, while differing from the more global feature basis approach described in this chapter, can produce nonlinear boundaries in the original feature space as illustrated for some particular examples in Fig. 6.18.

With the k-NN approach there is no training phase to the classification scheme. We simply use the training data directly to classify any new point $\mathbf{x}_{\text{new}}$ by taking the average of the labels of its k-nearest neighbors. That is, we create the label y_{new} for a point $\mathbf{x}_{\text{new}}$ by simply calculating

$$y_{\text{new}} = \text{sign}\left(\sum_{i \in \Omega} y_i\right), \tag{6.35}$$

where Ω is the set of indices of the k closest training points to $\mathbf{x}_{\text{new}}$. To avoid tie votes (i.e., a value of zero above) typically the number of neighbors k is chosen to be odd (however, in practice the value of k is typically set via cross-validation).

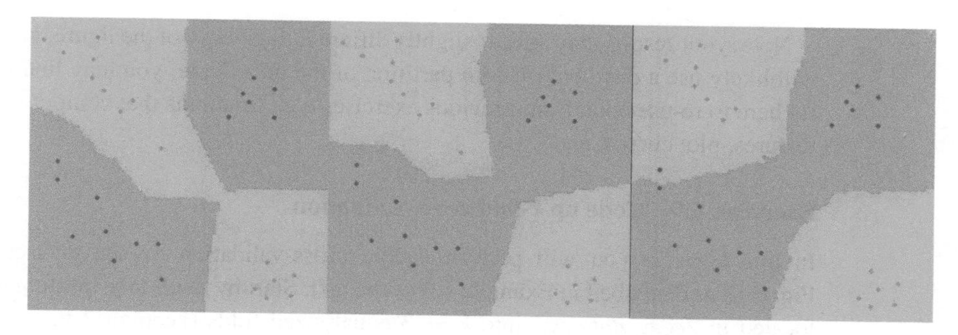

Fig. 6.18 The k-NN classifier applied to a two class dataset (the blue and red points) where (left panel) $k = 1$, (middle panel) $k = 5$, and (right panel) $k = 10$. All points in the space have been colored according to the rule given in Equation (6.35) where the red and blue classes have labels $+1$ and -1 respectively.

Code up the k-NN algorithm and reproduce the results shown in Fig. 6.18 using the dataset located in *knn_data.csv*.

Section 6.2 exercises

Exercises 6.7 One-versus-all using a polynomial basis

In this exercise you will reproduce the one-versus-all classification on the $C = 3$ class dataset shown in the bottom panels of Fig. 6.10 using polynomial features. Note that for this exercise you will need to have completed the poly_features module for polynomial features described in Exercise 6.3.

a) Place the *poly_features* module in the wrapper *ova_fixed_basis* and use the dataset *bullseye_data.csv*, both of which may be downloaded from the book website. After installing the module try running the wrapper to reproduce the results shown in Fig. 6.10.

b) Try adjusting the degree D in the wrapper and run it several times. Is there a value of D other than $D = 2$ that seems to produce a good fit to the data?

Section 6.3 exercises

Exercises 6.8 Code up hold out cross-validation

In this exercise you will perform hold out cross-validation on the dataset shown in Fig. 6.12 as described in Example 6.5 of the text. Start by randomly splitting the dataset, located in *2eggs_data.csv*, into $k = 3$ equal sized folds (keeping 2 folds as training, and 1 fold as testing data). Use the polynomial basis features with M in the range $M = 2, 5, 9, 14, 20, 27, 35, 44$ (or likewise D in the range $D = 1, 2, \ldots, 8$) and produce a graph showing the training and testing error for each D like the one shown in Fig. 6.12, as well as the best (i.e., the lowest test error) model fit to the data.

Note: your results may appear slightly different than those of the figure, given that you will likely use a different random partition of the data. Note: you may find it very useful here to re-use code from previous exercises e.g., functions that compute polynomial features, plot curves, etc.

Exercises 6.9 Code up *k*-fold cross-validation

In this exercise you will perform k-fold cross-validation on the dataset shown in Fig. 6.13 as described in Example 6.6 of the text. Start by randomly splitting the dataset, located in *2eggs_data.csv*, into $k = 3$ equal sized folds (keeping 2 folds as training, and 1 fold as testing data). Use the polynomial basis features with M in the range $M = 2, 5, 9, 14, 20, 27, 35, 44$ (or likewise D in the range $D = 1, 2, \ldots, 8$) and produce a graph showing the training and testing error for each D like the one shown in Fig. 6.13, as well as the best (i.e., the lowest average test error) model fit to the data.

Note: your results may appear slightly different than those of the figure given that you will likely use a different random partition of the data. Note: you may find it very useful here to re-use code from previous exercises e.g., functions that compute polynomial features, plot curves, etc.

7 Kernels, backpropagation, and regularized cross-validation

This chapter is essentially an appendix of technical material critically relevant to the ideas described in the previous two chapters, consisting of three sections each of which may be read independently of the other two. The first describes *fixed feature kernels*, which is a method of representing fixed basis features so that they scale more gracefully when applied to vector valued input. In the second we provide an organized set of computations of the gradient for any cost function employing multilayer neural network basis features for performing gradient descent, commonly referred to as the *backpropagation algorithm* when such deep network basis features are used. Finally, in Section 7.3 we describe a slight variation of the cross-validation technique discussed in previous chapters, called *regularized cross-validation*, that is more appropriate for fixed feature kernels, multilayer network features, as well as the softmax multiclass classification (using either fixed or neural network basis features).

7.1 Fixed feature kernels

A serious practical issue presents itself when applying fixed basis features to vector valued input: even with a moderate sized input dimension N, the corresponding dimension M of the transformed features grows rapidly with N and quickly becomes prohibitively large in terms of storage and computation. For example, the precise number M of non-bias features/feature weights of a degree D polynomial of an input with dimension N is $M = \binom{N+D}{D} - 1 = \frac{(N+D)!}{D!N!} - 1$. Even if the input dimension is of reasonably small size, for instance $N = 100$ or $N = 500$, then just the associated degree $D = 5$ polynomial feature map of these input dimensions has dimension $M = 96560645$ and $M = 268\,318\,178\,226$ respectively! In the latter case we cannot even hold the feature vectors in memory on a modern computer.[1]

This crucial issue, of not being able to effectively store high dimensional fixed basis feature transformations, motivates the search for more efficient representations of fixed bases. Here we introduce *kernels* or *kernelized representations* of fixed feature transformations, which are clever ways of constructing them that do not require explicit construction of the fixed features themselves. Kernels allow us to avoid this

[1] The corresponding number of transformed features with a Fourier basis/map is even more gargantuan: the degree D Fourier feature map of arbitrary input dimension N has $(2D + 1)^N$ associated/feature weights. When $D = 5$ and $N = 80$ this is 11^{80}, a number larger than current estimates of the number of atoms in the visible universe (around 10^{80} atoms)!

combinatorial storage problem and use fixed features with vector input (at the cost, as we will see, of scaling poorly with the size of a data-set). Additionally they provide a way of generating new fixed feature maps defined solely through such a kernelized representation.

7.1.1 The fundamental theorem of linear algebra

Before discussing the concept of kernelization, it will be helpful to first recall a useful fact, generally referred to as the *fundamental theorem of linear algebra*. This is a simple statement about how to deconstruct an M length vector $\mathbf{w} \in \mathbb{R}^M$ over the columns of a given matrix.

Recall that a set of M-dimensional vectors $\left\{\mathbf{f}_p\right\}_{p=1}^{P}$ spans a subspace of dimension P, where $P \leq M$, and that any vector $\mathbf{w}$ in this subspace can be written as some linear combination of the vectors as

$$\mathbf{w} = \sum_{p=1}^{P} \mathbf{f}_p z_p, \tag{7.1}$$

where z_p are weights associated with $\mathbf{w}$. By stacking the vectors $\mathbf{f}_p$ column-wise into an $M \times P$ matrix $\mathbf{F}$ and the z_p together into a $P \times 1$ vector $\mathbf{z}$ this relationship can be written more compactly as

$$\mathbf{w} = \mathbf{F}\mathbf{z}. \tag{7.2}$$

As illustrated in Fig. 7.1, any vector $\mathbf{w} \in \mathbb{R}^M$ can then be decomposed into two pieces: the portion of $\mathbf{w}$ belonging to the subspace spanned by the columns of $\mathbf{F}$ and an orthogonal component $\mathbf{r}$. Formally this decomposition is written as

$$\mathbf{w} = \mathbf{F}\mathbf{z} + \mathbf{r}. \tag{7.3}$$

Note that $\mathbf{r}$ being orthogonal to the span of $\mathbf{F}$'s columns means formally that $\mathbf{F}^T\mathbf{r} = \mathbf{0}_{P \times 1}$.

As we will now see this simple statement is the key to representing fixed basis features more effectively (when used to transform vector valued input for use) with every cost function discussed in this book.

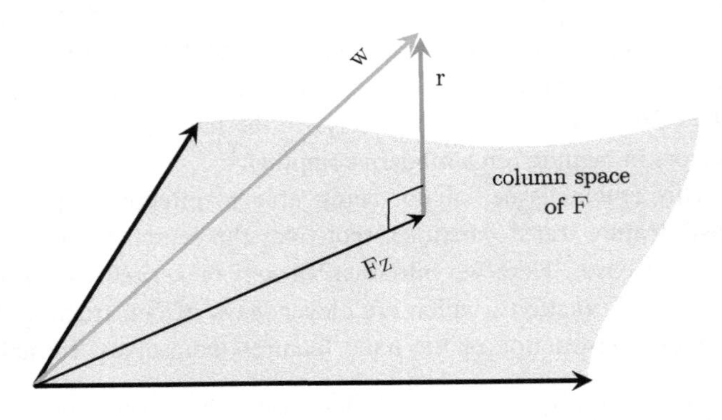

Fig. 7.1 An illustration of the fundamental theorem of linear algebra which states that any vector $\mathbf{w}$ in an M-dimensional space can be decomposed as $\mathbf{w} = \mathbf{F}\mathbf{z} + \mathbf{r}$. Here the vector $\mathbf{F}\mathbf{z}$ belongs in the subspace determined by the columns of the matrix $\mathbf{F}$ and $\mathbf{r}$ is orthogonal to this subspace.

7.1.2 Kernelizing cost functions

Suppose that we have a dataset of P points $\left\{\left(\mathbf{x}_p, y_p\right)\right\}_{p=1}^{P}$ where each input $\mathbf{x}_p$ has dimension N. Recall from Section 5.2 that when employing any fixed feature basis we learn proper parameters by minimizing the Least Squares regression cost,

$$g\left(b, \mathbf{w}\right) = \sum_{p=1}^{P} \left(b + \mathbf{f}_p^T \mathbf{w} - y_p\right)^2, \tag{7.4}$$

where we have used the vector notation $\mathbf{f}_p = \left[\begin{array}{cccc} f_1\left(\mathbf{x}_p\right) & f_2\left(\mathbf{x}_p\right) & \cdots & f_M\left(\mathbf{x}_p\right) \end{array}\right]^T$ to denote the M fixed basis feature transformations of the input $\mathbf{x}_p$. Denote by $\mathbf{F}$ the $M \times P$ matrix $\mathbf{F}$ formed by stacking the vectors $\mathbf{f}_p$ column-wise. Now, employing the fundamental theorem of linear algebra discussed in the previous section we may write $\mathbf{w}$ here as

$$\mathbf{w} = \mathbf{F}\mathbf{z} + \mathbf{r}, \tag{7.5}$$

where $\mathbf{r}$ satisfies $\mathbf{F}^T\mathbf{r} = \mathbf{0}_{P\times 1}$. Plugging this representation of $\mathbf{w}$ back into the cost function then gives

$$\sum_{p=1}^{P} \left(b + \mathbf{f}_p^T \left(\mathbf{F}\mathbf{z} + \mathbf{r}\right) - y_p\right)^2 = \sum_{p=1}^{P} \left(b + \mathbf{f}_p^T \mathbf{F}\mathbf{z} - y_p\right)^2. \tag{7.6}$$

Finally, denoting the symmetric matrix $\mathbf{H} = \mathbf{F}^T\mathbf{F}$ (and where $\mathbf{h}_p = \mathbf{F}^T\mathbf{f}_p$ is the pth column of this matrix), referred to as a fixed basis *kernel matrix*, our original cost function becomes equivalently

$$g\left(b, \mathbf{z}\right) = \sum_{p=1}^{P} \left(b + \mathbf{h}_p^T \mathbf{z} - y_p\right)^2. \tag{7.7}$$

Note that we have changed the arguments of the cost function from $g\left(b, \mathbf{w}\right)$ to $g\left(b, \mathbf{z}\right)$ due to our substitution of $\mathbf{w}$. The original problem of minimizing the Least Squares cost may now be written equivalently in this *kernelized form* as

$$\underset{b,\,\mathbf{z}}{\text{minimize}} \sum_{p=1}^{P} \left(b + \mathbf{h}_p^T \mathbf{z} - y_p\right)^2. \tag{7.8}$$

Using precisely the same argument given here we may *kernelize* all of the cost functions discussed in this book including: the softmax cost/logistic regression classifier, the squared margin-perceptron/soft-margin SVMs, the multiclass softmax cost function, as well as any ℓ_2 regularized version of these models. We show both the original and kernelized forms of these formulae in Table 7.1 for easy reference.

7.1.3 The value of kernelization

The real value of kernelizing any cost function is that for many fixed feature maps, including polynomials and Fourier features, the kernel matrix $\mathbf{H}$ may be constructed *without* first building the matrix $\mathbf{F}$, that is we need not construct it explicitly as $\mathbf{H} = \mathbf{F}^T\mathbf{F}$,

Table 7.1 Cost functions and their kernelized versions. Note that the ℓ_2 regularizer can be added to any cost function in the middle column and the resulting kernelized form of the sum will be the sum of the kernelized cost and the kernelized regularizer. For example, the kernelized form of the regularized Least Squares problem $\sum_{p=1}^{P}\left(b+\mathbf{f}_p^T\mathbf{w}-y_p\right)^2+\lambda\left\|\mathbf{w}\right\|_2^2$ is $\sum_{p=1}^{P}\left(b+\mathbf{h}_p^T\mathbf{z}-y_p\right)^2+\lambda\mathbf{z}^T\mathbf{H}\mathbf{z}$.

Cost function	Original version	Kernelized version
Least Squares	$\sum_{p=1}^{P}\left(b+\mathbf{f}_p^T\mathbf{w}-y_p\right)^2$	$\sum_{p=1}^{P}\left(b+\mathbf{h}_p^T\mathbf{z}-y_p\right)^2$
Softmax cost/logistic regression	$\sum_{p=1}^{P}\log\left(1+e^{-y_p\left(b+\mathbf{f}_p^T\mathbf{w}\right)}\right)$	$\sum_{p=1}^{P}\log\left(1+e^{-y_p\left(b+\mathbf{h}_p^T\mathbf{z}\right)}\right)$
Squared margin/ soft-margin SVMs	$\sum_{p=1}^{P}\max^2\left(0,1-y_p\left(b+\mathbf{f}_p^T\mathbf{w}\right)\right)$	$\sum_{p=1}^{P}\max^2\left(0,1-y_p\left(b+\mathbf{h}_p^T\mathbf{z}\right)\right)$
Multiclass softmax	$\sum_{c=1}^{C}\sum_{p\in\Omega_c}\log\left(1+\sum_{\substack{j=1\\j\neq c}}^{C}e^{\left(b_j-b_c\right)+\mathbf{f}_p^T\left(\mathbf{w}_j-\mathbf{w}_c\right)}\right)$	$\sum_{c=1}^{C}\sum_{p\in\Omega_c}\log\left(1+\sum_{\substack{j=1\\j\neq c}}^{C}e^{\left(b_j-b_c\right)+\mathbf{h}_p^T\left(\mathbf{z}_j-\mathbf{z}_c\right)}\right)$
ℓ_2-regularizer	$\lambda\left\|\mathbf{w}\right\|_2^2$	$\lambda\mathbf{z}^T\mathbf{H}\mathbf{z}$

but this matrix may be constructed entry-wise via simple formulae. In fact, as we will see, thinking about constructing kernel matrices in this way leads to the construction of fixed feature bases by defining the kernel matrix first (that is, not by beginning with an explicit feature transformation). As we see in the next section this can be done for both degree D polynomial and Fourier feature bases, as well as many other fixed maps. This is highly advantageous since recall, as discussed in the introduction to this section, that even with moderate sized input dimension N the dimension of a fixed feature transformation M will likely be gargantuan, so large that we may not even be able to store the matrix $\mathbf{F}$ let alone compute with it.

However, note that the non-bias optimization variable from the original to kernelized form has changed from $\mathbf{w}$, which had dimension M in Equation (7.4), to $\mathbf{z}$, which has dimension P in the kernelized version shown in Equation (7.7). This is precisely how the dimension of the non-bias optimization variable changes with kernelized cost functions as well, like those shown in Table 7.1.

While it is true that for large datasets (that is large values of P, e.g., in the thousands or tens of thousands) the minimization of a kernelized cost function becomes more challenging, the main obstacle is storing the $P\times P$ kernel matrix itself, which for large values of P is difficult or even impossible to do completely. For example, with $P=10\,000$ the corresponding kernel matrix will be of size $10\,000 \times 10\,000$, with 10^8 values to store, far more than a modern computer can store all at once. Moreover, the amount of computation required to perform, e.g. gradient descent, grows dramatically with the size of a kernel matrix due to its explosive size.

Common ways of dealing with these issues for large datasets include: 1) using advanced first order methods such as stochastic gradient descent, discussed in Chapter 8, so that only a small number of the kernelized points are dealt with at a time when optimizing; 2) reducing the dimension of data using techniques like those discussed in Chapter 9 and hence avoiding the need for kernelized versions of fixed bases; 3) using the explicit structure of certain problems (see e.g., [22, 49]); and 4) employing the tools from function approximation to avoid explicit construction of the kernel matrix [64, 68, 69].

7.1.4 Examples of kernels

Here we present a list of examples of kernels for popular fixed feature transformations that may be built without first constructing the explicit feature transformation itself. While these are the most commonly used kernels in practice, the reader can see e.g., [20, 51] for a more exhaustive list of kernels and their properties.

Example 7.1 The polynomial kernel

Consider the following second degree polynomial mapping from $N = 2$ to $M = 5$ dimensional space given by

$$\mathbf{f}\left(\begin{bmatrix} x_1 \\ x_2 \end{bmatrix}\right) = \begin{bmatrix} \sqrt{2}x_1 & \sqrt{2}x_2 & x_1^2 & \sqrt{2}x_1x_2 & x_2^2 \end{bmatrix}^T. \tag{7.9}$$

This is entirely equivalent to a standard degree 2 polynomial, as the $\sqrt{2}$ attached to several of the terms can be absorbed by their associated weights when taking the corresponding weighted sum $\sum_{m=1}^{5} f_m(\mathbf{x})\, w_m$. Denoting briefly by $\mathbf{u} = \mathbf{x}_i$ and $\mathbf{v} = \mathbf{x}_j$ the ith and jth input data points respectively, the (i, j)th element of the kernel matrix for a degree 2 polynomial $\mathbf{H} = \mathbf{F}^T\mathbf{F}$ may be written as

$$\begin{aligned}
\mathbf{H}_{ij} &= \begin{bmatrix} \sqrt{2}u_1 & \sqrt{2}u_2 & u_1^2 & \sqrt{2}u_1u_2 & u_2^2 \end{bmatrix} \begin{bmatrix} \sqrt{2}v_1 \\ \sqrt{2}v_2 \\ v_1^2 \\ \sqrt{2}v_1v_2 \\ v_2^2 \end{bmatrix} \\
&= \left(1 + 2u_1v_1 + 2u_2v_2 + u_1^2v_1^2 + 2u_1u_2v_1v_2 + u_2^2v_2^2\right) - 1 \\
&= (1 + u_1v_1 + u_2v_2)^2 - 1 = \left(1 + \mathbf{u}^T\mathbf{v}\right)^2 - 1.
\end{aligned} \tag{7.10}$$

In short, the *polynomial kernel* matrix $\mathbf{H}$ may be built without first constructing the explicit features in Equation (7.9), and may be simply defined entry-wise as

$$\mathbf{H}_{ij} = \left(1 + \mathbf{x}_i^T \mathbf{x}_j\right)^2 - 1. \tag{7.11}$$

Again note that with the polynomial kernel defined above we only require access to the original input data, not the explicit polynomial features themselves.

Although the kernel construction rule in (7.11) was derived specifically for $N = 2$ and a degree two polynomial, one can show that a polynomial kernel can be defined entry-wise for general N and degree D analogously as

$$\mathbf{H}_{ij} = \left(1 + \mathbf{x}_i^T \mathbf{x}_j\right)^D - 1. \tag{7.12}$$

Example 7.2 The Fourier kernel

Recall from Example 5.1 the degree D Fourier feature transformation from $N = 1$ to $M = 2D$ dimensional space, with corresponding transformed feature vector given as

$$\mathbf{f}_p = \left[\ \sqrt{2}\cos\left(2\pi x_p\right) \quad \sqrt{2}\sin\left(2\pi x_p\right) \quad \cdots \quad \sqrt{2}\cos\left(2D\pi x_p\right) \quad \sqrt{2}\sin\left(2D\pi x_p\right)\ \right]^T. \tag{7.13}$$

For a dataset of P points the corresponding (i,j)th element of the corresponding kernel matrix $\mathbf{H}$ can be written as

$$\mathbf{H}_{ij} = \mathbf{f}_i^T \mathbf{f}_j = 2 \sum_{m=1}^{D} \cos\left(2\pi m x_i\right) \cos\left(2\pi m x_j\right) + \sin\left(2\pi m x_i\right) \sin\left(2\pi m x_j\right). \tag{7.14}$$

Using trigonometric identities one can show (see Section 7.5.2) that this may equivalently be written as

$$\mathbf{H}_{ij} = \frac{\sin\left((2D+1)\,\pi\,\left(x_i - x_j\right)\right)}{\sin\left(\pi\,\left(x_i - x_j\right)\right)} - 1. \tag{7.15}$$

Note that whenever $x_i - x_j$ is integer valued the term $\frac{\sin\left((2D+1)\pi\left(x_i - x_j\right)\right)}{\sin\left(\pi\left(x_i - x_j\right)\right)}$ is not technically defined. In these cases it is simply replaced by its associated limit which, regardless of the integer value $x_i - x_j$, is always equal to $2D + 1$ meaning that $\mathbf{H}_{ij} = 2D$.

Moreover, for general N-dimensional input the corresponding kernel can be written similarly entry-wise as

$$\mathbf{H}_{ij} = \prod_{n=1}^{N} \frac{\sin\left((2D+1)\,\pi\,\left(x_{in} - x_{jn}\right)\right)}{\sin\left(\pi\,\left(x_{in} - x_{jn}\right)\right)} - 1. \tag{7.16}$$

As with the 1-dimensional version, whenever $x_{in} - x_{jn}$ is integer valued the associated term $\frac{\sin\left((2D+1)\pi\left(x_{in} - x_{jn}\right)\right)}{\sin\left(\pi\left(x_{in} - x_{jn}\right)\right)}$ in the product is replaced by its limit which, regardless of the value of $x_{in} - x_{jn}$, is always equal to $2D + 1$. See Section 7.5.3 for further details.

With this formula we can compute the degree D Fourier features for arbitrary N-dimensional input vectors without calculating the enormous number (see footnote 1) of basis features explicitly.

Example 7.3 Kernel representation of radial basis function (RBF) features

Another popular choice of kernel is the *radial basis function* (RBF) kernel which is typically defined explicitly as a kernel matrix over the input data as

$$\mathbf{H}_{ij} = e^{-\beta \left\Vert \mathbf{x}_i - \mathbf{x}_j \right\Vert_2^2}. \tag{7.17}$$

Here the kernel parameter β is tuned to the data in practice via cross-validation.

While the RBF kernel is typically defined directly as above, it can be traced back to an explicit fixed feature basis as with the polynomial and Fourier kernels, i.e., we have that

$$\mathbf{H}_{ij} = \mathbf{f}_i^T \mathbf{f}_j, \tag{7.18}$$

where $\mathbf{f}_i$ is the fixed feature transformation of the input $\mathbf{x}_i$ based on a fixed basis. While the length of a feature transformation corresponding to a degree D polynomial/Fourier kernel matrix can be extremely large (as discussed in the introduction to this section), with the RBF kernel the associated feature transformation is always *infinite* dimensional. For example, when $N = 1$ the feature vector $\mathbf{f}_i$ takes the form $\mathbf{f}_i = \begin{bmatrix} f_1(x_i) & f_2(x_i) & f_3(x_i) & \cdots \end{bmatrix}^T$, where the mth fixed basis feature is defined as

$$f_m(x_i) = e^{-\beta x_i^2} \sqrt{\frac{(2\beta)^{m-1}}{(m-1)!}} x_i^{m-1} \quad \text{for all } m \geq 1. \tag{7.19}$$

When $N > 1$ the corresponding feature vector takes on an analogous form (and is also infinite in length), but regardless of the input dimension it would be impossible to even construct and store a single $\mathbf{f}_i$ let alone such transformations of the entire dataset.

7.1.5 Kernels as similarity matrices

The polynomial, Fourier, and RBF kernel matrices introduced earlier are all *similarity matrices*, essentially encoding how close or similar a collection of data points are to one another, with points in proximity to one another receiving a high value and those far apart receiving a low value. In this sense all three kernels discussed here, and hence all three corresponding fixed feature bases, define some kind of similarity between data points $\mathbf{x}_i$ and $\mathbf{x}_j$ from different geometric perspectives.

In Fig. 7.2 we compare these three kernels geometrically by fixing a point $\mathbf{x}_p = \begin{bmatrix} 0.5 & 0.5 \end{bmatrix}^T$ and plotting $\mathbf{H}\left(\mathbf{x}, \mathbf{x}_p\right)$ over the range $\mathbf{x} \in [0,1]^2$, producing a color-coded surface showing how each kernel treats points near $\mathbf{x}_p$. Analyzing this figure we can judge more generally how the three kernels define "similarity" between points.

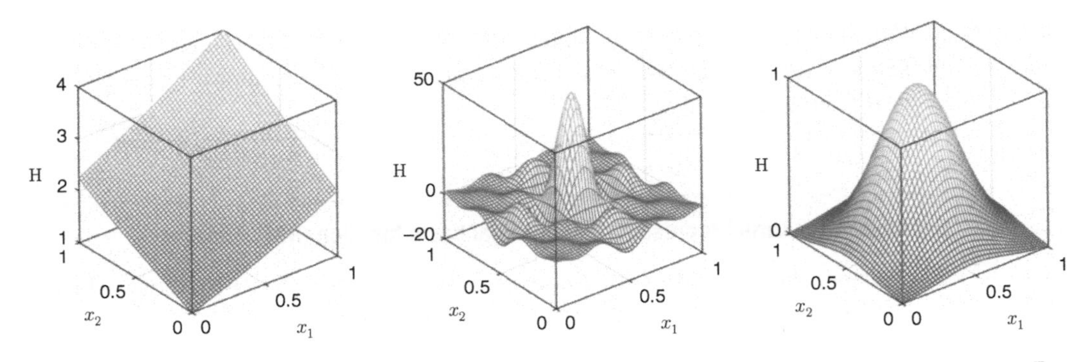

Fig. 7.2 Surfaces generated by polynomial, Fourier, and RBF kernels centered at $\mathbf{x}_p = \begin{bmatrix} 0.5 & 0.5 \end{bmatrix}^T$ with the surfaces color-coded based on their similarity to $\mathbf{x}_p$. (left panel) A degree 2 polynomial kernel, (middle panel) degree 3 Fourier kernel, and (right panel) RBF kernel with $\beta = 10$. See text for further details.

Firstly, we can see that a polynomial kernel treats data points $\mathbf{x}_i$ and $\mathbf{x}_j$ similarly if their inner product is high or, in other words, they highly correlate with each other. Likewise the points are treated as dissimilar when they are orthogonal to one another. On the other hand, the Fourier kernel treats points as similar if they lie close together, but their similarity differs like a "sinc" function as their distance from each other grows. Finally an RBF kernel provides a smooth similarity between points. If they are close to each other in a Euclidean sense they are highly similar; however, once the distance between them passes a certain threshold they are deemed rapidly dissimilar.

7.2 The backpropagation algorithm

In this section we provide details for applying gradient descent, commonly referred to as the *backpropagation algorithm*, to any cost function employing a multilayer neural network feature basis. The term "backpropagation" is often used because, as we will see, there is a natural movement or propagation of computation in calculating the gradient of such a cost function *backward* through a sum of neural network basis features. However, one should not think of this as somehow a special version of gradient descent, it is just the standard gradient descent procedure we have used throughout the text applied to a more complicated (cost) function.

While computing the gradient of such a function only requires the use of careful bookkeeping as well as repeated use of the chain rule, these calculations can easily be incorrect because of human error due to their tedious nature. Because of this we provide explicit calculations for general two and three layer hidden networks, and will write them assuming arbitrary cost and activation functions. Since there is no useful compact formula to express the derivatives associated with an arbitrary layered neural network, the reader can extend the pattern for computing two and three layer networks shown here if employing deeper networks.

Variations of gradient descent, including stochastic gradient descent as well as gradient descent with momentum (see Sections 8.3 and 7.2.3 for further details), are also commonly used to minimize a cost function employing neural network features as they tend to speed up convergence in practice.

7.2.1 Computing the gradient of a two layer network cost function

Let g be any cost function for regression or classification described in this book. Note that, over a dataset of P points $\left\{\left(\mathbf{x}_p, y_p\right)\right\}_{p=1}^{P}$, each can be decomposed over the P points as

$$g = \sum_{p=1}^{P} h\left(b + \mathbf{x}_p^T \mathbf{w}\right); \tag{7.20}$$

e.g., if g is the Least Squares cost for regression or the softmax cost for classification then $h\left(b + \mathbf{x}_p^T \mathbf{w}\right) = \left(b + \mathbf{x}_p^T \mathbf{w} - y_p\right)^2$ and $h\left(b + \mathbf{x}_p^T \mathbf{w}\right) = \log\left(1 + e^{-y_p\left(b + \mathbf{x}_p^T \mathbf{w}\right)}\right)$ respectively. In what follows we will compute derivatives of $h\left(b + \mathbf{x}_p^T \mathbf{w}\right)$, which may then be added up to give corresponding derivatives of g.

Substituting an M_2 two-layer neural network feature $\mathbf{f}_p$ map of $\mathbf{x}_p$, whose mth coordinate takes the form

$$f_m\left(\mathbf{x}_p\right) = a\left(c_m^{(1)} + \sum_{m_2=1}^{M_2} a\left(c_{m_2}^{(2)} + \sum_{n=1}^{N} x_{p,n} v_{n,m_2}^{(2)}\right) v_{m_2,m}^{(1)}\right), \tag{7.21}$$

or more compactly all together we can write

$$\mathbf{f}_p = a\left(\mathbf{c}^{(1)} + \mathbf{V}^{T(1)} a\left(\mathbf{c}^{(2)} + \mathbf{V}^{T(2)} \mathbf{x}_p\right)\right), \tag{7.22}$$

where we slightly abuse notation and say that the activation $a(\cdot)$ applies the function to each coordinate of its input. With this network map, our cost summand is given as $h\left(b + \mathbf{f}_p^T \mathbf{w}\right)$. Here we have stacked the parameters of the first layer into the $M_1 \times 1$ vector $\mathbf{c}^{(1)}$ and $M_2 \times M_1$ matrix $\mathbf{V}^{(1)}$, and those of the second layer into the $M_2 \times 1$ vector $\mathbf{c}^{(2)}$ and $N \times M_2$ matrix $\mathbf{V}^{(2)}$ respectively.

Because we will need to employ the chain rule many times, in order to more effectively compute the gradient of this summand it will be helpful to introduce notation for the argument or *residual* of each layer of the network, as well as the result of each layer after passing through the activation function. Firstly, we will write the arguments at each layer recursively as

$$\begin{aligned} r &= b + \mathbf{w}^T a\left(\mathbf{r}^{(1)}\right) \\ \mathbf{r}^{(1)} &= \mathbf{c}^{(1)} + \mathbf{V}^{T(1)} a\left(\mathbf{r}^{(2)}\right) \\ \mathbf{r}^{(2)} &= \mathbf{c}^{(2)} + \mathbf{V}^{T(2)} \mathbf{x}_p. \end{aligned} \tag{7.23}$$

Note that the first argument r is a scalar, while the latter two $\mathbf{r}^{(1)}$ and $\mathbf{r}^{(2)}$ are M_1 and M_2 length vectors respectively. Correspondingly, we can write $\mathbf{a}^{(1)} = a\left(\mathbf{r}^{(1)}\right)$ and $\mathbf{a}^{(2)} = a\left(\mathbf{r}^{(2)}\right)$, the result of the first layer and second layer argument passed through the activation function respectively (note these are also M_1 and M_2 length vectors).

With this notation we have in particular $h(r) = h\left(b + \mathbf{f}_p^T\mathbf{w}\right)$, and we can write the derivatives of the parameters $(b, \mathbf{w})$ via the chain rule as

$$\begin{aligned}\frac{\partial h}{\partial b} &= \frac{\partial h}{\partial r}\frac{\partial r}{\partial b}\\ \nabla_{\mathbf{w}} h &= \frac{\partial h}{\partial r}\nabla_{\mathbf{w}} r.\end{aligned} \tag{7.24}$$

Each derivative on the right hand side above may be calculated in closed form, e.g., $\frac{\partial r^{(0)}}{\partial b} = 1$ and $\nabla_{\mathbf{w}} r = \mathbf{a}^{(1)}$, and if $h(t) = \log\left(1 + e^{-t}\right)$ is the softmax cost summand then $\frac{\partial h}{\partial r} = h'(r) = \sigma(-r)$. Computing derivatives of the first and second hidden layers' parameters similarly yields, via applying the chain rule multiple times,

$$\begin{aligned}\frac{\partial h}{\partial c_i^{(1)}} &= \frac{\partial h}{\partial r}\frac{\partial r}{\partial a_i^{(1)}}\frac{\partial a_i^{(1)}}{\partial r_i^{(1)}}\frac{\partial r_i^{(1)}}{\partial c_i^{(1)}}\\ \nabla_{\mathbf{v}_i^{(1)}} h &= \frac{\partial h}{\partial r}\frac{\partial r}{\partial a_i^{(1)}}\frac{\partial a_i^{(1)}}{\partial r_i^{(1)}}\nabla_{\mathbf{v}_i^{(1)}} r_i^{(1)}\\ \frac{\partial h}{\partial c_i^{(2)}} &= \frac{\partial h}{\partial r}\left(\sum_{n_1=1}^{M_1}\frac{\partial r}{\partial a_{n_1}^{(1)}}\frac{\partial a_{n_1}^{(1)}}{\partial r_{n_1}^{(1)}}\frac{\partial r_{n_1}^{(1)}}{\partial a_i^{(2)}}\right)\frac{\partial a_i^{(2)}}{\partial r_i^{(2)}}\frac{\partial r_i^{(2)}}{\partial c_i^{(2)}}\\ \nabla_{\mathbf{v}_i^{(2)}} h &= \frac{\partial h}{\partial r}\left(\sum_{n_1=1}^{L_1}\frac{\partial r}{\partial a_{n_1}^{(1)}}\frac{\partial a_{n_1}^{(1)}}{\partial r_{n_1}^{(1)}}\frac{\partial r_{n_1}^{(1)}}{\partial a_i^{(2)}}\right)\frac{\partial a_i^{(2)}}{\partial r_i^{(2)}}\nabla_{\mathbf{v}_i^{(2)}} r_i^{(2)}\end{aligned} \tag{7.25}$$

Note that due to the decision to denote the residuals in each layer, these derivatives take a predictable form, consisting of repeating pairs of partial derivatives: "partial of a function with respect to its residual" and "partial of the residual with respect to a parameter or the following layer activation." Again each individual derivative on the right hand side of the equalities may be computed in closed form given a choice of cost summand h and activation function a. As we have already seen, $\frac{\partial h}{\partial r} = h'(r)$, and as for the remainder of the derivatives we have

$$\begin{array}{llll}\frac{\partial r}{\partial a_i^{(1)}} = w_i & \frac{\partial a_i^{(1)}}{\partial r_i^{(1)}} = a'\left(r_i^{(1)}\right) & \frac{\partial r_i^{(1)}}{\partial c_i^{(1)}} = 1 & \nabla_{\mathbf{v}_i^{(1)}} r_i^{(1)} = \mathbf{a}^{(2)}\\ \frac{\partial r_{n_1}^{(1)}}{\partial a_i^{(2)}} = v_{n_1,i}^{(2)} & \frac{\partial a_i^{(2)}}{\partial r_i^{(2)}} = a'\left(r_i^{(2)}\right) & \frac{\partial r_i^{(2)}}{\partial c_i^{(2)}} = 1 & \nabla_{\mathbf{v}_i^{(2)}} r_i^{(2)} = \mathbf{x}_p.\end{array} \tag{7.26}$$

Note that due to the recursive nature of the arguments, as shown in Equation (7.23), these are typically precomputed (that is, prior to computing derivatives) and, in particular, must be computed in a *forward* manner from inner (i.e., closer to the input) to outer (i.e. farther from the input) layers of the network sequentially. Conversely, as we can see above in Equation (7.25), the propagation of derivative calculations is performed *backwards*. In other words, first the outer layer derivatives in Equation (7.24) are computed, then derivatives of layer one are constructed, and then finally layer two. This pattern, of computing the residuals and derivatives in a forward and backward manner respectively, holds more generally when employing an arbitrary number of layers in a neural network.

As mentioned in the introduction, gradient descent is often referred to as backpropagation because the residuals having been computed in a forward fashion, we can then compute the output residual r which is propagated backwards as we compute the gradients layer by layer.

7.2.2 Three layer neural network gradient calculations

To reiterate the points made in computing two-layer derivatives as shown previously, we briefly show mirrored results for a three-layer network so that the reader can be more comfortable with the notation, as well as the pattern of partial derivatives. Again to make the use of the chain rule more predictable, we define residuals at each layer of the network (in complete similarity to Equation (7.23)) as

$$
\begin{aligned}
r &= b + \mathbf{w}^T a\left(\mathbf{r}^{(1)}\right) \\
\mathbf{r}^{(1)} &= \mathbf{c}^{(1)} + \mathbf{V}^{T(1)} a\left(\mathbf{r}^{(2)}\right) \\
\mathbf{r}^{(2)} &= \mathbf{c}^{(2)} + \mathbf{V}^{T(2)} a\left(\mathbf{r}^{(3)}\right) \\
\mathbf{r}^{(3)} &= \mathbf{c}^{(3)} + \mathbf{V}^{T(3)} \mathbf{x}_p
\end{aligned}
\tag{7.27}
$$

where again we slightly abuse notation and say that the activation $a(\cdot)$ applies the function to each coordinate of its input. The forms of the derivatives of the bias and feature weights are given precisely as shown in Equation (7.24), and the first two layer-wise derivatives precisely as shown in Equation (7.25). Just note that the form of the residuals and of the total summand h have changed, since we now have three layers in the network. Using the chain rule the third-layer derivatives can then be computed as

$$
\begin{aligned}
\frac{\partial h}{\partial c_i^{(3)}} &= \frac{\partial h}{\partial r} \sum_{n_1=1}^{M_1} \frac{\partial r}{\partial a_{n_1}^{(1)}} \frac{\partial a_{n_1}^{(1)}}{\partial r_{n_1}^{(1)}} \left(\sum_{n_2=1}^{M_2} \frac{\partial r_{n_1}^{(1)}}{\partial a_{n_2}^{(2)}} \frac{\partial a_{n_2}^{(2)}}{\partial r_{n_2}^{(2)}} \frac{\partial r_{n_2}^{(2)}}{\partial a_i^{(3)}} \right) \frac{\partial a_i^{(3)}}{\partial r_i^{(3)}} \frac{\partial r_i^{(3)}}{\partial c_i^{(3)}} \\
\nabla_{\mathbf{v}_i^{(3)}} h &= \frac{\partial h}{\partial r} \sum_{n_1=1}^{L_1} \frac{\partial r}{\partial a_{n_1}^{(1)}} \frac{\partial a_{n_1}^{(1)}}{\partial r_{n_1}^{(1)}} \left(\sum_{n_2=1}^{L_2} \frac{\partial r_{n_1}^{(1)}}{\partial a_{n_2}^{(2)}} \frac{\partial a_{n_2}^{(2)}}{\partial r_{n_2}^{(2)}} \frac{\partial r_{n_2}^{(2)}}{\partial a_i^{(3)}} \right) \frac{\partial a_i^{(3)}}{\partial r_i^{(3)}} \nabla_{\mathbf{v}_i^{(3)}} r_i^{(3)}
\end{aligned}
\tag{7.28}
$$

where, as with the previous layers, all derivatives on the right hand side can be computed in closed form. Also, as with the previous two-layer case, again the residuals are computed first in a forward manner (from the inner to outer layer of the network), while the derivatives are naturally computed in a backward manner given their structure.

Example 7.4 Comparison of different networks on a toy classification dataset

Here we give a simple example comparing neural network feature maps with one, two, and three hidden layers on the toy classification dataset shown in the middle panel of Fig. 6.9. In particular we use $M_1 = M_2 = M_3 = 10$ units in each hidden layer of each network respectively. In Fig. 7.3 we show the objective value per iteration of gradient descent, for illustrative purposes showing iterations 10 through 50. In this figure we

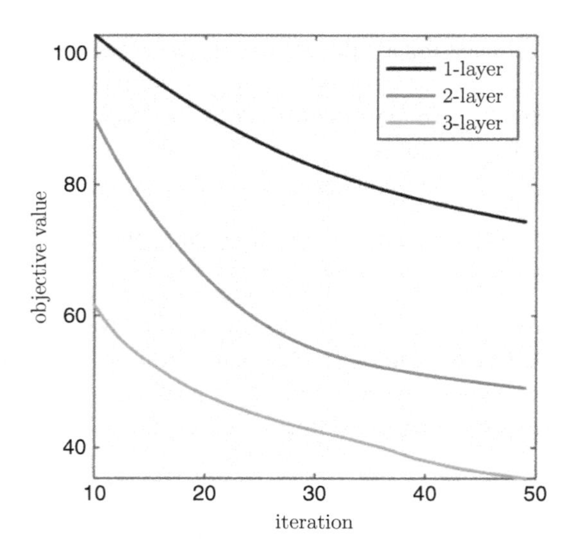

Fig. 7.3 Objective value resulting from the first 50 iterations of gradient descent applied to minimizing the softmax cost employing one/two/three hidden layer neural network features for the dataset shown in the middle panel of Fig. 6.9 (see text for further details). This plot simply reflects the fact that as we increase the number of layers of a neural network each basis feature becomes more flexible, allowing for faster fitting to nonlinearly separable data. However, as described in the context of both regression and classification, overfitting becomes more of a potential problem as we increase the flexibility of a feature map and thus cross-validation is absolutely required when using deep net features.

can see that the deeper networks provide stronger fits faster to the dataset, with the three layer network providing the best progress throughout the range of iterations shown. This is true more generally since, as described in Section 5.1.4, as we increase the number of layers in a network each basis feature becomes more flexible. However, as discussed in Sections 5.3 and 6.4, such flexibility in representation can lead deeper networks to overfit a dataset and therefore cross-validation must be employed with deep networks (typically performed via the regularization approach detailed in Section 7.3).

7.2.3 Gradient descent with momentum

A practical problem that occurs in minimizing some cost functions g, especially in higher dimensions, is that the gradient descent steps tend to *zig-zag* towards a solution as illustrated in two dimensions in Fig. 7.4. In this hypothetical example we illustrate the use of an adaptive step length rule (see Section 8.2) in order to exaggerate the real problem in higher dimensions.

This problem motivates the concept of an old and simple heuristic known as the *momentum term*, which is added to the standard gradient descent step applied to both convex and non-convex functions. The momentum term is a simple weighted difference

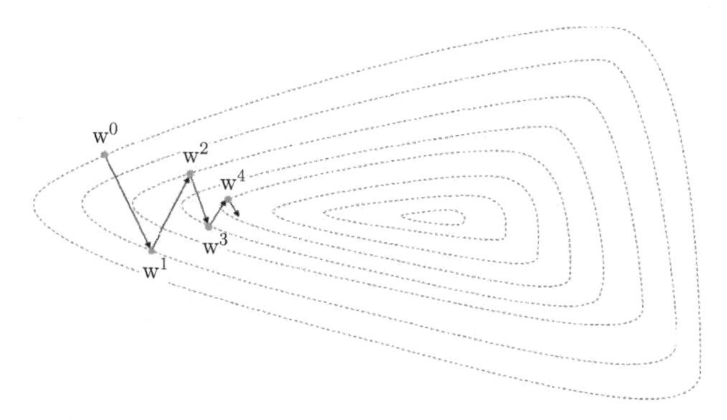

Fig. 7.4 A figurative illustration of gradient steps toward the minimum of a function in two dimensions. Note that the gradient step directions are perpendicular to the contours of the surface shown with dashed ellipses.

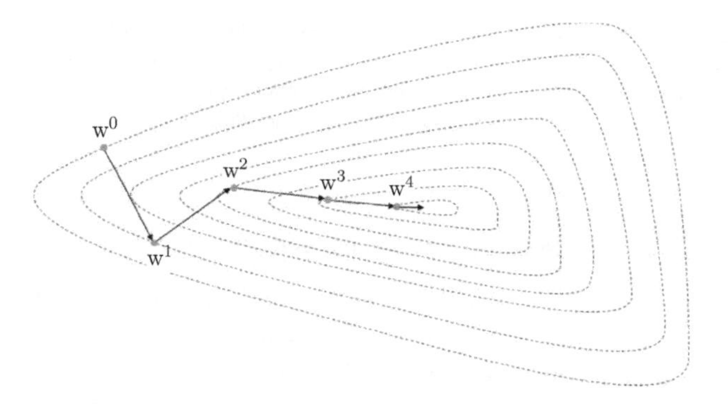

Fig. 7.5 A figurative illustration of momentum-adjusted gradient descent steps toward the minimum of the same function shown in Fig. 7.4. The addition of the momentum term averages out the zig-zagging inherent in standard gradient descent steps.

of the subsequent kth and $(k-1)$th gradient steps, i.e., $\beta\left(\mathbf{w}^k - \mathbf{w}^{k-1}\right)$ for some $\beta > 0$, and is designed to even out the zig-zagging effect of the gradient descent iterates. Hypothetical steps from a gradient descent scheme with momentum are illustrated in Fig. 7.5.

Adding this term to the $(k+1)$th gradient step gives the combined update of

$$\mathbf{w}^{k+1} = \mathbf{w}^k - \alpha_k \nabla g\left(\mathbf{w}^k\right) + \beta\left(\mathbf{w}^k - \mathbf{w}^{k-1}\right), \tag{7.29}$$

where β can be adjusted as well at each iteration if desired. When tuned properly the adjusted gradient descent step with momentum is known empirically to significantly improve the convergence of gradient descent (see e.g., [66]).

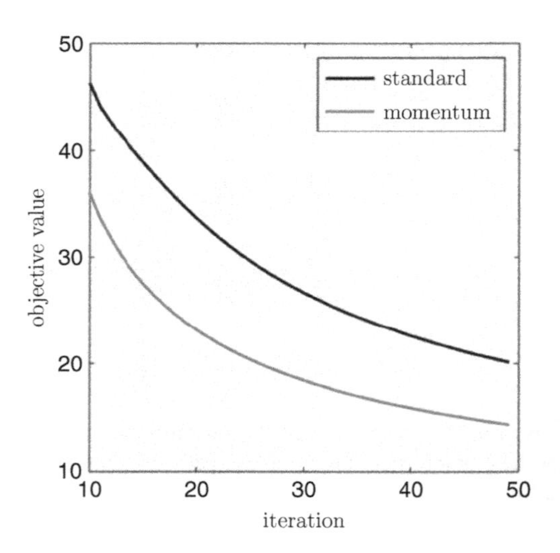

Fig. 7.6 The cost function value of the first $k = 10$–50 iterations of standard and momentum gradient descent procedures (shown in black and magenta respectively) on a classification dataset, averaged over five runs. By tuning the momentum parameter we can create faster converging gradient descent schemes. See text for further details.

Example 7.5 Multilayer neural network example

In this example we use the softmax cost with a three-layer neural network feature map, using $M_1 = M_2 = M_3 = 10$ hidden units in all three layers and the rectified linear unit activation function, and the classification dataset first shown in the left panel of Fig. 4.3. We compare standard gradient descent to the momentum version shown in Equation (7.29) by averaging the results of five runs of each algorithm (using the same random initialization for each run for both algorithms).

Shown in Fig. 7.6 is the average objective function value for the first 50 iterations of both algorithms. β has been tuned[2] such that the momentum gradient descent runs converge significantly faster than do the standard runs.

7.3 Cross-validation via ℓ_2 regularization

In this section we describe the concept of cross-validation via ℓ_2 regularization, a variation of the direct comparison method for cross-validation described in previous chapters. This method provides a much more useful framework for performing cross-validation in a variety of important circumstances. These include:

- **Fixed/kernelized feature bases of high dimensional input.** The difference in dimension M, or the number of basis elements, of fixed feature representations having

[2] For both we used a constant fixed step size of $\alpha = 10^{-3}$ and a momentum weight of $\beta = 0.5$ for all runs.

subsequent degrees D and $D+1$ becomes extremely large as we increase the input dimension N. Therefore comparison of fixed features of various degrees becomes far too coarse: because there is an increasing number of possible feature configurations between degrees D and $D+1$ we do not try.[3] Furthermore, by kernelizing such fixed features we lose direct access to individual elements of the original basis, making it impossible to make comparisons at a finer scale. Moreover for fixed bases defined explicitly at the level of a kernel (e.g., the RBF kernel in Example 7.3), regardless of input dimension N, we cannot even practically compare different length feature vectors, as they are all infinite in dimension.

- **Multilayer neural networks.** As the number of layers/activation units of a neural network feature basis is increased each corresponding entry of the feature vector/basis element used, having many layers of internal parameters, becomes increasingly flexible. This makes simply testing the effectiveness of M versus $M+1$ dimensional neural network features increasingly coarse, meaning again that there is an increasing number of feature configurations we do not test, as the number of layers/activation units employed is increased.
- **Multiclass softmax classification.** Both of the previously described problems are exacerbated in the case of multiclass classification, particularly when using the multiclass softmax classifier.

We begin by discussing regularized cross-validation with regression followed by a mirrored section for classification. Note that throughout the remainder of this section we will work with P input/output pairs $\left\{\left(\mathbf{x}_p, y_p\right)\right\}_{p=1}^{P}$, where the input $\mathbf{x}_p$ is of dimension N and y_p are either continuous in the case of regression, or discrete in the case of classification. Further we will assume that either M-dimensional fixed or neural network basis features with arbitrary number of layers have been taken of each input $\mathbf{x}_p$, denoted as $\mathbf{f}_p = \left[\begin{array}{cccc} f_1\left(\mathbf{x}_p\right) & f_2\left(\mathbf{x}_p\right) & \cdots & f_M\left(\mathbf{x}_p\right)\end{array}\right]^T$, and Θ is the set of possible internal parameters of the feature basis elements, which is empty in the case of fixed/kernelized features.

7.3.1 ℓ_2 regularization and cross-validation

Regularized approaches to cross-validation can be understood as following from the same simple observation that motivated the original direct approach described in Sections 5.3 and 6.4. However, instead of trying to determine the proper number M of basis features in order to avoid the pitfall of overfitting, the regularization approach first fixes M at a reasonably high value and adds a second term (the regularizer) to the associated cost function. This additional regularizer constrains the weights, prohibiting the cost function itself from achieving too small values (i.e., creating an overfitting model) over the entire dataset.

[3] Recall that, for example, the number of non-constant basis elements in a degree D polynomial is given as $M = \frac{(N+D)!}{D!N!} - 1$. This means, for example, that with $N = 500$ dimensional input there are 20 958 500 more basis features in a degree $D = 3$ than a degree $D = 2$ polynomial.

While many choices of regularizer function are available, by far the most commonly used in practice is the squared ℓ_2 norm of the weights, which we have seen previously used to "convexify" non-convex functions (see Section 3.3.2) in order to aid in their minimization, as well as a mathematical way of encoding the margin length in the soft-margin SVM classifier (see Section 4.3.3).

7.3.2 Regularized k-fold cross-validation for regression

Formally to regularize the Least Squares cost function using the approach described previously, we add to it a weighted version of the squared ℓ_2 norm of all the weights, giving the ℓ_2 regularized Least Squares cost function

$$g\left(b,\mathbf{w},\Theta\right)=\sum_{p=1}^{P}\left(b+\mathbf{f}_p^T\mathbf{w}-y_p\right)^2+\lambda\left(\left\|\mathbf{w}\right\|_2^2+\sum_{\theta\in\Theta}\theta^2\right). \tag{7.30}$$

Note that since the definition of the squared ℓ_2 norm gives $\left\|\mathbf{a}\right\|_2^2=\sum_{m=1}^{M}a_m^2$, the final term $\sum_{\theta\in\Theta}\theta^2$ is equivalent to first forming a single long column vector containing all internal parameters in Θ, and then taking the squared ℓ_2 norm of the result. Also note that we do not regularize the bias b since we are only concerned with mitigating the impact of the features themselves on our final model. Finally note that when using a kernelized fixed map, as described in Section 7.1.2, the parameter set is empty and the above is written as

$$g\left(b,\mathbf{z}\right)=\sum_{p=1}^{P}\left(b+\mathbf{h}_p^T\mathbf{z}-y_p\right)^2+\lambda\mathbf{z}^T\mathbf{H}\mathbf{z}. \tag{7.31}$$

In either case, the parameter $\lambda\geq 0$ controls the strength of each term, the Least Squares cost, and regularizer, in the final sum. For example, if $\lambda=0$ we have our original cost again, but if on the other hand λ is set very high then the regularizer drowns out the cost function.

Now, the precise value of λ is determined by employing the cross-validation framework described in Section 5.3, where instead of trying various values for M we (having fixed the feature dimension at M) try a discrete set of values in some range $\lambda\in\left[\lambda_{\min},\lambda_{\max}\right]$. In other words, we split our dataset of P points into k non-overlapping portions, for each fold by forming $k-1$ portions of the data into a training set and leaving a single (distinct for each fold) portion as a test set.

Fitting to a single fold training set for one choice of λ, we minimize (7.30) over the training set by solving

$$\underset{b,\mathbf{w},\Theta}{\text{minimize}}\ \sum_{p\in\Omega_{\text{train}}}\left(b+\mathbf{f}_p^T\mathbf{w}-y_p\right)^2+\lambda\left(\left\|\mathbf{w}\right\|_2^2+\sum_{\theta\in\Theta}\theta^2\right), \tag{7.32}$$

where we once again denote via index sets those points belonging to this fold's training and testing sets as

$$\begin{aligned}\Omega_{\text{train}} &= \left\{p \mid \left(\mathbf{x}_p, y_p\right) \text{ belongs to the training set}\right\} \\ \Omega_{\text{test}} &= \left\{p \mid \left(\mathbf{x}_p, y_p\right) \text{ belongs to the testing set}\right\}.\end{aligned} \tag{7.33}$$

Denoting an optimal solution to the above problem as $\left(b_\lambda^\star, \mathbf{w}_\lambda^\star, \Theta_\lambda^\star\right)$, we then compute the training and testing errors on a single fold for one choice of λ as

$$\begin{aligned}\text{Training error} &= \frac{1}{|\Omega_{\text{train}}|}\sum_{p\in\Omega_{\text{train}}}\left(b_\lambda^\star + \mathbf{f}_p^T\mathbf{w}_\lambda^\star - y_p\right)^2 \\ \text{Testing error} &= \frac{1}{|\Omega_{\text{test}}|}\sum_{p\in\Omega_{\text{test}}}\left(b_\lambda^\star + \mathbf{f}_p^T\mathbf{w}_\lambda^\star - y_p\right)^2,\end{aligned} \tag{7.34}$$

where the notation $|\Omega_{\text{train}}|$ and $|\Omega_{\text{test}}|$ denotes the cardinality or number of indices in the training and testing sets respectively.

To perform k-fold cross-validation we then execute these calculations over all k folds and average the results for each value of λ. We then pick the value $\lambda^\star$ providing the lowest average testing error, and fit the final model to the entire dataset by solving

$$\underset{b,\,\mathbf{w},\Theta}{\text{minimize}} \sum_{p=1}^{P}\left(b + \mathbf{f}_p^T\mathbf{w} - y_p\right)^2 + \lambda^\star\left(\|\mathbf{w}\|_2^2 + \sum_{\theta\in\Theta}\theta^2\right) \tag{7.35}$$

7.3.3 Regularized cross-validation for classification

As with the case of regression, we may regularize any cost function like e.g., the softmax cost with the ℓ_2 norm squared of all feature weights as

$$g\left(b, \mathbf{w}, \Theta\right) = \sum_{p=1}^{P}\log\left(1 + e^{-y_p\left(b + \mathbf{f}_p^T\mathbf{w}\right)}\right) + \lambda\left(\|\mathbf{w}\|_2^2 + \sum_{\theta\in\Theta}\theta^2\right), \tag{7.36}$$

where again if a kernelized fixed feature map is used we may write the above as

$$g\left(b, \mathbf{z}\right) = \sum_{p=1}^{P}\log\left(1 + e^{-y_p\left(b + \mathbf{h}_p^T\mathbf{z}\right)}\right) + \lambda\mathbf{z}^T\mathbf{H}\mathbf{z}. \tag{7.37}$$

Following the same format as with regression, to determine a proper value of $\lambda \geq 0$ we perform k-fold cross-validation for a discrete set of values in a range of $\lambda \in [\lambda_{\min}, \lambda_{\max}]$ and choose the value providing the lowest average testing error.

To determine the training/testing error for one value of λ on a single fold we first fit to one fold's training set, solving

$$\underset{b,\,\mathbf{w},\Theta}{\text{minimize}} \sum_{p\in\Omega_{\text{train}}}\log\left(1 + e^{-y_p\left(b + \mathbf{f}_p^T\mathbf{w}\right)}\right) + \lambda\left(\|\mathbf{w}\|_2^2 + \sum_{\theta\in\Theta}\theta^2\right), \tag{7.38}$$

and computing the training and testing errors on this fold:

$$\begin{aligned}\text{Training error} &= \frac{1}{|\Omega_{\text{train}}|}\sum_{p\in\Omega_{\text{train}}}\max\left(0,\ \text{sign}\left(-y_p\left(b_\lambda^\star+\mathbf{x}_p^T\mathbf{w}_\lambda^\star\right)\right)\right)\\ \text{Testing error} &= \frac{1}{|\Omega_{\text{test}}|}\sum_{p\in\Omega_{\text{test}}}\max\left(0,\ \text{sign}\left(-y_p\left(b_\lambda^\star+\mathbf{x}_p^T\mathbf{w}_\lambda^\star\right)\right)\right)\end{aligned} \tag{7.39}$$

where once again Ω_{train} and Ω_{test} denote the training and testing sets on this fold. Averaging these values over all k folds we then pick the $\lambda^\star$ with lowest average testing error, and fit the corresponding model to the entire dataset as

$$\underset{b,\,\mathbf{w},\Theta}{\text{minimize}}\ \sum_{p=1}^{P}\log\left(1+e^{-y_p\left(b+\mathbf{f}_p^T\mathbf{w}\right)}\right)+\lambda^\star\left(\|\mathbf{w}\|_2^2+\sum_{\theta\in\Theta}\theta^2\right). \tag{7.40}$$

7.4 Summary

In the first section of this chapter we described how kernel representations are used to overcome the serious scaling issue of fixed feature bases with vector valued input. Furthermore, we have seen how new kinds of fixed bases can be defined directly through a kernelized representation. We have also showed how every machine learning cost function discussed in this book may be kernelized (permitting the use of any fixed basis kernel).

In Section 7.2 we gave careful derivations of the gradient when using multilayer network features. As we saw, this requires very careful bookkeeping, as well as repeated use of the chain rule.

In the final section we detailed a variation of cross-validation based on the ℓ_2 regularizer. This approach is founded on the same principles that led to the direct approach described in previous chapters, but here k-fold cross-validation is used to determine the proper value of the penalty parameter on the regularizer (instead of determining the best number of basis features to use). This regularized approach to cross-validation is a much more effective way of properly fitting regression/classification models employing either kernelized fixed feature or deep net feature bases.

7.5 Further kernel calculations

7.5.1 Kernelizing various cost functions

Here we derive the kernelization of the three core classification models: softmax cost/logistic regression, soft-margin SVMs, and the multiclass softmax classifier. Although we will only describe how to kernelize the ℓ_2 regularizer along with the SVM model, precisely the same argument can be made in combination with any other machine learning model shown in Table 7.1. As with the derivation for Least Squares regression shown in Section 7.1.2, here the main tool for kernelizing these models is again the fundamental theorem of linear algebra described in Section 7.1.1.

Throughout this section we will suppose that an arbitrary M-dimensional fixed feature vector has been taken of the input of P points $\left\{\left(\mathbf{x}_p, y_p\right)\right\}_{p=1}^{P}$ giving feature vectors $\mathbf{f}_p = \left[\begin{array}{cccc} f_1\left(\mathbf{x}_p\right) & f_2\left(\mathbf{x}_p\right) & \cdots & f_M\left(\mathbf{x}_p\right)\end{array}\right]^T$ for each $\mathbf{x}_p$.

Example 7.6 Kernelizing two-class softmax classification/logistic regression

Recall that the softmax perceptron cost function used with fixed feature mapped input is given as

$$g\left(b,\mathbf{w}\right) = \sum_{p=1}^{P} \log\left(1 + e^{-y_p\left(b+\mathbf{f}_p^T\mathbf{w}\right)}\right). \tag{7.41}$$

Using the fundamental theorem of linear algebra for any $\mathbf{w}$ we can then write $\mathbf{w} = \mathbf{F}\mathbf{z}+\mathbf{r}$ where $\mathbf{F}^T\mathbf{r} = \mathbf{0}_{P\times 1}$. Making this substitution into the above and simplifying gives

$$g\left(b,\mathbf{z}\right) = \sum_{p=1}^{P} \log\left(1 + e^{-y_p\left(b+\mathbf{f}_p^T\mathbf{F}\mathbf{z}\right)}\right), \tag{7.42}$$

and denoting the kernel matrix $\mathbf{H} = \mathbf{F}^T\mathbf{F}$ (where $\mathbf{h}_p = \mathbf{F}^T\mathbf{f}_p$ is the pth column of $\mathbf{H}$) we can then write the above in kernelized form as

$$g\left(b,\mathbf{z}\right) = \sum_{p=1}^{P} \log\left(1 + e^{-y_p\left(b+\mathbf{h}_p^T\mathbf{z}\right)}\right). \tag{7.43}$$

This is the kernelized form of logistic regression shown in Table 7.1.

Example 7.7 Kernelizing soft-margin SVM/regularized margin-perceptron

Recall the soft-margin SVM cost/regularized margin-perceptron cost:

$$g\left(b,\mathbf{w}\right) = \sum_{p=1}^{P} \max^2\left(0, 1 - y_p\left(b + \mathbf{f}_p^T\mathbf{w}\right)\right) + \lambda\left\Vert \mathbf{w}\right\Vert_2^2 \tag{7.44}$$

Applying the fundamental theorem of linear algebra we may then write $\mathbf{w}$ as $\mathbf{w} = \mathbf{F}\mathbf{z} + \mathbf{r}$ where $\mathbf{F}^T\mathbf{r} = \mathbf{0}_{P\times 1}$. Substituting into the cost and noting that $\mathbf{w}^T\mathbf{w} = \left(\mathbf{F}\mathbf{z}+\mathbf{r}\right)^T\left(\mathbf{F}\mathbf{z}+\mathbf{r}\right) = \mathbf{z}^T\mathbf{F}^T\mathbf{F}\mathbf{z} + \mathbf{r}^T\mathbf{r} = \mathbf{z}^T\mathbf{H}\mathbf{z} + \left\Vert\mathbf{r}\right\Vert_2^2$, denoting $\mathbf{H} = \mathbf{F}^T\mathbf{F}$ as the kernel matrix we may rewrite the above equivalently as

$$g\left(b,\mathbf{z},\mathbf{r}\right) = \sum_{p=1}^{P} \max^2\left(0, 1 - y_p\left(b + \mathbf{h}_p^T\mathbf{z}\right)\right) + \lambda\mathbf{z}^T\mathbf{H}\mathbf{z} + \lambda\left\Vert\mathbf{r}\right\Vert_2^2. \tag{7.45}$$

Note that since we are aiming to minimize the quantity above over $\left(b,\mathbf{z},\mathbf{r}\right)$, and since the only term with $\mathbf{r}$ remaining is $\left\Vert\mathbf{r}\right\Vert_2^2$, the optimal value of $\mathbf{r}$ is zero, for otherwise the value of the cost function would be larger than necessary. Therefore we can ignore $\mathbf{r}$ and write the cost function above in kernelized form as

$$g\left(b,\mathbf{z}\right)=\sum_{p=1}^{P}\max^{2}\left(0,1-y_{p}\left(b+\mathbf{h}_{p}^{T}\mathbf{z}\right)\right)+\lambda\mathbf{z}^{T}\mathbf{H}\mathbf{z},\tag{7.46}$$

as originally shown in Table 7.1.

Example 7.8 Kernelizing the multiclass softmax loss

Recall that the multiclass softmax cost function is written as

$$g\left(b_{1},\ldots,b_{C},\mathbf{w}_{1},\ldots,\mathbf{w}_{C}\right)=\sum_{c=1}^{C}\sum_{p\in\Omega_{c}}\log\left(1+\sum_{\substack{j=1\\j\neq c}}^{C}e^{\left(b_{j}-b_{c}\right)+\mathbf{f}_{p}^{T}\left(\mathbf{w}_{j}-\mathbf{w}_{c}\right)}\right).\tag{7.47}$$

Rewriting each $\mathbf{w}_j$ as $\mathbf{w}_j=\mathbf{F}\mathbf{z}_j+\mathbf{r}_j$, where $\mathbf{F}^T\mathbf{r}_j=\mathbf{0}_{P\times 1}$ for all j, we can rewrite each $\mathbf{f}_p^T\left(\mathbf{w}_j-\mathbf{w}_c\right)$ term as $\mathbf{f}_p^T\left(\mathbf{w}_j-\mathbf{w}_c\right)=\mathbf{f}_p^T\left(\mathbf{F}\left(\mathbf{z}_j-\mathbf{z}_c\right)+\left(\mathbf{r}_j-\mathbf{r}_c\right)\right)=\mathbf{f}_p^T\mathbf{F}\left(\mathbf{z}_j-\mathbf{z}_c\right)$. And denoting $\mathbf{H}=\mathbf{F}^T\mathbf{F}$ the kernel matrix we have that $\mathbf{f}_p^T\left(\mathbf{w}_j-\mathbf{w}_c\right)=\mathbf{h}_p^T\left(\mathbf{z}_j-\mathbf{z}_c\right)$ and so the cost may be written equivalently (kernelized) as

$$g\left(b_{1},\ldots,b_{C},\mathbf{z}_{1},\ldots,\mathbf{z}_{C}\right)=\sum_{c=1}^{C}\sum_{p\in\Omega_{c}}\log\left(1+\sum_{\substack{j=1\\j\neq c}}^{C}e^{\left(b_{j}-b_{c}\right)+\mathbf{h}_{p}^{T}\left(\mathbf{z}_{j}-\mathbf{z}_{c}\right)}\right),\tag{7.48}$$

as shown in Table 7.1.

7.5.2 Fourier kernel calculations – scalar input

From Example 7.2 the (i,j)th element of the kernel matrix $\mathbf{H}$ is given as

$$\mathbf{H}_{ij}=2\sum_{m=1}^{D}\cos\left(2\pi mx_{i}\right)\cos\left(2\pi mx_{j}\right)+\sin\left(2\pi mx_{i}\right)\sin\left(2\pi mx_{j}\right).\tag{7.49}$$

Writing this using the complex exponential notation (see Exercise 5.5), we have equivalently

$$\mathbf{H}_{ij}=\sum_{m=-D}^{D}e^{2\pi im\left(x_{i}-x_{j}\right)}-1.\tag{7.50}$$

If x_i-x_j is an integer then $e^{2\pi im\left(x_i-x_j\right)}=1$ and so clearly the above sums to $2D$. Supposing this is not the case, examining the summation alone we may write

$$\sum_{m=-D}^{D} e^{2\pi im(x_i-x_j)} = e^{-2\pi iD(x_i-x_j)} \sum_{m=0}^{2D} e^{2\pi im(x_i-x_j)}. \tag{7.51}$$

Now the sum on the right hand side above is a geometric series, thus we have the above is equal to

$$e^{-2\pi iD(x_i-x_j)} \frac{1-e^{2\pi i(x_i-x_j)(2D+1)}}{1-e^{2\pi i(x_i-x_j)}} = \frac{\sin\left((2D+1)\,\pi\,\left(x_i-x_j\right)\right)}{\sin\left(\pi\,\left(x_i-x_j\right)\right)}, \tag{7.52}$$

where final equality follows from the definition of the complex exponential. Because in the limit as t approaches any integer value $\frac{\sin((2D+1)\pi t)}{\sin(\pi t)} = 2D+1$, which one can show using L'Hospital's rule from basic calculus, we may therefore generally write in conclusion that

$$\mathbf{H}_{ij} = \frac{\sin\left((2D+1)\,\pi\,\left(x_i-x_j\right)\right)}{\sin\left(\pi\,\left(x_i-x_j\right)\right)} - 1, \tag{7.53}$$

where at integer values of the input it is defined by the associated limit.

7.5.3 Fourier kernel calculations – vector input

Like the multidimensional polynomial basis element (see footnote 5 in the previous chapter) with the complex exponential notation for a general N-dimensional input, each Fourier basis element takes the form $f_{\mathbf{m}}(\mathbf{x}) = e^{2\pi im_1x_1}e^{2\pi im_2x_2}\cdots e^{2\pi im_Nx_N} = e^{2\pi i\mathbf{m}^T\mathbf{x}}$ where $\mathbf{m} = \begin{bmatrix} m_1 & m_2 & \cdots & m_N \end{bmatrix}^T$, a product of 1-dimensional basis elements. Further a "degree D" sum contains all such basis elements where $-D \leq m_1, m_2, \cdots, m_N \leq D$, and one may deduce that there are $M = (2D+1)^N - 1$ non-constant basis elements in this sum.

The corresponding (i,j)th entry of the kernel matrix in this instance takes the form

$$\mathbf{H}_{ij} = \mathbf{f}_i^T\overline{\mathbf{f}_j} = \left(\sum_{-D\leq m_1,m_2,\cdots,m_N\leq D} e^{2\pi i\mathbf{m}^T(\mathbf{x}_i-\mathbf{x}_j)}\right) - 1. \tag{7.54}$$

Since $e^{a+b} = e^ae^b$ we may write each summand above as $e^{2\pi i\mathbf{m}^T(\mathbf{x}_i-\mathbf{x}_j)} = \prod_{n=1}^{N} e^{2\pi im_n(x_{in}-x_{jn})}$ and the entire summation as

$$\sum_{-D\leq m_1,m_2,\cdots,m_N\leq D} \prod_{n=1}^{N} e^{2\pi im_n(x_{in}-x_{jn})}. \tag{7.55}$$

Finally, one can show that the above can be written simply as

$$\sum_{-D \le m_1, m_2, \cdots, m_N \le D} \prod_{n=1}^{N} e^{2\pi i m_n (x_{in} - x_{jn})} = \prod_{n=1}^{N} \left(\sum_{m=-D}^{D} e^{2\pi i m (x_{in} - x_{jn})} \right). \tag{7.56}$$

Since we already have that $\sum_{m=-D}^{D} e^{2\pi i m (x_{in} - x_{jn})} = \frac{\sin((2D+1)\pi (x_{in} - x_{jn}))}{\sin(\pi (x_{in} - x_{jn}))}$, the (i, j)th entry of the kernel matrix can easily be calculated as

$$\mathbf{H}_{ij} = \prod_{n=1}^{N} \frac{\sin\left((2D+1)\pi\left(x_{in} - x_{jn}\right)\right)}{\sin\left(\pi\left(x_{in} - x_{jn}\right)\right)} - 1. \tag{7.57}$$

Part III

Methods for large scale machine learning

Overview of Part III

In the following two chapters we describe two general sets of tools for dealing with large scale data. First, in Chapter 8, we introduce advanced gradient methods which extend the standard gradient descent scheme first described in Chapter 2. These techniques help us to deal with very large datasets directly by making use of more efficient algorithms. In Chapter 9, on the other hand, we introduce general techniques for significantly reducing the size of datasets prior to performing regression or classification.

8 Advanced gradient schemes

In Chapter 2 we introduced two canonical approaches to unconstrained minimization, namely, gradient descent and Newton's method. In the current chapter we add to that discussion by fully describing two popular step length rules, both of which provide mathematically provable convergence to a stationary point for the gradient descent algorithm. We then describe *stochastic* (or *iterative*) *gradient descent*, an important extension of the original gradient descent scheme that helps scale the algorithm to very large datasets.

8.1 Fixed step length rules for gradient descent

In the following two sections we discuss two of the most popular ways of automatically determining proper step lengths for each step of a gradient descent run, which we refer to as *step length rules*. In particular we discuss two commonly used rules which guarantee, mathematically speaking, convergence of the gradient descent algorithm to a stationary point: *fixed* and *adaptive* step length rules each of which has practical strengths and weaknesses. While typically providing a conservative (i.e., small) step length that is kept fixed for all iterations, the fixed step length rule discussed first provides both a convenient choice for many of the cost functions described in this book, as well as a benchmark by which to easily test larger fixed values. On the other hand, with the adaptive rule discussed second we adaptively compute the step length at each gradient descent step by using local information from the part of the cost function near the current step. This typically produces larger steps in practice than a fixed rule, meaning fewer steps are necessary for convergence, but the determination of each step requires computation.

8.1.1 Gradient descent and simple quadratic surrogates

Recall from Section 2.2.4 how the second order Taylor series expansion of a cost function g centered at a point $\mathbf{w}^0$,

$$g\left(\mathbf{w}^0\right) + \nabla g\left(\mathbf{w}^0\right)^T\left(\mathbf{w} - \mathbf{w}^0\right) + \frac{1}{2}\left(\mathbf{w} - \mathbf{w}^0\right)^T \nabla^2 g\left(\mathbf{w}^0\right)\left(\mathbf{w} - \mathbf{w}^0\right), \qquad (8.1)$$

leads to a well-defined descent step known as Newton's method. This is indeed the most natural quadratic approximation to g at $\mathbf{w}^0$ that is available to us. However, as detailed

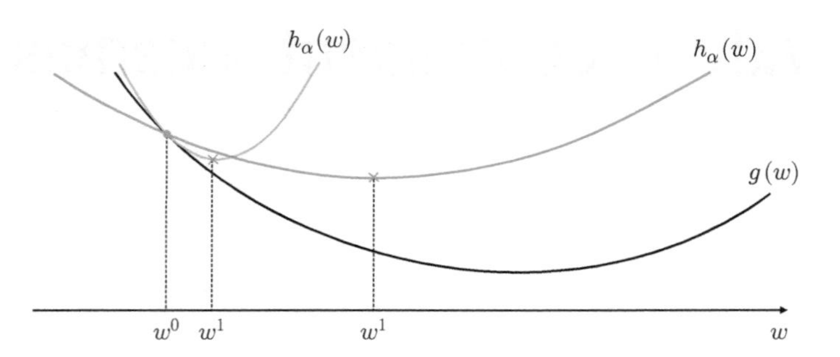

Fig. 8.1 Two quadratic functions approximating the function g around $\mathbf{w}^0$ given by (8.2). The value of α is larger with the red quadratic than with the blue one.

in that section, there are potential difficulties in storing and even calculating the Hessian matrix $\nabla^2 g\left(\mathbf{w}^0\right)$ for large scale problems. Still the idea of minimizing a function g by "hopping down" the stationary points of quadratic approximations (also referred to as *surrogates*), as opposed to the linear approximations/surrogates as employed by gradient descent, is a pleasing one with great intuitive appeal. So a natural question is: can we replace the Hessian with a simpler quadratic term and produce an effective descent algorithm?

For example, we may consider the following simple quadratic function:

$$h_\alpha\left(\mathbf{w}\right) = g\left(\mathbf{w}^0\right) + \nabla g\left(\mathbf{w}^0\right)^T\left(\mathbf{w} - \mathbf{w}^0\right) + \frac{1}{2\alpha}\left\|\mathbf{w} - \mathbf{w}^0\right\|_2^2, \tag{8.2}$$

where $\alpha > 0$. This is just the second order Taylor series of g around $\mathbf{w}^0$ where we have replaced the Hessian $\nabla^2 g\left(\mathbf{w}^0\right)$ with the simple diagonal matrix $\frac{1}{\alpha}\mathbf{I}_{N\times N}$. This kind of quadratic is illustrated in Fig. 8.1 for two values of α. Note that the larger the α the wider the associated quadratic becomes. Also, when $\mathbf{w} = \mathbf{w}^0$ the last two terms on the right hand side of (8.2) disappear and we have $h_\alpha\left(\mathbf{w}^0\right) = g\left(\mathbf{w}^0\right)$.

Our simple quadratic surrogate h_α has a unique global minimum which may be found by checking the first order condition (see Section 2.1.2) by setting its gradient to zero,

$$\nabla h_\alpha\left(\mathbf{w}\right) = \nabla g\left(\mathbf{w}^0\right) + \frac{1}{\alpha}\left(\mathbf{w} - \mathbf{w}^0\right) = \mathbf{0}, \tag{8.3}$$

and solving for $\mathbf{w}$. Doing this we can compute the minimizer of h_α, which we call $\mathbf{w}^1$, as

$$\mathbf{w}^1 = \mathbf{w}^0 - \alpha\nabla g\left(\mathbf{w}^0\right). \tag{8.4}$$

Note the minimizer of the quadratic in Equation (*8.2*) is precisely a gradient descent step at $\mathbf{w}^0$ with a step length of α. Therefore our attempt at replacing the Hessian with a very simple quadratic, and locating the minimum of that quadratic, does not lead to a new descent method but to the familiar gradient descent step. If we continue taking steps in this manner the kth update is found as the minimum of the simple quadratic surrogate associated with the previous update $\mathbf{w}^{k-1}$,

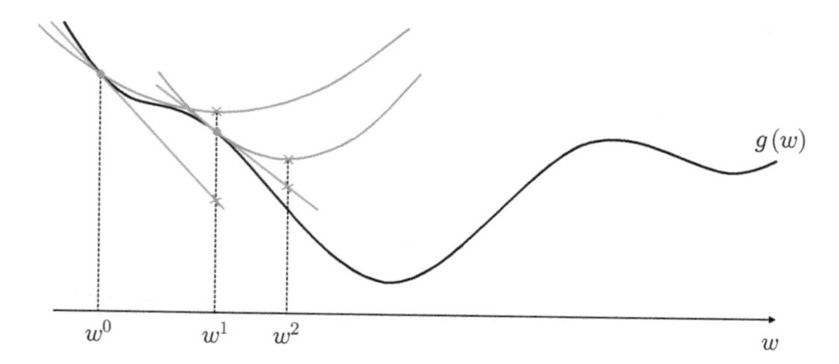

Fig. 8.2 Gradient descent can be viewed simultaneously as using either linear or simple quadratic surrogates to find a stationary point of g. At each step the associated step length defines both how far along the linear surrogate we move before hopping back onto the function g, and at the same time the width of the simple quadratic surrogate which we minimize to reach the same point on g.

$$h_{\alpha}(\mathbf{w}) = g\left(\mathbf{w}^{k-1}\right) + \nabla g\left(\mathbf{w}^{k-1}\right)^T\left(\mathbf{w} - \mathbf{w}^{k-1}\right) + \frac{1}{2\alpha}\left\|\mathbf{w} - \mathbf{w}^{k-1}\right\|_2^2, \tag{8.5}$$

where the minimum is given as the kth gradient descent step

$$\mathbf{w}^k = \mathbf{w}^{k-1} - \alpha\nabla g\left(\mathbf{w}^{k-1}\right). \tag{8.6}$$

Therefore, as illustrated in Fig. 8.2, we can interpret gradient descent as a method that uses linear surrogates or simultaneously one that uses simple fixed curvature quadratic surrogates to locate a stationary point of g. The chosen step length at the kth iteration then determines how far along the linear surrogate we move, or equivalently the width of the quadratic we minimize, in order to reach the next point on g.

Using the simple quadratic perspective of gradient descent we can naturally wonder if, akin to the operation of Newton's method (see e.g., Fig. 2.11), for a given cost function g we can design a step length rule such that we can "hop down" the minima of the associated quadratic surrogates to reach a stationary point of g. As we describe in the remainder of this and the next section, we absolutely can.

8.1.2 Functions with bounded curvature and optimally conservative step length rules

What would happen if we chose a small enough step length α so that the curvature of the associated quadratic surrogate, which would be fixed at each step of gradient descent, matched the greatest curvature of the underlying cost function itself? As illustrated in Fig. 8.3, this would force not only the minimum of each quadratic surrogate to lie *above* the cost function, but the entire quadratic surrogate itself.[1] While this is a conservative choice of step length (and by conservative, we mean small) we refer to it as "optimally conservative" because we can actually compute the maximum curvature of every cost

[1] It is easy to show that the simple quadratic surrogate with $\alpha = \frac{1}{L}$, where L is defined as in Equation (8.50), around $\mathbf{w}^{k-1}$ given by

$$h_{\frac{1}{L}}(\mathbf{w}) = g\left(\mathbf{w}^{k-1}\right) + \nabla g\left(\mathbf{w}^{k-1}\right)^T\left(\mathbf{w} - \mathbf{w}^{k-1}\right) + \frac{L}{2}\left\|\mathbf{w} - \mathbf{w}^{k-1}\right\|_2^2, \tag{8.7}$$

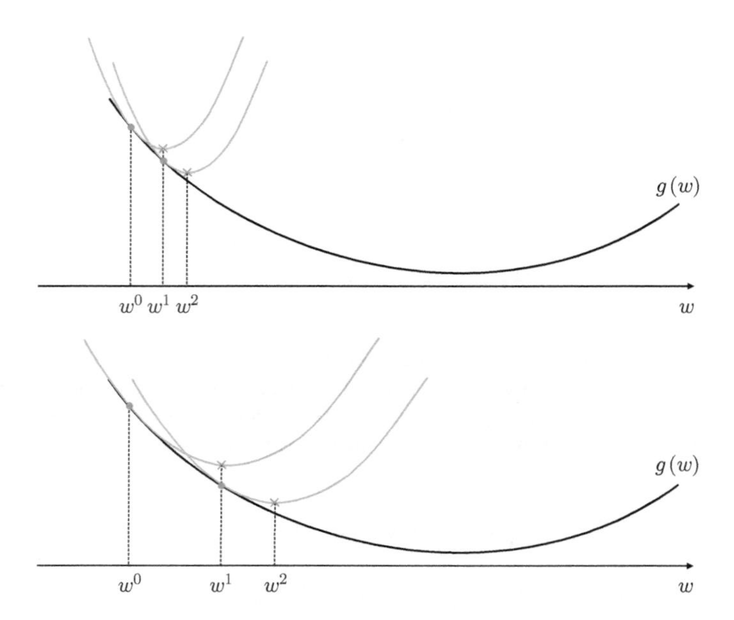

Fig. 8.3 (top panel) Too conservative a fixed step length leads to smaller descent steps. (bottom panel) Another conservative fixed step length where the curvature of its associated quadratic just matches the greatest curvature of the hypothetical cost function while still lying entirely above the function. Such a step length is referred to as "optimally conservative." Note that while the underlying cost here is drawn convex this applies to non-convex cost functions as well, whose greatest curvature could be negative, i.e., on a concave part of the function.

function described in this book (or a reasonable approximation of it) analytically. Therefore this choice of step length can be very convenient in practice and, moreover, using this step length the gradient descent procedure is guaranteed (mathematically speaking) to converge to a stationary point of g (see Section 8.4.1 for further details).

To define this step length formally, recall from Section 2.1.3 that the curvature of a function g is encoded in its *second derivative* when g takes in scalar input w, and more generally its matrix of second derivates or *Hessian* when g takes in vector valued input $\mathbf{w}$. More formally, if g has globally bounded curvature then there must exist an $L > 0$ that bounds its second derivative above and below in the case of a scalar input function

indeed lies completely above the function g at all points as shown in Fig. 8.3. Writing out the *first order Taylor's formula* for g centered at $\mathbf{w}^{k-1}$, we have

$$g(\mathbf{w}) = g\left(\mathbf{w}^{k-1}\right) + \nabla g\left(\mathbf{w}^{k-1}\right)^T\left(\mathbf{w} - \mathbf{w}^{k-1}\right) + \frac{1}{2}\left(\mathbf{w} - \mathbf{w}^{k-1}\right)^T \nabla^2 g(\mathbf{c})\left(\mathbf{w} - \mathbf{w}^{k-1}\right), \tag{8.8}$$

where $\mathbf{c}$ is a point on the line segment connecting $\mathbf{w}$ and $\mathbf{w}^{k-1}$. Since $\nabla^2 g \preceq L\mathbf{I}_{N\times N}$ we can bound the right hand side of (8.8) by replacing $\nabla^2 g(\mathbf{c})$ with $L\mathbf{I}_{N\times N}$, giving

$$g(\mathbf{w}) \leq g\left(\mathbf{w}^{k-1}\right) + \nabla g\left(\mathbf{w}^{k-1}\right)^T\left(\mathbf{w} - \mathbf{w}^{k-1}\right) + \frac{L}{2}\|\mathbf{w} - \mathbf{w}^{k-1}\|_2^2 = h_{\frac{1}{L}}(\mathbf{w}), \tag{8.9}$$

which indeed holds for all $\mathbf{w}$.

Table 8.1 Common cost functions and corresponding Lipschitz constants for each cost where the optimally conservative fixed step length rule is given as $\alpha = \frac{1}{L}$. Note that the regularizer can be added to any cost function in the middle column and the resulting Lipschitz constant is the sum of the two Lipschitz constants listed here.

Cost function	Form of cost function	Lipschitz constant
Least Squares regression	$\sum_{p=1}^{P}\left(\tilde{\mathbf{x}}_p^T\tilde{\mathbf{w}} - y_p\right)^2$	$L = 2\left\Vert\tilde{\mathbf{X}}\right\Vert_2^2$
Softmax cost/logistic regression	$\sum_{p=1}^{P}\log\left(1 + e^{-y_p\tilde{\mathbf{x}}_p^T\tilde{\mathbf{w}}}\right)$	$L = \frac{1}{4}\left\Vert\tilde{\mathbf{X}}\right\Vert_2^2$
Squared margin/soft-margin SVMs	$\sum_{p=1}^{P}\max^2\left(0, 1 - y_p\tilde{\mathbf{x}}_p^T\tilde{\mathbf{w}}\right)$	$L = 2\left\Vert\tilde{\mathbf{X}}\right\Vert_2^2$
Multiclass softmax	$\sum_{c=1}^{C}\sum_{p\in\Omega_c}\log\left(1 + \sum_{\substack{j=1\\ j\neq c}}^{C} e^{\tilde{\mathbf{x}}_p^T\left(\tilde{\mathbf{w}}_j - \tilde{\mathbf{w}}_c\right)}\right)$	$L = \frac{C}{4}\left\Vert\tilde{\mathbf{X}}\right\Vert_2^2$
ℓ_2-regularizer	$\lambda\left\Vert\mathbf{w}\right\Vert_2^2$	$L = 2\lambda$

$$-L \leq \frac{\partial^2}{\partial w^2} g(w) \leq L, \tag{8.10}$$

or bounds the eigenvalues of its Hessian in the general case of vector input

$$-L\mathbf{I}_{N\times N} \preceq \nabla^2 g(\mathbf{w}) \preceq L\mathbf{I}_{N\times N}. \tag{8.11}$$

For square symmetric matrices $\mathbf{A}$ and $\mathbf{B}$ the notation $\mathbf{A} \preceq \mathbf{B}$ is shorthand for saying that each eigenvalue of $\mathbf{A}$ is smaller than or equal to the corresponding eigenvalue of $\mathbf{B}$. When described in this mathematical way, functions satisfying the above for finite values of L are said to have "bounded curvature" or equivalently to have a "Lipschitz continuous gradient[2]" with Lipschitz constant L.

As mentioned, all of the cost functions discussed in this book have computable maximum curvature (or some estimation of it) including the Least Squares costs, the squared margin hinge and soft-margin SVM costs, as well as two-class and multiclass soft-margin perceptrons. For convenience, in Table 8.1 we provide a complete list of Lipschitz constants for these cost functions (one can find associated calculations producing these constants in Section 8.5). Note here[3] that we write each cost function using the compact vector notation commonly used throughout the book (see e.g., Examples 4.1 and 4.2). Also recall that the notation $\left\Vert\tilde{\mathbf{X}}\right\Vert_2^2$ refers to the so-called "spectral norm" of

[2] In rare instances where g is only once differentiable but not twice (e.g., for the squared margin cost), it is said to have a Lipschitz continuous gradient if $\frac{\left\Vert\nabla g(\mathbf{w}) - \nabla g(\mathbf{v})\right\Vert_2}{\left\Vert\mathbf{w}-\mathbf{v}\right\Vert_2} \leq L$ for any $\mathbf{v}$ and $\mathbf{w}$ in its domain.

[3] Also note that the results shown here can be easily generalized to pair with fixed basis feature transformations, but while cost functions paired with (deep) neural network features also typically have bounded curvature, explicitly computing Lipschitz constants for them becomes very challenging as the number of layers increases due to the difficulty in gradient/Hessian computations (as one can see by noting the difficulty in simply computing the gradient in such instances, as in Section 7.2). Therefore the Lipschitz constants reported here do not extend to the use of multilayer network basis features.

the matrix $\tilde{\mathbf{X}}$ and denotes the largest eigenvalue of $\tilde{\mathbf{X}}\tilde{\mathbf{X}}^T$ (which is always the largest eigenvalue of $\tilde{\mathbf{X}}^T\tilde{\mathbf{X}}$ as well).

8.1.3 How to use the conservative fixed step length rule

The conservatively optimal Lipschitz constant step length rule will always work "right out of the box" to produce a convergent gradient descent scheme, therefore it can be a very convenient rule to use in practice. However, as its name implies and as described previously, it is indeed a conservative rule by nature.[4]

Therefore in practice, if one has the resources, one should use the rule as a benchmark to search for larger convergence-forcing fixed step length rules. In other words, with the Lipschitz constant step length $\alpha = \frac{1}{L}$ calculated one can easily test larger step lengths of the form $\alpha = t \cdot \frac{1}{L}$ for any constant $t > 1$.

> The conservatively optimal step length rule is convenient both because it works "right out of the box," and because it provides a benchmark for trying larger fixed step length values in practice.

Depending on both the cost function and dataset, values of t ranging from 1 to large values like 100 can work well in practice. For convenience we give the complete gradient descent algorithm with this sort of fixed step length in Algorithm 8.1.

Algorithm 8.1 Gradient descent with fixed step length based on a conservatively optimal fixed base.

Input: function g with Lipschitz continuous gradient, and initial point $\mathbf{w}^0$
$k = 1$
Find the smallest L such that $-L\mathbf{I} \preceq \nabla^2 g(\mathbf{w}) \preceq L\mathbf{I}$ for all $\mathbf{w}$ in the domain of g
Choose a constant $t \geq 1$
Set $\alpha = t \cdot \frac{1}{L}$
Repeat until stopping condition is met:
$\quad \mathbf{w}^k = \mathbf{w}^{k-1} - \alpha \nabla g\left(\mathbf{w}^{k-1}\right)$
$\quad k \leftarrow k + 1$

Example 8.1 Conservative fixed rate base comparisons

In Fig. 8.4 we show the result of employing several fixed step length rules using the conservatively optimal Lipschitz base for minimizing the softmax cost/logistic regression,

[4] Moreover, in many instances the Lipschitz constants shown in Table 8.1 are themselves conservative estimates of the true maximum curvature of a cost function (that is, a *larger* than ideal estimate) due to necessary mathematical inequalities involved in their derivation (see Section 8.5 for details).

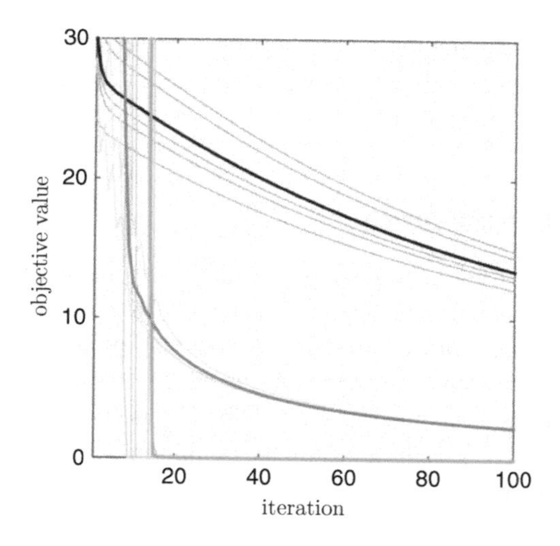

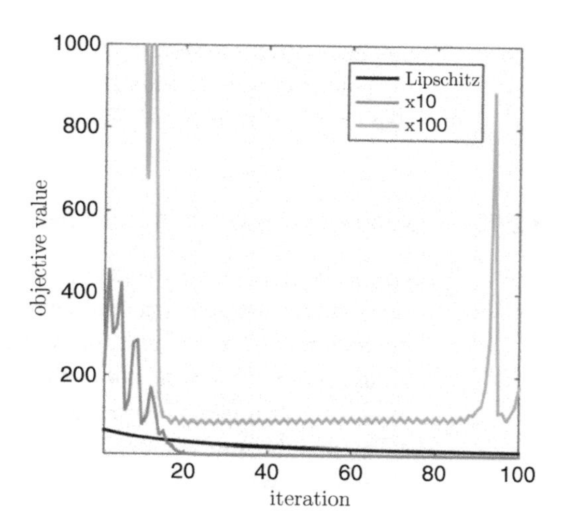

Fig. 8.4 The objective value resulting from the first 100 iterations of gradient descent applied to minimizing the softmax cost over two simple two-class datasets (see text for further details). Three constant step size rules were employed, with five runs for each (shown in lighter colors) as well as their average (shown in darker versions of the matching color): the gradient Lipschitz constant $\alpha = \frac{1}{L}$ (guaranteed to force convergence, shown in black) was used as a base, along with $\alpha = 10 \cdot \frac{1}{L}$ and $\alpha = 100 \cdot \frac{1}{L}$ (shown in magenta and green respectively). For the first dataset (left panel) both of the larger step lengths produce faster convergence than the base, with the largest providing extremely rapid convergence for this dataset. With the second dataset (right panel) the medium step length produces the best result, with the largest step length producing a divergent sequence of gradient steps.

employing both two-class datasets of $P = 100$ points each first shown in Fig. 4.3. In particular, gradient descent was run using three fixed step length rules: the Lipschitz step length $\alpha = \frac{1}{L}$ (where L is as shown in Table 8.1), as well as $\alpha = 10 \cdot \frac{1}{L}$ and $\alpha = 100 \cdot \frac{1}{L}$. Shown in the figure (left panel) is the objective or cost function value for the first 100 iterations of five runs with each step length (shown in light black, magenta, and green respectively), as well as their average shown in bolder versions of the colors.

With the first dataset (left panel), both fixed step length rules larger than the Lipschitz base produce more rapid convergence, with the step length 100 times that of the Lipschitz base producing extremely rapid convergence. Conversely, with the second dataset (right panel), only the medium step length rule produces more rapid convergence than the Lipschitz base, with the 100 times rate producing a divergent sequence of steps.

8.2 Adaptive step length rules for gradient descent

We have just seen how gradient descent can be thought of as a geometric minimization technique that, akin to Newton's method, works by hopping down the global minima of simple quadratic surrogates towards a function's minimum. In this section we continue this geometric intuition in order to develop a commonly used adaptive step length rule for gradient descent, which is a convenient and well-performing alternative to the fixed step length rule previously described.

8.2.1 Adaptive step length rule via backtracking line search

Using the quadratic surrogate perspective of gradient descent, we can now construct a very simple yet powerful and generally applicable method for adaptively determining the appropriate step length for gradient descent at each iteration. Recall that in the previous section we saw how the kth gradient descent step is given as the global minimizer of the simple quadratic surrogate h_α given in Equation (8.5). Note that if α is chosen in a way that the minimum of h_α lies above $g\left(\mathbf{w}^k\right)$ we have, using the definition of h_α and plugging in $\mathbf{w}^k = \mathbf{w}^{k-1} - \alpha\nabla g\left(\mathbf{w}^{k-1}\right)$ for $\mathbf{w}$,

$$g\left(\mathbf{w}^k\right) \leq g\left(\mathbf{w}^{k-1}\right) + \nabla g\left(\mathbf{w}^{k-1}\right)^T\left(\mathbf{w}^k - \mathbf{w}^{k-1}\right) + \frac{1}{2\alpha}\left\|\mathbf{w}^k - \mathbf{w}^{k-1}\right\|_2^2. \tag{8.12}$$

Simplifying[5] the right hand side gives

$$g\left(\mathbf{w}^k\right) \leq g\left(\mathbf{w}^{k-1}\right) - \frac{\alpha}{2}\left\|\nabla g\left(\mathbf{w}^{k-1}\right)\right\|_2^2. \tag{8.13}$$

Note that as long as we have not reached a minimum of g, the term $\frac{\alpha}{2}\left\|\nabla g\left(\mathbf{w}^{k-1}\right)\right\|_2^2$ is always positive and we have descent at each step $g\left(\mathbf{w}^k\right) < g\left(\mathbf{w}^{k-1}\right)$. While this conclusion was based on our assumption that the global minimum of h_α lay above g, we can in fact conclude the converse as well. That is, if the inequality in (8.13) holds for an $\alpha > 0$ then the minimum of the associated quadratic h_α lies above g, and the related gradient descent step decreases the objective value, i.e., $g\left(\mathbf{w}^k\right) < g\left(\mathbf{w}^{k-1}\right)$.

Therefore the inequality in (8.13) can be used as a tool, referred to as *backtracking line search*, for finding an appropriate step length α at each step in performing gradient descent (which leads to a provably convergent sequence of gradient steps to a stationary point of g, see Section 8.4.2 for further details). That is, we can test a range of decreasing values for the learning rate until we find one that satisfies the inequality, or equivalently a simple quadratic surrogate whose minimum lies above the corresponding point on g, as illustrated in Fig. 8.5.

One common way of performing this search is to initialize a step length $\alpha > 0$ and check that the desired inequality,

$$g\left(\mathbf{w}^{k-1} - \alpha\nabla g\left(\mathbf{w}^{k-1}\right)\right) \leq g\left(\mathbf{w}^{k-1}\right) - \frac{\alpha}{2}\left\|\nabla g\left(\mathbf{w}^{k-1}\right)\right\|_2^2, \tag{8.14}$$

holds. If it does not, then we multiply α by some number $t \in (0, 1)$, set $\alpha \longleftarrow t\alpha$, and try again until the inequality is satisfied. Note that the larger t is set the more fine grained

[5] Making the substitution $\mathbf{w}^k = \mathbf{w}^{k-1} - \alpha\nabla g\left(\mathbf{w}^{k-1}\right)$ the right hand side becomes $g\left(\mathbf{w}^{k-1}\right)$ $-\alpha\left\|\nabla g\left(\mathbf{w}^{k-1}\right)\right\|_2^2 + \frac{\alpha}{2}\left\|\nabla g\left(\mathbf{w}^{k-1}\right)\right\|_2^2 = g\left(\mathbf{w}^{k-1}\right) - \frac{\alpha}{2}\left\|\nabla g\left(\mathbf{w}^{k-1}\right)\right\|_2^2$.

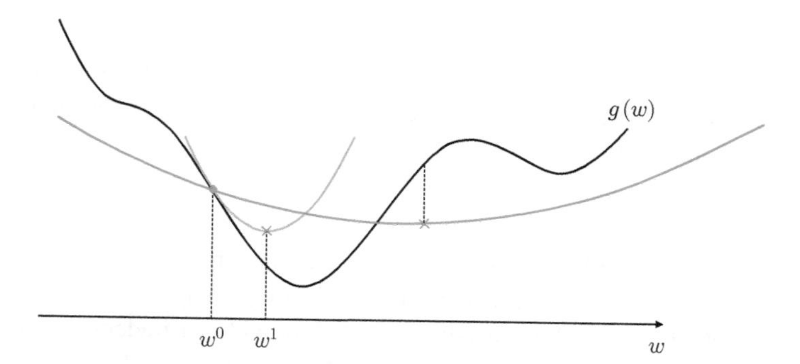

Fig. 8.5 A geometric illustration of backtracking line search. We begin with a relatively large initial guess for the step length, which generates the larger red quadratic, whose minimum may not lie above g. The guess is then adjusted downwards until the minimum of the associated quadratic (in blue) lies above the function.

the search will be. Also the terms $g\left(\mathbf{w}^{k-1}\right)$ and $\left\|\nabla g\left(\mathbf{w}^{k-1}\right)\right\|_2^2$ in (8.14) need only be computed a single time, making the procedure very efficient, and the same initial α and t can be used at each gradient descent step.

Furthermore, this sequence can be shown to be mathematically convergent to a stationary point of g, as detailed in Section 8.4.2. For convenience the backtracking line search rule is summarized in Algorithm 8.2.

Algorithm 8.2 Gradient descent with backtracking line search

Input: starting point $\mathbf{w}^0$, damping factor $t \in (0, 1)$, and initial $\alpha > 0$

$k = 1$

Repeat until stopping condition is met:

$\alpha_k = \alpha$

While $g\left(\mathbf{w}^{k-1} - \alpha_k \nabla g\left(\mathbf{w}^{k-1}\right)\right) > g\left(\mathbf{w}^{k-1}\right) - \frac{\alpha_k}{2}\left\|\nabla g\left(\mathbf{w}^{k-1}\right)\right\|_2^2$

$\alpha_k \longleftarrow t\alpha_k$

End while

$\mathbf{w}^k = \mathbf{w}^{k-1} - \alpha_k \nabla g\left(\mathbf{w}^{k-1}\right)$

$k \leftarrow k + 1$

8.2.2 How to use the adaptive step length rule

Like the optimally conservative fixed step length, backtracking line search is a convenient rule for determining step lengths at each iteration of gradient descent that works right out of the box. Furthermore, because each step length is determined using the local curvature of g, the backtracking step length will typically be superior (i.e., larger) than that of the conservative fixed step length described in Section 8.1. However each gradient step using backtracking line search, compared to using the fixed step length rule, typically includes higher computational cost due to the search for a proper step length.

> The adaptive step length rule works right out of the box, and tends to produce large step lengths at each iteration. However, each step length must be actively computed.

Due to this tradeoff it is difficult to judge universally which rule (conservative or adaptive) works best in practice, and both are widely used. The choice of diminishing parameter $t \in (0, 1)$ in Algorithm 8.2 provides a tradeoff between computation and step size with backtracking line search. The larger the diminishing parameter is set the more careful is the search for each step length (leading to more computation) but the larger will be the final step length chosen, while the converse holds for smaller values of t.

Example 8.2 Simple comparison of adaptive and optimal conservative step length rules

In Fig. 8.6 we show the result of 100 iterations of gradient descent using the backtracking line search step size rule as well as the optimally conservative step length rule discussed in Section 8.1. The dataset used here is a two-class dataset consisting of $P = 10\,000$ points (see Example 8.4 for further details), and the cost function used

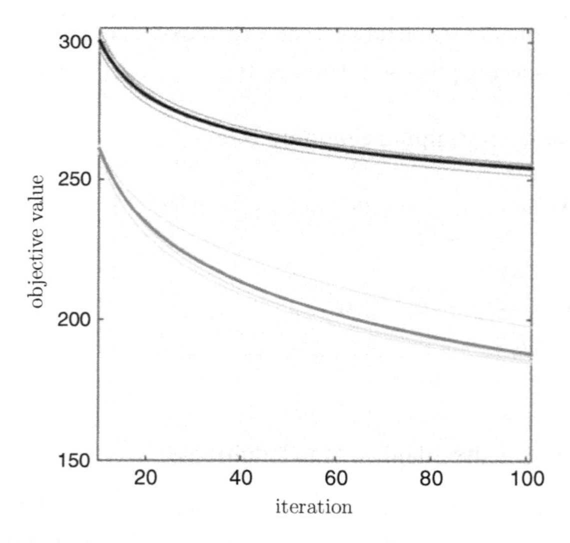

Fig. 8.6 The convergence of gradient descent using conservative fixed and backtracking line search rules on a two class classification dataset (see text for further details). Shown here are the objective values per iteration for each rule, with runs using the backtracking line search and fixed step lengths shown in magenta and black respectively (lighter colored curves indicate values over a single run of gradient descent, with the two darker colored curves showing the respective average value over five runs). As expected the backtracking runs typically display greater decrease per iteration than runs employing the fixed step length rule.

was the softmax cost. Gradient descent was run five times with each step length rule, and the objective value per iteration is shown in the figure (with the backtracking runs in magenta and the runs with fixed step length in black). As expected the adaptive backtracking line search rule, due to its larger individual step lengths, leads to more rapid decrease in the cost function value per iteration.

8.3 Stochastic gradient descent

As the size of a dataset grows, storing it in active memory in order to just compute a gradient becomes challenging if not impossible, making the standard gradient descent scheme particularly ineffective in practice for large datasets. In this section we introduce an extension of gradient descent, known as *stochastic* (or *iterative*) *gradient descent* (see footnote 8), which not only completely overcomes the memory issue but, for large datasets, is also significantly more effective computationally speaking in practice than the standard gradient method. In particular, the stochastic gradient descent provides one of the best ways of dealing with the large datasets often employed in image/audio/text-based learning tasks when paired with both non-convex cost functions (e.g., any regression/classification cost employing a neural network feature map), as well as convex costs where storage is an issue or when the feature space of a dataset is too large to employ Newton's method.

Throughout the section we will discuss minimization of cost functions over a dataset of P points $\left\{\left(\mathbf{x}_p, y_p\right)\right\}_{p=1}^{P}$, where the y_p are continuous in the case of regression, or $y_p \in \{-1, 1\}$ in the case of two class classification (although the method described here can also be applied to multiclass classification as well), and for simplicity we will make use of the compact optimization notation $\tilde{\mathbf{x}}_p = \begin{bmatrix} 1 \\ \mathbf{x}_p \end{bmatrix}$ $\tilde{\mathbf{w}} = \begin{bmatrix} b \\ \mathbf{w} \end{bmatrix}$ introduced in e.g., Examples 4.1 and 4.2.

Finally, note that in what follows one may replace each input data point with any M-dimensional fixed or neural network feature vector (as in Chapters 5 and 6) with no adjustment to the general ideas described here.

8.3.1 Decomposing the gradient

As we have seen, the cost function of a predictive model is written as a sum of individual costs over each data point as

$$g\left(\tilde{\mathbf{w}}\right) = \sum_{p=1}^{P} h\left(\tilde{\mathbf{w}}, \tilde{\mathbf{x}}_p\right). \tag{8.15}$$

For example, the Least Squares regression and softmax perceptron costs each take this form as

$$g\left(\tilde{\mathbf{w}}\right) = \sum_{p=1}^{P} \left(\tilde{\mathbf{x}}_p^T \tilde{\mathbf{w}} - y_p\right)^2, \qquad g\left(\tilde{\mathbf{w}}\right) = \sum_{p=1}^{P} \log\left(1 + e^{-y_p \tilde{\mathbf{x}}_p^T \tilde{\mathbf{w}}}\right), \tag{8.16}$$

where $h\left(\tilde{\mathbf{w}},\tilde{\mathbf{x}}_p\right) = \left(\tilde{\mathbf{x}}_p^T\tilde{\mathbf{w}} - y_p\right)^2$ and $h\left(\tilde{\mathbf{w}},\tilde{\mathbf{x}}_p\right) = \log\left(1 + e^{-y_p\tilde{\mathbf{x}}_p^T\tilde{\mathbf{w}}}\right)$ respectively. Because of this, when we minimize a predictive modeling cost via gradient descent, the gradient itself is a summation of the gradients of each of the P summands. For example, the gradient of the softmax cost may be written as

$$\nabla g\left(\tilde{\mathbf{w}}\right) = \sum_{p=1}^{P} \nabla h\left(\tilde{\mathbf{w}},\tilde{\mathbf{x}}_p\right) = -\sum_{p=1}^{P} \sigma\left(-y_p\tilde{\mathbf{x}}_p^T\tilde{\mathbf{w}}\right) y_p\tilde{\mathbf{x}}_p, \tag{8.17}$$

where in this instance $\nabla h\left(\tilde{\mathbf{w}},\tilde{\mathbf{x}}_p\right) = -\sigma\left(-y_p\tilde{\mathbf{x}}_p^T\tilde{\mathbf{w}}\right) y_p\tilde{\mathbf{x}}_p$ with $\sigma\left(\cdot\right)$ being the logistic sigmoid function (as first defined in Section 3.3.1). When minimizing any predictive modeling cost via gradient descent we can therefore think about the kth gradient descent step in terms of these individual gradients as

$$\tilde{\mathbf{w}}^k = \tilde{\mathbf{w}}^{k-1} - \alpha_k \nabla g\left(\tilde{\mathbf{w}}^{k-1}\right) = \tilde{\mathbf{w}}^{k-1} - \alpha_k \sum_{p=1}^{P} \nabla h\left(\tilde{\mathbf{w}}^{k-1},\tilde{\mathbf{x}}_p\right), \tag{8.18}$$

where α_k is an appropriately chosen step length such as those discussed in the previous sections. As described in the introduction to this section, for large datasets/values of P this gradient can be difficult or even impossible to produce given memory limitations.

Given this memory issue and the fact that the gradient decomposes over each data point, it is natural to ask if, in place of a single gradient step over the entire dataset, whether or not we can instead take a sequence of P shorter gradient steps in each data point individually. In other words, instead of taking a single full gradient step as in Equation (8.18), at the kth iteration of the procedure, will taking P smaller gradient steps in each data point similarly lead to a properly convergent method (that is, a method convergent to a stationary point of the cost function $g\left(\tilde{\mathbf{w}}\right)$)? If this were the case then we could resolve the memory problem discussed in the introduction to this section, as data would need only to be loaded into active memory a single point at a time.

Indeed with the appropriate choice of step length this procedure, called stochastic gradient descent, is provably convergent (for a formal proof see Section 8.4.3).

8.3.2 The stochastic gradient descent iteration

More formally, the analog of the kth iteration of the full gradient scheme shown in Equation (8.18) consists of P sequential point-wise gradient steps written as

$$\tilde{\mathbf{w}}^{k,p} = \tilde{\mathbf{w}}^{k,p-1} - \alpha_k \nabla h\left(\tilde{\mathbf{w}}^{k,p-1},\tilde{\mathbf{x}}_p\right) \qquad p = 1 \ldots P. \tag{8.19}$$

In analogy with the kth full gradient step, here we have used the double superscript $\tilde{\mathbf{w}}^{k,p}$ which reads "the pth individual gradient step of the kth stochastic gradient descent iteration." In this notation the initial point of the kth iteration is written as $\tilde{\mathbf{w}}^{k,0}$, the corresponding sequence of P individual gradient steps as $\left\{\tilde{\mathbf{w}}^{k,1}, \tilde{\mathbf{w}}^{k,2} \ldots, \tilde{\mathbf{w}}^{k,P}\right\}$, and the final output of the kth iteration (i.e., the Pth stochastic step) as $\tilde{\mathbf{w}}^{k,P} = \tilde{\mathbf{w}}^{k+1,0}$. After completing the kth iteration we perform the $(k+1)$th iteration by cycling through the data in precisely the same order, taking individual gradient steps for $p = 1 \ldots P$.

To reaffirm the vocabulary being used here, with the standard gradient descent we use "step" and "iteration" interchangeably, i.e., each iteration consists of one full gradient step in all P data points simultaneously as shown in Equation (8.18). Conversely, with the stochastic method we refer to a single "iteration" as consisting of all P individual gradient steps, one in each data point, executed sequentially for $p = 1 \ldots P$ as shown in Equation (8.19).

Example 8.3 Comparing stochastic and standard gradient descent on a simulated dataset

As an example, in Fig. 8.7 we show the result of applying 25 iterations of the standard gradient descent method with a fixed conservative step length (discussed in Section 8.1) shown in black, and an adaptively chosen one at each iteration (discussed in Section 8.2) shown in magenta, as well as the stochastic gradient descent method shown in green (we discuss the choice of step length for the iterative gradient method in the next section).

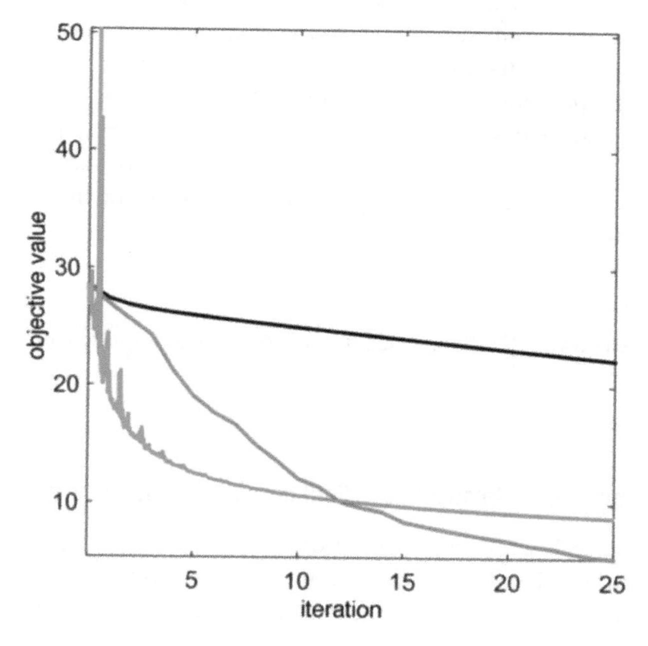

Fig. 8.7 The objective value of the first 25 iterations of stochastic gradient descent (shown in green), compared with standard gradient descent with conservative fixed step length (black) and adaptively chosen step length (magenta) (all three algorithms used the same initialization). Each iteration of the stochastic gradient method consists of P sequential gradient steps, one in each of the data points, as shown in Equation (8.19). Note how the stochastic method converges rapidly in the first few iterations, outperforming both standard gradient runs, when far from a point of convergence. See text for further details.

The softmax cost function is minimized here in order to perform logistic regression on the dataset[6] of $P = 100$ points first shown in the left panel of Fig. 4.3.

In Fig. 8.7 we show the objective value at all iterations of the algorithm. For the stochastic gradient method in each case we show these values for each step of the algorithm over all 25 of its iterations, for a total of 2500 individual steps (since $P = 100$ and 25 iterations were performed). Interestingly, we can see that while the stochastic gradient method has roughly the same computational cost as the standard gradient method, it actually outperforms the fixed step length run and, at least for the first 12 iterations, the adaptive run as well.

8.3.3 The value of stochastic gradient descent

The result of the previous example is indicative of a major computational advantage of stochastic gradient descent: when far from a point of convergence the stochastic method tends in practice to progress much faster towards a solution compared to standard gradient descent schemes. Because moderately accurate solutions (provided by a moderate amount of minimization of a cost function) tend to perform reasonably well in machine learning applications, and because with large datasets a random initialization will tend to lie far from a convergent point, substantial *empirical* evidence (see e.g., [18, 21] and references therein) suggests that stochastic gradient descent is often far more effective in practice (than standard gradient descent) for dealing with large datasets.[7] In many such instances even a single iteration (i.e., one gradient step through each point of the dataset) can provide a good solution.

Stochastic gradient descent tends to work extremely well with large datasets.

Example 8.4 Stochastic gradient descent performed on a large dataset

In Fig. 8.8 we show the result of 100 iterations of gradient descent on a $P = 10\,000$ two class classification dataset on which the softmax cost was used to perform logistic regression. This data was generated for the task of face detection (see Example 1.4)

[6] When applying stochastic gradient descent it is common practice to first randomize the order of the data prior to running the algorithm. In practice this has been found to improve convergence of the method, and is done with the data in this example.

[7] Note that due to the manner in which MATLAB/Octave currently performs loops an implementation of stochastic gradient descent can run slowly. However, in other programming languages like C++ or Python this is not a problem.

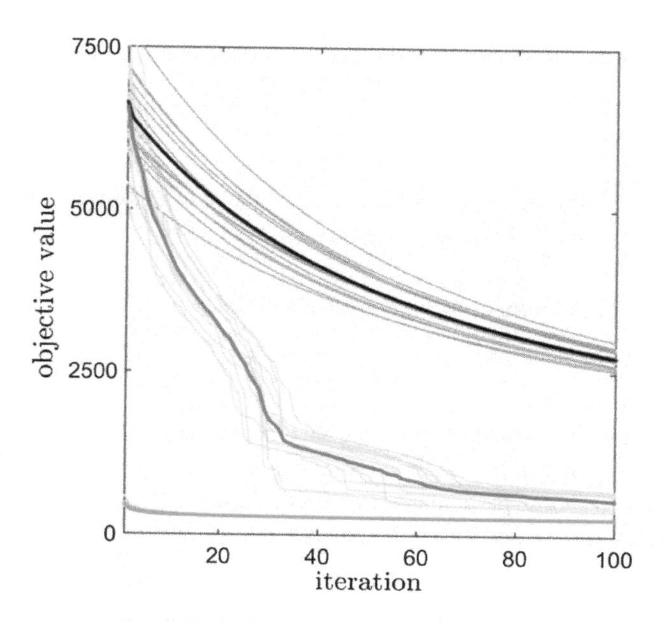

Fig. 8.8 The objective value of the first 100 iterations of stochastic gradient descent (shown in green), compared with standard gradient descent with conservatively optimal fixed step length (black) and adaptively chosen step length (magenta). In this instance a real two class classification dataset is used consisting of $P = 10\,000$ points (see text for further details). 15 runs of each method are shown as the lighter colored curves, with their averages shown in bold. Here the stochastic gradient descent scheme is massively superior to both standard gradient methods in terms of the rapidity of its convergence.

and consists of 3000 facial images, the remainder consisting of examples of non-face images.

Shown in the figure is the objective or cost function value of two runs of the standard gradient descent scheme, the first using the conservatively optimal fixed step length and the second using the adaptive rule, along with the corresponding cost function value of the stochastic gradient procedure. Here the results of 15 runs of each method are shown in lighter colors, where in each instance a shared random initialization is used by all three methods, and the average over all runs of each method is highlighted as a darker curve. We can see that the stochastic gradient descent scheme is massively superior on this large dataset in terms of the rapidity of its convergence when compared to the standard gradient method.

8.3.4 Step length rules for stochastic gradient descent

By slightly extending the convergence-forcing mechanism used in determining a step size rule for standard gradient descent one can conclude that a *diminishing* step size can similarly guarantee mathematically the convergence of the stochastic gradient method (see Section 8.4.3 for a formal derivation). More precisely, a step size rule satisfying the following two requirements is guaranteed to cause the stochastic gradient descent procedure to converge to a stationary point:

(1) The step size must diminish as the number of iterations increases: $\alpha_k \longrightarrow 0$ as $k \longrightarrow \infty$.

(2) The sum of the step sizes is not finite: i.e., $\sum_{k=1}^{\infty} \alpha_k = \infty$.

Common choices of step size in practice with the iterative gradient method include $\alpha_k = \frac{1}{k}$ and $\alpha_k = \frac{1}{\sqrt{k}}$, or variations of these (see e.g., [21] for further information about how to choose particular variations of these step lengths in practice). The former of these rules, $\alpha_k = \frac{1}{k}$, was used in both examples shown in Fig. 8.7 and 8.8.

8.3.5 How to use the stochastic gradient method in practice

Even though a diminishing step length mathematically ensures convergence, like the optimal conservative fixed step length rules discussed for standard gradient descent in Section 8.1, the diminishing step length rule for the stochastic gradient method is *conservative in nature*. Again as with the standard gradient this does not reduce the utility of the diminishing step length rule, it is indeed very useful as it can always be counted on to work right out of the box in practice, and it is therefore commonly used.

However, be aware that in practice one may successfully use other step length rules such as fixed step lengths that, while they do not ensure convergence theoretically, work very well in practice. For example, fixed step lengths (tuned properly on a given dataset) are also commonly used in practice with the stochastic gradient method.

Example 8.5 Comparison of diminishing versus (tuned) fixed step lengths on a large dataset

In Fig. 8.9 we show several runs of the stochastic gradient method to minimize the softmax cost for the dataset of $P = 10\,000$ two-class data points first described in Example 8.4. In particular, we compare the result of $k = 100$ total iterations using three distinct step-size rules: two diminishing step sizes, $\alpha_k = \frac{1}{k}$ and $\alpha_k = \frac{1}{\sqrt{k}}$, and the constant step size of $\alpha_k = \frac{1}{3}$ for all k. We run stochastic gradient descent with each step size rule five times each, in each instance providing the same random initialization to each version of the algorithm. As can be seen in the figure, where we show the objective value at each iteration of the stochastic method for runs of all three step length choices, the run with $\alpha_k = \frac{1}{\sqrt{k}}$ provides better performance than $\alpha_k = \frac{1}{k}$, but the constant step size $\alpha_k = \frac{1}{3}$ runs outperform both diminishing rules overall.

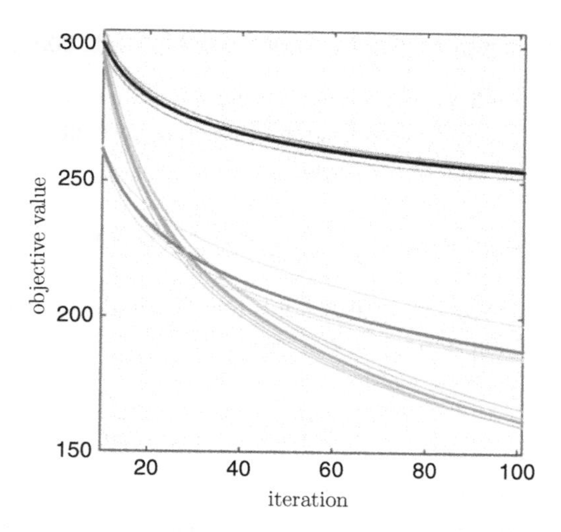

Fig. 8.9 The objective value of the first $k = 1$–100 iterations of three versions of stochastic gradient descent with diminishing step-size rules, $\alpha_k = \frac{1}{k}$ and $\alpha_k = \frac{1}{\sqrt{k}}$, and a constant step size $\alpha_k = \frac{1}{3}$ for all k (shown in light black, magenta, and green respectively, with the average of the runs shown in bold of each color). The two-class classification dataset used here consists of $P = 10\,000$ data points (see text for further details). In this instance the constant step size-runs tend to outperform those governed by the provably convergent diminishing step-size rules.

8.4 Convergence proofs for gradient descent schemes

To set the stage for the material of this section, it will be helpful to briefly point out the specific set of mild conditions satisfied by all of the cost functions we aim to minimize in this book, as these conditions are relied upon explicitly in the upcoming convergence proofs. These three basic conditions are listed below:

① They have piecewise-differentiable first derivative.

② They are bounded from below, i.e., they never take on values at $-\infty$.

③ They have bounded curvature.

While the first condition is specific to the set of cost functions we discuss, in particular so that we include the squared margin perceptron which is not smooth and indeed has piecewise differentiable first derivative while the other costs are completely smooth, the latter two conditions are very common assumptions made in the study of mathematical optimization more generally.

8.4.1 Convergence of gradient descent with Lipschitz constant fixed step length

With the gradient of g being Lipschitz continuous with constant L, from Section 8.1 we know that at the kth iteration of gradient descent we have a corresponding quadratic upper bound on g of the form

$$g(\mathbf{w}) \leq g\left(\mathbf{w}^{k-1}\right) + \nabla g\left(\mathbf{w}^{k-1}\right)^T\left(\mathbf{w} - \mathbf{w}^{k-1}\right) + \frac{L}{2}\|\mathbf{w} - \mathbf{w}^{k-1}\|_2^2, \tag{8.20}$$

where indeed this inequality holds for all $\mathbf{w}$ in the domain of g. Now plugging the form of the gradient step $\mathbf{w}^k = \mathbf{w}^{k-1} - \frac{1}{L}\nabla g\left(\mathbf{w}^{k-1}\right)$ into the above and simplifying gives

$$g\left(\mathbf{w}^k\right) \leq g\left(\mathbf{w}^{k-1}\right) - \frac{1}{2L}\|\nabla g\left(\mathbf{w}^{k-1}\right)\|_2^2, \tag{8.21}$$

which, since $\|\nabla g\left(\mathbf{w}^{k-1}\right)\|_2^2 \geq 0$, indeed shows that the sequence of gradient steps with conservative fixed step length is decreasing. To show that it converges to a stationary point where the gradient vanishes we subtract off $g\left(\mathbf{w}^{k-1}\right)$ from both sides of the above, and sum the result over $k = 1 \ldots K$ giving

$$\sum_{k=1}^{K} g\left(\mathbf{w}^k\right) - g\left(\mathbf{w}^{k-1}\right) = g\left(\mathbf{w}^K\right) - g\left(\mathbf{w}^0\right) \leq -\frac{1}{2L}\sum_{k=1}^{K}\left\|\nabla g\left(\mathbf{w}^{k-1}\right)\right\|_2^2. \tag{8.22}$$

Note importantly here that since g is bounded below so too is $g\left(\mathbf{w}^K\right)$ for all K, and this implies that, taking $K \longrightarrow \infty$, we *must* have that

$$\sum_{k=1}^{\infty}\left\|\nabla g\left(\mathbf{w}^{k-1}\right)\right\|_2^2 < \infty. \tag{8.23}$$

If this were not the case then we would contradict the assumption that g has a finite lower bound, since Equation (8.22) would say that $g\left(\mathbf{w}^K\right)$ would be negative infinity! Hence the fact that the infinite sum above must be finite implies that as $k \longrightarrow \infty$ we have that

$$\left\|\nabla g\left(\mathbf{w}^{k-1}\right)\right\|_2^2 \longrightarrow 0, \tag{8.24}$$

or that the sequence of gradient descent steps with step length determined by the Lipschitz constant of the gradient of g produces a vanishing gradient. Or, in other words, that this sequence indeed converges to a stationary point of g.

Note that we could have made the same argument above using any fixed step length smaller than $\frac{1}{L}$ as well.

8.4.2 Convergence of gradient descent with backtracking line search

With the assumption that g has bounded curvature, stated formally in Section 8.1.2 that g has a Lipschitz continuous gradient with some constant L (even if we cannot calculate

L explicitly), it follows that with a fixed choice of initial step length $\alpha > 0$ and $t \in (0, 1)$ for all gradient descent steps, we can always find an integer n_0 such that

$$t^{n_0}\alpha \leq \frac{1}{L}. \tag{8.25}$$

Thus we always have the lower bound on an adaptively chosen step length $\hat{t} = t^{n_0}\alpha$, meaning formally that the backtracking found step length at the kth gradient descent step will always be larger than this lower bound, i.e.,

$$\alpha_k \geq \hat{t} > 0. \tag{8.26}$$

Now, recall from Section 8.2 that by running the backtracking procedure at the kth gradient step we produce a step length α_k that ensures the associated quadratic upper bound

$$g\left(\mathbf{w}\right) \leq g\left(\mathbf{w}^{k-1}\right) + \nabla g\left(\mathbf{w}^{k-1}\right)^T\left(\mathbf{w} - \mathbf{w}^{k-1}\right) + \frac{1}{2\alpha_k}\|\mathbf{w} - \mathbf{w}^{k-1}\|_2^2 \tag{8.27}$$

holds for all $\mathbf{w}$ in the domain of g. Plugging the gradient step $\mathbf{w}^k = \mathbf{w}^{k-1} - \alpha_k \nabla g\left(\mathbf{w}^{k-1}\right)$ into the above and simplifying we have equivalently that

$$g\left(\mathbf{w}^k\right) \leq g\left(\mathbf{w}^{k-1}\right) - \frac{\alpha_k}{2}\left\|\nabla g\left(\mathbf{w}^{k-1}\right)\right\|_2^2, \tag{8.28}$$

which indeed shows that that step produces decrease in the cost function. To show that the sequence of gradient steps converges to a stationary point of g we first subtract off $g\left(\mathbf{w}^{k-1}\right)$ and sum the above over $k = 1 \ldots K$ which gives

$$\sum_{k=1}^{K} g\left(\mathbf{w}^k\right) - g\left(\mathbf{w}^{k-1}\right) = g\left(\mathbf{w}^K\right) - g\left(\mathbf{w}^0\right) \leq -\frac{1}{2}\sum_{k=1}^{K}\alpha_k\left\|\nabla g\left(\mathbf{w}^{k-1}\right)\right\|_2^2. \tag{8.29}$$

Since g is bounded below so too is $g\left(\mathbf{w}^K\right)$ for all K, and therefore taking $K \longrightarrow \infty$ the above says that we must have

$$\sum_{k=1}^{\infty}\alpha_k\left\|\nabla g\left(\mathbf{w}^{k-1}\right)\right\|_2^2 < \infty. \tag{8.30}$$

Now, we know from Equation (8.26) that since each $\alpha_k \geq \hat{t} > 0$ for all k, this implies that we must have that

$$\sum_{k=1}^{\infty}\alpha_k = \infty. \tag{8.31}$$

And this is just fine, because in order for Equation (8.30) to hold under this condition we *must* have that

$$\left\|\nabla g\left(\mathbf{w}^{k-1}\right)\right\|_2^2 \longrightarrow 0, \tag{8.32}$$

as $k \longrightarrow \infty$, for otherwise Equation (8.30) could not be true. This shows that the sequence of gradient steps determined by backtracking line search converges to a stationary point of g.

8.4.3 Convergence of the stochastic gradient method

To understand what kind of step length rule we will need to mathematically force the stochastic gradient method to converge we first relate the kth iteration of stochastic gradient descent to a full standard gradient step. This is accomplished by unraveling the definition of the gradient iteration given in Equation (8.19), and writing out the kth iteration (note here that we ignore the bias term for ease of exposition) as

$$\begin{aligned}\mathbf{w}^{k,0} &= \mathbf{w}^{k-1,P} = \mathbf{w}^{k-1,P-1} - \alpha_k \nabla h\left(\mathbf{w}^{k-1,P-1}, \mathbf{x}_P\right) \\ &= \ldots = \mathbf{w}^{k-1,0} - \alpha_k \sum_{p=1}^{P} \nabla h\left(\mathbf{w}^{k-1,p-1}, \mathbf{x}_p\right).\end{aligned} \tag{8.33}$$

Next, by adding and subtracting the full gradient at $\mathbf{w}^{k-1,0}$, that is $\nabla g\left(\mathbf{w}^{k-1,0}\right) = \sum_{p=1}^{P} \nabla h\left(\mathbf{w}^{k-1,0}, \mathbf{x}_p\right)$, weighted by the step length α_k, and by referring to

$$\epsilon_k = \sum_{p=2}^{P} \left(\nabla h\left(\mathbf{w}^{k-1,p-1}, \mathbf{x}_p\right) - \nabla h\left(\mathbf{w}^{k-1,0}, \mathbf{x}_p\right)\right), \tag{8.34}$$

the gradient iteration can be rewritten equivalently as

$$\mathbf{w}^{k,0} = \mathbf{w}^{k-1,0} - \alpha_k \left(\nabla g\left(\mathbf{w}^{k-1,0}\right) + \epsilon_k\right). \tag{8.35}$$

In other words, the above expresses the kth gradient iteration as a standard gradient step, with the additional "error" term ϵ_k. Since we have expressed the kth iteration of the stochastic gradient method in this manner, to simplify notation from here on we remove the redundant second superscript, writing the above more simply as

$$\mathbf{w}^{k} = \mathbf{w}^{k-1} - \alpha_k \left(\nabla g\left(\mathbf{w}^{k-1}\right) + \epsilon_k\right). \tag{8.36}$$

This can be thought of as a "noisy" gradient step.[8] As we have seen previously, a properly designed step size α_k that forces the gradient $\nabla g\left(\mathbf{w}^{k-1}\right)$ to vanish for large k is that $\sum_{k=1}^{\infty} \alpha_k = \infty$, where each $\alpha_k \leq \frac{1}{L}$. However, in order for the stochastic gradient method to converge to a stationary point of g we will also need the error ϵ_k to vanish.

By analyzing ϵ_k one can show[9] that the norm of the kth error term $\|\epsilon_k\|_2$ is in fact bounded by a constant proportional to the corresponding step length α_k, i.e.,

$$\|\epsilon_k\|_2 \leq \alpha_k S \tag{8.41}$$

[8] The mathematical form of Equation (8.36), as a noisy gradient step, arises more generally in the case where the gradient of a function cannot be effectively computed (i.e., the only gradient calculation available is polluted by noise). In this more general instance, ϵ_k is typically modeled using a random variable, in which case the step defined in Equation (8.36) is referred to as a *stochastic gradient descent*. Because of this similarity in mathematical form the iterative gradient method is also often referred to as stochastic gradient descent, although the error ϵ_k is not random but given explicitly by Equation (8.34).

[9] Using the definition of the error, the triangle inequality, and the fact that each h has Lipschitz continuous gradient with constant L_h, we have that

$$\|\epsilon_k\|_2 \leq \sum_{p=2}^{P} \left\|\left(\nabla h\left(\mathbf{w}^{k-1,p-1}, \mathbf{x}_p\right) - \nabla h\left(\mathbf{w}^{k-1,0}, \mathbf{x}_p\right)\right)\right\|_2 \leq L_h \sum_{p=2}^{P} \left\|\mathbf{w}^{k-1,p-1} - \mathbf{w}^{k-1,0}\right\|_2. \tag{8.37}$$

for some constant S. Therefore, by adding the condition to the step size scheme that $\alpha_k \longrightarrow 0$ as $k \longrightarrow \infty$ we can force the error term $\|\epsilon_k\| \longrightarrow 0$ as well. Altogether then our conditions on the step size for convergence of the stochastic gradient method include that $\sum_{k=1}^{\infty} \alpha_k = \infty$ (which forces the gradient to vanish) and $\alpha_k \longrightarrow 0$ as k grows large (which forces the error to vanish). Any diminishing sequence like e.g., $\alpha_k = \frac{1}{k}$ or $\alpha_k = \frac{1}{\sqrt{k}}$ etc., satisfies such requirements. However, these two kinds of step-size rules are commonly used in practice as they balance our desire to cause both the gradient and error term to vanish (while slower or faster diminishing step size favors one term over the other practically speaking).

8.4.4 Convergence rate of gradient descent for convex functions with fixed step length

Suppose, in addition to g having bounded curvature with Lipschitz constant L, that g is also convex. As illustrated in Fig. 8.10, this implies that for any $\mathbf{w}$, $g(\mathbf{w})$ is majorized by (or lies underneath) a quadratic and minorized by (or lies over) a linear function. With convexity we can, in addition to assuring convergence of gradient descent, more easily calculate a convergence rate of gradient descent.

Now that we know the sequence is decreasing, all that is left is to make sure that the sequence converges to a global minimum of g, say $\mathbf{w}^{\star}$. We would like to show that $g\left(\mathbf{w}^k\right)$ converges to $g(\mathbf{w}^{\star}) > -\infty$ as k increases. Looking at the quadratic upper bound on $g\left(\mathbf{w}^i\right)$ centered at $\mathbf{w}^{i-1}$, as in (8.21), we have

$$g\left(\mathbf{w}^k\right) \leq g\left(\mathbf{w}^{k-1}\right) - \frac{1}{2L}\|\nabla g\left(\mathbf{w}^{k-1}\right)\|_2^2. \tag{8.42}$$

By using the definition of the stochastic gradient step we can roll back each $\mathbf{w}^{k-1,p-1}$ to $\mathbf{w}^{k-1,0}$ by expanding first $\mathbf{w}^{k-1,p-1} = \mathbf{w}^{k-1,p-2} - \alpha_k \nabla h\left(\mathbf{w}^{k-1,p-2}, \mathbf{x}_{p-1}\right)$ and then doing the same to $\mathbf{w}^{k-1,p-2}$, etc., giving

$$\mathbf{w}^{k-1,p-1} - \mathbf{w}^{k-1,0} = -\alpha_k \sum_{t=1}^{p-1} \nabla h\left(\mathbf{w}^{k-1,t-1}, \mathbf{x}_t\right). \tag{8.38}$$

Substituting this into the right hand side of Equation (8.37) for each $p = 1 \ldots P$ then gives the bound

$$\|\epsilon_k\|_2 \leq L_h \sum_{p=1}^{P} \left\| \alpha_k \sum_{t=1}^{p} \nabla h\left(\mathbf{w}^{k-1,t-1}, \mathbf{x}_t\right) \right\|_2 \leq L_h \alpha_k \sum_{p=1}^{P} \sum_{t=1}^{p} \left\| \nabla h\left(\mathbf{w}^{k-1,t-1}, \mathbf{x}_t\right) \right\|_2, \tag{8.39}$$

where the right hand side follows by the triangle inequality and the definition of the norm $\|\alpha \mathbf{z}\|_2 = \alpha \|\mathbf{z}\|_2$ when $\alpha \geq 0$. Finally, because we have chosen α_k such that the gradient $\nabla g\left(\mathbf{w}^{k-1}\right)$ will vanish, or in other words that $\left\{\mathbf{w}^{k-1}\right\}_{k=1}^{\infty}$ converges, and because each set of points $\left\{\mathbf{w}^{k-1,t-1}\right\}_{t=1}^{P}$ lies in a neighborhood of $\mathbf{w}^{k-1}$ for each k, we can say that the individual gradients $\left\| \nabla h\left(\mathbf{w}^{k-1,t-1}, \mathbf{x}_t\right) \right\|_2$ are bounded. Hence we can say that there is some fixed constant S such that $L_h \sum_{p=1}^{P} \sum_{t=1}^{p} \left\| \nabla h\left(\mathbf{w}^{k-1,t-1}, \mathbf{x}_t\right) \right\|_2 \leq S$ and therefore conclude that

$$\|\epsilon_k\|_2 \leq \alpha_k S. \tag{8.40}$$

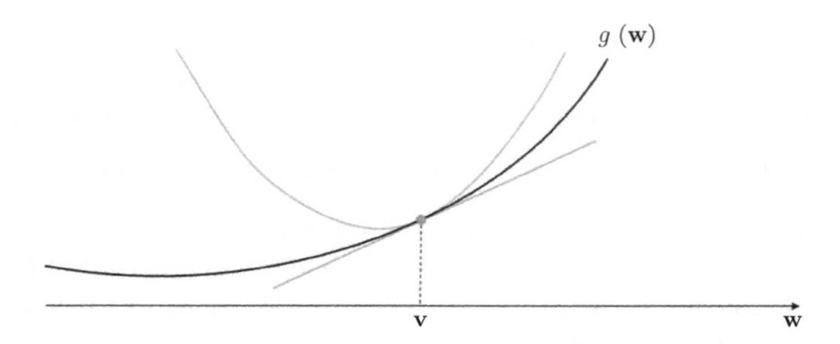

Fig. 8.10 The conservatively optimal quadratic form (in blue) majorizes the convex function g at $\mathbf{v}$, which (since it is convex) is also minorized by a linear function (in green) at each $\mathbf{v}$.

The first order definition of convexity at $\mathbf{w}^{\star}$gives the inequality

$$g(\mathbf{w}) + \nabla g(\mathbf{w})\left(\mathbf{w}^{\star} - \mathbf{w}\right) \leq g\left(\mathbf{w}^{\star}\right). \tag{8.43}$$

Plugging in $\mathbf{w} = \mathbf{w}^{i-1}$ into the right side and rearranging gives

$$g\left(\mathbf{w}^{k-1}\right) \leq g\left(\mathbf{w}^{\star}\right) + \nabla g\left(\mathbf{w}^{k-1}\right)^{T}\left(\mathbf{w}^{k-1} - \mathbf{w}^{\star}\right). \tag{8.44}$$

Now we use this upper bound for $g\left(\mathbf{w}^{i-1}\right)$ on the right hand side of (8.42) to obtain

$$g\left(\mathbf{w}^{k}\right) \leq g\left(\mathbf{w}^{\star}\right) + \nabla g\left(\mathbf{w}^{k-1}\right)^{T}\left(\mathbf{w}^{k-1} - \mathbf{w}^{\star}\right) - \frac{1}{2L}\|\nabla g\left(\mathbf{w}^{k-1}\right)\|_2^2. \tag{8.45}$$

Bringing $g\left(\mathbf{w}^{\star}\right)$ to the left hand side, and using the fact that $\mathbf{w}^{k} = \mathbf{w}^{k-1} - \frac{1}{L}\nabla g\left(\mathbf{w}^{k-1}\right)$ implies that $\nabla g\left(\mathbf{w}^{k-1}\right) = L\left(\mathbf{w}^{k-1} - \mathbf{w}^{k}\right)$, simple rearrangement gives that the above is equivalent to

$$g\left(\mathbf{w}^{k}\right) - g\left(\mathbf{w}^{\star}\right) \leq \frac{L}{2}\left(\|\mathbf{w}^{k-1} - \mathbf{w}^{\star}\|_2^2 - \|\mathbf{w}^{k} - \mathbf{w}^{\star}\|_2^2\right). \tag{8.46}$$

Now averaging both sides over the first k steps gives the corresponding inequality

$$\frac{1}{k}\sum_{i=1}^{k}\left(g\left(\mathbf{w}^{i}\right) - g\left(\mathbf{w}^{\star}\right)\right) \leq \frac{L}{2k}\left(\|\mathbf{w}^{0} - \mathbf{w}^{\star}\|_2^2 - \|\mathbf{w}^{k} - \mathbf{w}^{\star}\|_2^2\right). \tag{8.47}$$

Note that because $\left\{g\left(\mathbf{w}^{i}\right)\right\}_{i=1}^{k}$ is decreasing, the left hand side itself has lower bound given by $g\left(\mathbf{w}^{k}\right) - g\left(\mathbf{w}^{\star}\right)$, and the right side has upper bound given by $\frac{L}{2k}\|\mathbf{w}^{0} - \mathbf{w}^{\star}\|_2^2$. Therefore we have

$$g\left(\mathbf{w}^{k}\right) - g\left(\mathbf{w}^{\star}\right) \leq \frac{L}{2k}\|\mathbf{w}^{0} - \mathbf{w}^{\star}\|_2^2. \tag{8.48}$$

The right hand side goes to zero as $k \longrightarrow \infty$ with the rate of $\frac{1}{k}$. In other words, to get within $\frac{1}{k}$ of the global optimum takes in the order of k steps.

8.5 Calculation of computable Lipschitz constants

Here we provide the calculations associated with several of the Lipschitz constants reported in Table 8.1. In particular instances it will be convenient to express curvature in terms of the *curvature function,* which for a general cost function g taking input $\mathbf{w}$ is given as

$$\psi\left(\mathbf{z}\right) = \mathbf{z}^T \left(\nabla^2 g\left(\mathbf{w}\right)\right) \mathbf{z}. \tag{8.49}$$

As was the case with convexity (see Exercise 2.11), the greatest curvature given by the Lipschitz constant L can be defined in terms of the eigenvalues of $\nabla^2 g\left(\mathbf{w}\right)$ in Equation (8.11) or the curvature function above. In particular, the greatest curvature of g is equivalently given by the minimum and maximum values taken on by its associated curvature function on the unit sphere where $\|\mathbf{z}\|_2 = 1$. Formally this is the smallest nonnegative L such that

$$-L \leq \psi\left(\mathbf{z}\right) \leq L \quad \text{for any } \mathbf{z} \text{ where } \|\mathbf{z}\|_2 = 1 \tag{8.50}$$

holds over the domain of g. Here L is precisely the Lipschitz constant defined in Equation (8.11). One can show fairly easily that the two definitions of Lipschitz constant are indeed equivalent (see Exercise 2.11).

Example 8.6 Least Squares for linear regression

The Least Squares cost function for linear regression $g\left(\tilde{\mathbf{w}}\right) = \|\tilde{\mathbf{X}}^T\tilde{\mathbf{w}} - \mathbf{y}\|_2^2$ has an easily calculable gradient and Hessian, given respectively by $\nabla g\left(\tilde{\mathbf{w}}\right) = 2\tilde{\mathbf{X}}\left(\tilde{\mathbf{X}}^T\tilde{\mathbf{w}} - \mathbf{y}\right)$ and $\nabla^2 g\left(\tilde{\mathbf{w}}\right) = 2\tilde{\mathbf{X}}\tilde{\mathbf{X}}^T$. Because the Least Squares cost is convex we know that its Hessian is positive semidefinite, so we have that

$$\mathbf{0}_{N+1\times N+1} \preceq \nabla^2 g\left(\tilde{\mathbf{w}}\right) \preceq 2\left\|\tilde{\mathbf{X}}\right\|_2^2 \mathbf{I}_{N+1\times N+1}, \tag{8.51}$$

using the definition of greatest curvature given in Equation (8.50). Therefore the Lipschitz constant is given by the largest eigenvalue of this matrix,

$$L = 2\left\|\tilde{\mathbf{X}}\right\|_2^2. \tag{8.52}$$

Example 8.7 Two class softmax cost

As we saw in Exercise 4.4, the Hessian of the softmax perceptron/convex logistic regression cost function is convex with Hessian given as

$$\nabla^2 g\left(\tilde{\mathbf{w}}\right) = \tilde{\mathbf{X}}\text{diag}\left(\mathbf{r}\right)\tilde{\mathbf{X}}^T, \tag{8.53}$$

where $\mathbf{r} = \left[r_1 \ldots r_P\right]^T$ is defined entry-wise as $r_p = \sigma\left(-y_p\tilde{\mathbf{x}}_p^T\tilde{\mathbf{w}}\right)\left(1 - \sigma\left(-y_p\tilde{\mathbf{x}}_p^T\tilde{\mathbf{w}}\right)\right)$.

Using the fact that $0 < r_n \leq \frac{1}{4}$ and the curvature definition in Equation (8.50) we can say that the following holds for each $\mathbf{z}$ on the unit sphere:

$$\mathbf{z}^T \left(\tilde{\mathbf{X}}\text{diag}\left(\mathbf{r}\right)\tilde{\mathbf{X}}^T\right)\mathbf{z} \leq \frac{1}{4}\mathbf{z}^T\tilde{\mathbf{X}}\tilde{\mathbf{X}}^T\mathbf{z}. \tag{8.54}$$

Now, because the right hand side is maximized when $\mathbf{z}$ is the eigenvector of the matrix $\tilde{\mathbf{X}}\tilde{\mathbf{X}}^T$ associated to its largest eigenvalue, the maximum value attainable by the right hand side above is $\frac{1}{4}\left\|\tilde{\mathbf{X}}\right\|_2^2$ (see exercises). Therefore, altogether we have that

$$\mathbf{0}_{N\times N} \preceq \nabla^2 g\left(\tilde{\mathbf{w}}\right) \preceq \frac{1}{4}\left\|\tilde{\mathbf{X}}\right\|_2^2, \tag{8.55}$$

and therefore we may take as a Lipschitz constant

$$L = \frac{1}{4}\left\|\tilde{\mathbf{X}}\right\|_2^2. \tag{8.56}$$

Example 8.8 Squared margin hinge

The Hessian of the squared margin hinge function $g\left(\tilde{\mathbf{w}}\right) = \sum_{p=1}^{P}\max^2\left(0, 1 - y_p\tilde{\mathbf{x}}_p^T\tilde{\mathbf{w}}\right)$ can be written as

$$\nabla^2 g\left(\tilde{\mathbf{w}}\right) = 2\tilde{\mathbf{X}}_s\tilde{\mathbf{X}}_s^T. \tag{8.57}$$

Because the maximum eigenvalue of $\tilde{\mathbf{X}}_s\tilde{\mathbf{X}}_s^T$ is bounded above by that of the matrix $\tilde{\mathbf{X}}\tilde{\mathbf{X}}^T$ (see Exercise 8.5), a bound on the maximum eigenvalue of this matrix/the curvature of the cost g is given by

$$L = 2\left\|\tilde{\mathbf{X}}\right\|_2^2. \tag{8.58}$$

Example 8.9 Multiclass softmax

Since the multiclass softmax cost is convex we know that its Hessian, described blockwise in Exercise 4.18, must have all nonnegative eigenalues. Note that the maximum eigenvalue of its cth diagonal block $\nabla_{\tilde{\mathbf{w}}_c\tilde{\mathbf{w}}_c} g = \sum_{p=1}^{P}\frac{e^{\tilde{\mathbf{x}}_p^T\tilde{\mathbf{w}}_c}}{\sum_{d=1}^{C}e^{\tilde{\mathbf{x}}_p^T\tilde{\mathbf{w}}_d}}\left(1 - \frac{e^{\tilde{\mathbf{x}}_p^T\tilde{\mathbf{w}}_c}}{\sum_{d=1}^{C}e^{\tilde{\mathbf{x}}_p^T\tilde{\mathbf{w}}_d}}\right)\mathbf{x}_p\mathbf{x}_p^T$ is, in the same manner as the two class softmax cost in Example 8.7, given by $\frac{1}{4}\left\|\tilde{\mathbf{X}}\right\|_2^2$. Because the maximum eigenvalue of a symmetric block matrix with nonnegative eigenvalues is bounded above by the sum of the maximum eigenvalues of its diagonal blocks [28], an upper bound on the maximum eigenvalue of the Hessian is given by

$$L = \frac{C}{4}\left\|\tilde{\mathbf{X}}\right\|_2^2. \tag{8.59}$$

8.6 Summary

In the first two sections of this chapter we described two rigorous ways for determining step lengths for gradient descent. First in Section 8.1 we introduced the optimally conservative fixed step length rule (based on the notion of the maximum curvature of a cost function). This can be used to produce a fixed step length that, while conservative in nature, will guarantee convergence of gradient descent when applied to every cost function detailed in this book (see Table 8.1).

Next, in Section 8.2 we introduced an adaptive step length rule for gradient descent, another rule that works right out of the box for any cost function. As discussed in Section 8.2.2, the adaptive step length rule requires more computation at each step of gradient descent but typically takes considerably longer steps than a conservative fixed counterpart.

Finally, in Section 8.3 we introduced the stochastic gradient descent method. A natural extension of the standard gradient descent scheme, stochastic gradient descent is especially useful when dealing with large datasets (where, for example, loading the full dataset into memory is challenging or impossible). On large datasets the stochastic gradient method can converge extremely rapidly to a reasonable solution of a problem, even after only a single pass through the dataset.

8.7 Exercises

Section 8.1 exercises

Exercises 8.1 Code up backtracking line search for the squared margin cost

Code up an adaptive step length sub-function for the minimization of the squared margin perceptron and install it into the wrapper detailed in Exercise 4.7, replacing the fixed step length given there. Test your code by running the wrapper, and produce a plot of the objective value decrease at each iteration.

Exercises 8.2 Code up stochastic gradient descent

Reproduce a part of the experiment shown in Example 8.4 by comparing standard and stochastic gradient descent methods on a large two class classification dataset of $P = 10\,000$ points. Employ any cost function (e.g., softmax) to fit to this data and plot the cost value at each iteration from runs of each method, along with the average value of these runs (as in Fig. 8.8).

Exercises 8.3 Code up backtracking line search for the multiclass softmax cost

Code up an adaptive step length sub-function for the minimization of the multiclass softmax perceptron and install it into the wrapper detailed in Exercise 4.15, replacing the fixed step length given there. Test your code by running the wrapper, and produce a plot of the objective value decrease at each iteration.

Exercises 8.4 Alternative formal definition of Lipschitz gradient

An alternative to defining the Lipschitz constant by Equation (8.11) for functions f with Lipschitz continuous gradient is given by

$$\left\|\nabla f\left(\mathbf{x}\right)-\nabla f\left(\mathbf{y}\right)\right\|_2 \leq L\left\|\mathbf{x}-\mathbf{y}\right\|_2, \tag{8.60}$$

which follows from the limit definition of a derivative (in defining the Hessian of f). This definition is especially helpful to employ when f has only a single continuous derivative.

Suppose f has Lipschitz continuous gradient with constant J, and g is Lipschitz continuous with constant K, i.e.,

$$\left\|g\left(\mathbf{x}\right)-g\left(\mathbf{y}\right)\right\|_2 \leq K\left\|\mathbf{x}-\mathbf{y}\right\|_2 \tag{8.61}$$

for all $\mathbf{x}$, $\mathbf{y}$ in the domain of g. Using this definition of Lipschitz continuous gradient show that the composition $f\left(g\right)$ also has Lipschitz continuous gradient. What is the corresponding Lipschitz constant?

Exercises 8.5 Verifying Lipschitz constants

Let $\mathbf{a}$ and $\mathbf{b}$ be two $N \times 1$ column vectors.

a) Show that the maximum eigenvalue of $\mathbf{a}\mathbf{a}^T$ is less than or equal to the that of $\mathbf{a}\mathbf{a}^T + \mathbf{b}\mathbf{b}^T$.

b) Use the result of part a) to verify the bound reported in Example 8.8 on the maximum eigenvalue of the soft-margin SVM's Hessian.

c) Use the result of part a) to verify the bound on the maximum eigenvalue of the Hessian of the logistic regression function reported in Example 8.7.

9 Dimension reduction techniques

Large datasets, as well as data consisting of a large number of features, present computational problems in the training of predictive models. In this chapter we discuss several useful techniques for reducing the dimension of a given dataset, that is reducing the number of data points or number of features, often employed in order to make predictive learning methods scale to larger datasets. More specifically, we discuss widely used methods for reducing the *data dimension*, that is the number of data points, of a dataset including random subsampling and K-means clustering. We then detail a common way of reducing the *feature dimension*, or number features, of a dataset as explained in Fig. 9.1. A classical approach for feature dimension reduction, principal component analysis (PCA), while often used for general data analysis is a relatively poor tool for reducing the feature dimension of predictive modeling data. However, PCA presents a fundamental mathematical archetype, the *matrix factorization*, that provides a very useful way of organizing our thinking about a wide array of important learning models (including linear regression, K-means, recommender systems – introduced after detailing PCA in this chapter), all of which may be thought of as variations of the simple theme of matrix factorization.

9.1 Techniques for data dimension reduction

In this section we detail two common ways of reducing the data dimension of a dataset: random subsampling and K-means clustering.

9.1.1 Random subsampling

Random subsampling is a simple and intuitive way of reducing the data dimension of a dataset, and is often the first approach employed when performing regression/classification on datasets too large for available computational resources. Given a set of P points we keep a random subsample of $S < P$ of the entire set. Clearly the smaller we choose S the larger the chance we may loose an important structural characteristic of the underlying dataset (for example the geometry of the separating boundary between two classes of data). While there is no formula or hard rule for how large S should be, a simple guideline used in practice is to choose S as large as possible given the computational resources available so as to minimize this risk.

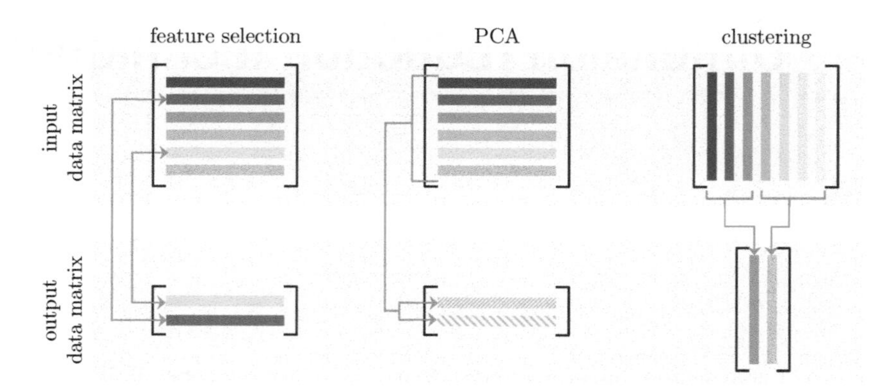

Fig. 9.1 Comparison of feature selection, PCA, and clustering as dimension reduction schemes on an arbitrary data matrix, like those we have discussed in previous chapters for predictive modeling, whose rows contain features and columns individual data points. The former two methods reduce the dimension of the feature space, or in other words the number of rows in a data matrix. However, the two methods work differently: while feature selection literally selects rows from the original matrix to keep, PCA uses the geometry of the feature space to produce a new data matrix based on a lower feature dimensional version of the data. K-means, on the other hand, reduces the dimension of the data/number of data points, or equivalently the number of columns in the input data matrix. It does so by finding a small number of new averaged representatives or "centroids" of the input data, forming a new data matrix whose fewer columns (which are not present in the original data matrix) are precisely these centroids.

9.1.2 K-means clustering

With K-means clustering we reduce the data dimension by finding suitable representatives or *centroids* for clusters, or groups, of data points. All members of each cluster are then represented by their cluster's respective centroid. Hence the problem of clustering is that of partitioning data into clusters of points with similar characteristics, and with K-means specifically this characteristic is geometric closeness in the feature space.[1] Figure 9.2 illustrates K-means clustering performed on a 2-D toy dataset with $P = 10$ data points, where in the right panel data points are clustered into $K = 3$ clusters.

With K-means we look to partition P data points, each of dimension N, into K clusters and find a representative centroid denoted for each cluster. For the moment we will assume that we know the location of these K cluster centroids, as illustrated figuratively in the toy example in the right panel of Fig. 9.2, in order to derive formally the desired relationship between the data and centroids. Once this is expressed clearly we will use it in order to form a learning problem for the accurate recovery of cluster centroids, dropping the unrealistic notion that we have pre-conceived knowledge of their location.

Denoting by $\mathbf{c}_k$ the centroid of the kth cluster and $\mathcal{S}_k$ the set of indices of the subset of those P data points, denoted $\mathbf{x}_1 \ldots \mathbf{x}_P$, belonging to this cluster, the desire that points in the kth cluster should lie close to its centroid may be written mathematically as

[1] Although it is possible to adopt a variety of different ways to define similarity between data points in the feature space (e.g., spectral clustering and subspace clustering), proximity in the Euclidean sense is the most popular measure for clustering data.

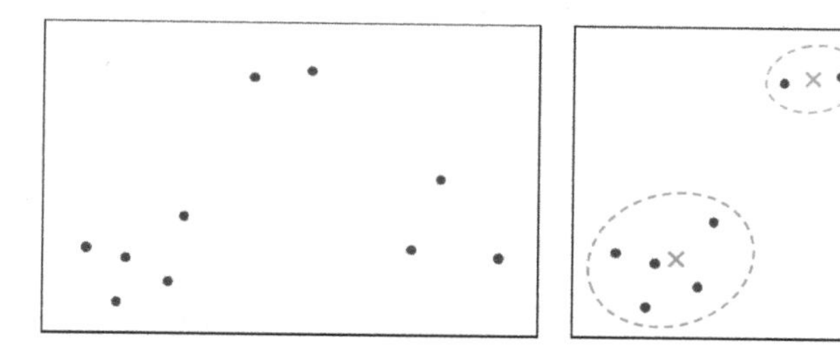

Fig. 9.2 (left) A 2-dimensional toy dataset with $P = 10$ data points. (right) Original data clustered into $K = 3$ clusters where each cluster centroid is marked by a $\times$ symbol. Points that are geometrically close to one another belong to the same cluster.

$$\mathbf{c}_k \approx \mathbf{x}_p \quad \text{for all } p \in \mathcal{S}_k, \tag{9.1}$$

for all $k = 1 \ldots K$. These desired relations can be written more conveniently by first stacking the centroids column-wise into the *centroid matrix* $\mathbf{C} = \begin{bmatrix} \mathbf{c}_1 & \mathbf{c}_2 & \cdots & \mathbf{c}_K \end{bmatrix}$. Then denoting by $\mathbf{e}_k$ the kth standard basis vector (that is a $K \times 1$ vector with a 1 in the kth slot and zeros elsewhere), we may write $\mathbf{C}\mathbf{e}_k = \mathbf{c}_k$, and hence the relations in Equation (9.1) may be written equivalently for each k as

$$\mathbf{C}\mathbf{e}_k \approx \mathbf{x}_p \quad \text{for all } p \in \mathcal{S}_k. \tag{9.2}$$

Next, to write these equations even more conveniently we stack the data column-wise into the *data matrix* $\mathbf{X} = \begin{bmatrix} \mathbf{x}_1 & \mathbf{x}_2 & \cdots & \mathbf{x}_P \end{bmatrix}$ and form a $K \times P$ *assignment matrix* $\mathbf{W}$. The pth column of this matrix, denoted as $\mathbf{w}_p$, is the standard basis vector associated with the cluster to which the pth point belongs: i.e., $\mathbf{w}_p = \mathbf{e}_k$ if $p \in \mathcal{S}_k$. With this $\mathbf{w}_p$ notation we may write each equation in (9.2) as $\mathbf{C}\mathbf{w}_p \approx \mathbf{x}_p$ for all $p \in \mathcal{S}_k$, or using matrix notation all K such relations simultaneously as

$$\mathbf{C}\mathbf{W} \approx \mathbf{X}. \tag{9.3}$$

Figure 9.3 illustrates the compactly written desired K-means relationship in (9.3) for the dataset shown in Fig. 9.2. Note that the location of the only nonzero entry in each column of the assignment matrix $\mathbf{W}$ determines the cluster membership of its corresponding data point in $\mathbf{X}$.

We now drop the assumption that we know the locations of cluster centroids and have knowledge of which points are assigned to them, i.e., the exact description of the centroid matrix $\mathbf{C}$ and assignment matrix $\mathbf{W}$. We want to *learn* the right values for these two matrices. Specifically, we know that the ideal $\mathbf{C}$ and $\mathbf{W}$ satisfy the compact relations described in Equation (9.3), i.e., that $\mathbf{C}\mathbf{W} \approx \mathbf{X}$ or in other words that $\|\mathbf{C}\mathbf{W} - \mathbf{X}\|_F^2$ is small, while $\mathbf{W}$ consists of properly chosen standard basis vectors relating the data points to their respective centroids. Thus we phrase a K-means optimization problem whose solution precisely satisfies these requirements as

$$\begin{aligned} &\underset{\mathbf{C},\mathbf{W}}{\text{minimize}} \;\; \|\mathbf{C}\mathbf{W} - \mathbf{X}\|_F^2 \\ &\text{subject to} \;\; \mathbf{w}_p \in \{\mathbf{e}_k\}_{k=1}^K \;\; p = 1, \ldots, P. \end{aligned} \tag{9.4}$$

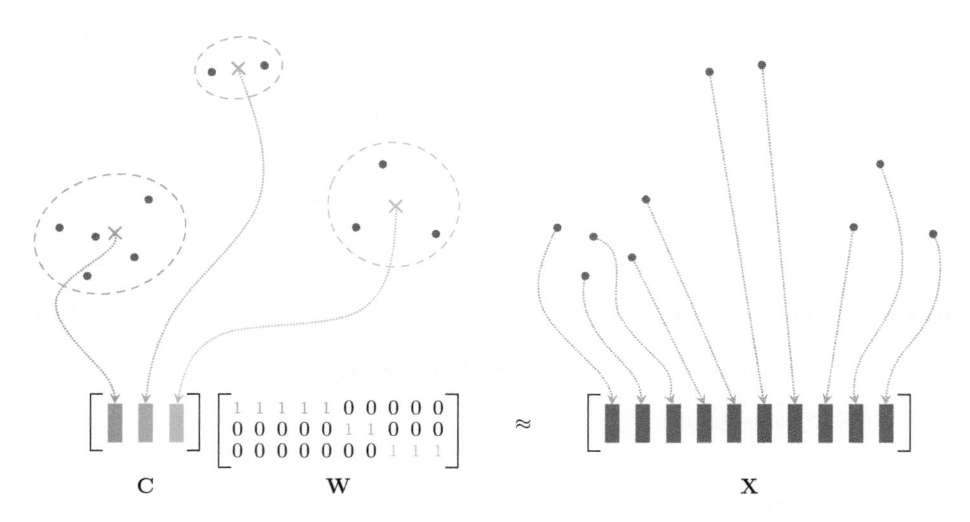

Fig. 9.3 K-means clustering relations described in a compact matrix form. Cluster centroids in $\mathbf{C}$ lie close to their corresponding cluster points in $\mathbf{X}$. The pth column of the assignment matrix $\mathbf{W}$ contains the standard basis vector corresponding to the data point's cluster centroid.

Note that the objective here is non-convex, and because we cannot minimize over both $\mathbf{C}$ and $\mathbf{W}$ simultaneously, it is solved via *alternating minimization*, that is by alternately minimizing the objective function in (9.4) over one of the variables ($\mathbf{C}$ or $\mathbf{W}$), while keeping the other variable fixed. We derive the steps corresponding to this procedure in the following section, and for convenience summarize the resulting simple procedure (often called the *K-means algorithm*) in Algorithm 9.1.

Algorithm 9.1 The K-means algorithm

Input(s): Data matrix $\mathbf{X}$, centroid matrix $\mathbf{C}$ initialized (e.g., randomly), and assignment matrix $\mathbf{W}$ initialized at zero

Output(s): Optimal centroid matrix $\mathbf{C}^{\star}$ and assignment matrix $\mathbf{W}^{\star}$

Repeat until convergence: (e.g., until $\mathbf{C}$ does not change)

(1) Update $\mathbf{W}$ (assign each data point to its closest centroid)

for $p = 1 \ldots P$

Set $\mathbf{w}_p = \mathbf{e}_{k^\star}$ where $k^\star = \underset{k=1\ldots K}{\operatorname{argmin}} \left\|\mathbf{c}_k - \mathbf{x}_p\right\|_2^2$

(2) Update $\mathbf{C}$ (assign each centroid the average of its current points)

for $k = 1 \ldots K$

Denote $\mathcal{S}_k$ the index set of points $\mathbf{x}_p$ currently assigned to the kth cluster

Set $\mathbf{c}_k = \frac{1}{|\mathcal{S}_k|} \sum_{p \in \mathcal{S}_k} \mathbf{x}_p$

Before showing alternating minimization derivation, note that because the objective in (9.4) is non-convex it is possible for the procedure to find non-global minima of the objective function. As with all non-convex problems, this depends on the initializations of our optimization variables (or in this instance just the $\mathbf{C}$ matrix initialization since the procedure begins by updating it independently of $\mathbf{W}$). The result of the algorithm

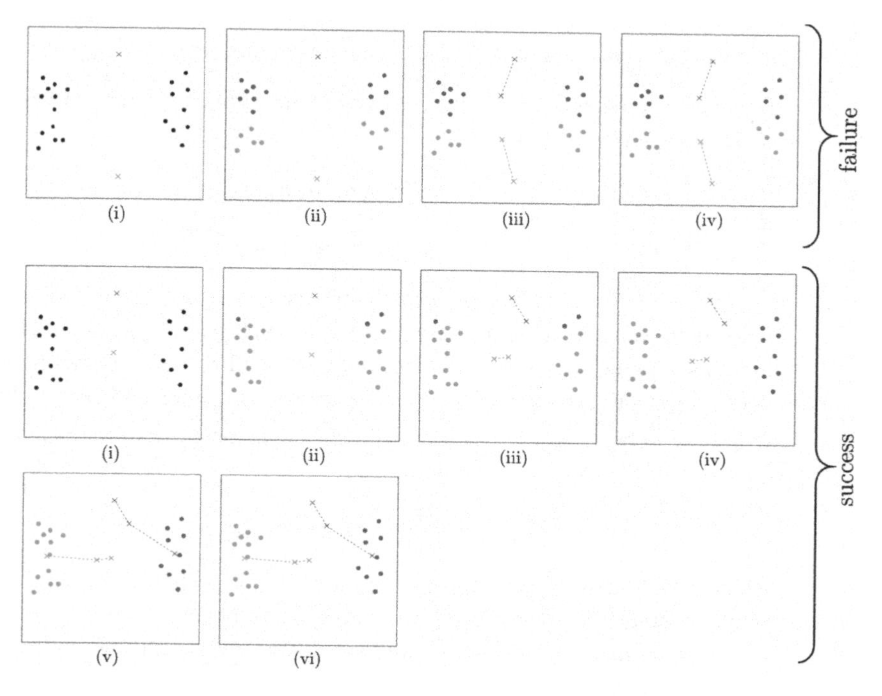

Fig. 9.4 Success or failure of K-means depends on the centroids' initialization. (top) (i) two centroids are initialized, (ii) cluster assignment is updated, (iii) centroid locations are updated, (iv) no change in the cluster assignment of the data points leads to stopping of the algorithm. (below) (i) two centroids are initialized with the red one being initialized differently, (ii) cluster assignment is updated, (iii) centroid locations are updated, (iv) cluster assignment is updated, (v) centroid locations are updated, (vi) no change in the cluster assignment of the data points leads to stopping of the algorithm.

reaching poor minima can have significant impact on the quality of the clusters learned. For example, in Fig. 9.4 we use a 2-D toy dataset with $K = 2$ clusters. With the initial centroid positions shown in the top panel, the K-means algorithm gets stuck in a local minimum and consequently fails to cluster the data properly. A different initialization for one of the centroids, however, leads to a successful clustering of the data, as shown in the lower panel of Fig. 9.4. To overcome the issue of non-convexity of K-means in practice we usually run the algorithm multiple times with different initializations, seeking out the lowest possible minimum of the objective, and the solution resulting in the smallest value of the objective function is selected as the final solution.

9.1.3 Optimization of the K-means problem

Over $\mathbf{W}$ the problem in Equation (9.4) reduces to

$$\begin{aligned} \underset{\mathbf{W}}{\text{minimize}} \quad & \|\mathbf{CW} - \mathbf{X}\|_F^2 \\ \text{subject to} \quad & \mathbf{w}_p \in \{\mathbf{e}_k\}_{k=1}^K \quad p = 1, \ldots, P. \end{aligned} \tag{9.5}$$

Noting that the objective in (9.5) can be equivalently written as $\sum_{p=1}^{P} \left\| \mathbf{C}\mathbf{w}_p - \mathbf{x}_p \right\|_2^2$ (again $\mathbf{C}$ is fixed) and that each $\mathbf{w}_p$ appears in only one summand, we can recover each column $\mathbf{w}_p$ independently by solving

$$\begin{aligned} & \underset{\mathbf{w}_p}{\text{minimize}} \ \left\| \mathbf{C}\mathbf{w}_p - \mathbf{x}_p \right\|_2^2 \\ & \text{subject to } \mathbf{w}_p \in \{\mathbf{e}_k\}_{k=1}^{K}, \end{aligned} \tag{9.6}$$

for each $p = 1, \ldots, P$. Note that this is precisely the problem of assigning a data point $\mathbf{x}_p$ to its closest centroid, i.e., finding k such that $\left\| \mathbf{c}_k - \mathbf{x}_p \right\|_2^2$ is smallest! We can see that this is precisely the problem above by unravelling our compact notation: given the constraint on $\mathbf{w}_p$, we have $\mathbf{C}\mathbf{w}_p = \mathbf{c}_k$ whenever $\mathbf{w}_p = \mathbf{e}_k$ and so the objective may be written as $\left\| \mathbf{C}\mathbf{w}_p - \mathbf{x}_p \right\|_2^2 = \left\| \mathbf{c}_k - \mathbf{x}_p \right\|_2^2$. Hence the problem of finding $\mathbf{w}_p$, or finding the closest centroid to $\mathbf{x}_p$, may be written as

$$\underset{k=1\ldots K}{\text{minimize}} \ \left\| \mathbf{c}_k - \mathbf{x}_p \right\|_2^2, \tag{9.7}$$

which can be solved by simply computing the objective for each k and finding the smallest value. Then for whichever $k^\star$ minimizes the above we may set $\mathbf{w}_p = \mathbf{e}_{k^\star}$.

Now minimizing (9.4) over $\mathbf{C}$, we have no constraints (they being applied only to $\mathbf{W}$) and have the problem

$$\underset{\mathbf{C}}{\text{minimize}} \ \|\mathbf{C}\mathbf{W} - \mathbf{X}\|_F^2. \tag{9.8}$$

Here we may use the first order condition: setting the derivative of $g(\mathbf{C}) = \|\mathbf{C}\mathbf{W} - \mathbf{X}\|_F^2$ to zero gives the linear system

$$\mathbf{C}\mathbf{W}\mathbf{W}^T = \mathbf{X}\mathbf{W}^T, \tag{9.9}$$

for the optimal $\mathbf{C}$ denoted as $\mathbf{C}^\star$. It is easy to show (see exercises) that $\mathbf{W}\mathbf{W}^T$ is a $K \times K$ diagonal matrix whose kth diagonal entry is equal to the number of data points assigned to the kth cluster, and that the kth column of $\mathbf{X}\mathbf{W}^T$ is the sum of all data points in the kth cluster. Hence each column of $\mathbf{C}^\star$ can therefore be calculated independently as

$$\mathbf{c}_k^\star = \frac{1}{|\mathcal{S}_k|} \sum_{p \in \mathcal{S}_k} \mathbf{x}_p \quad \forall k. \tag{9.10}$$

In other words, $\mathbf{c}_k^\star$ is the average of all data points in the kth cluster.

Finally, we repeat these alternating updates until neither matrix changes too much from iteration to iteration. We present the resulting simple column-wise updates for $\mathbf{C}$ and $\mathbf{W}$ in Algorithm 9.1.

9.2 Principal component analysis

As shown abstractly in Fig. 9.1, feature selection is a reasonable and general approach to lowering the dimension of the feature space when working on predictive modeling

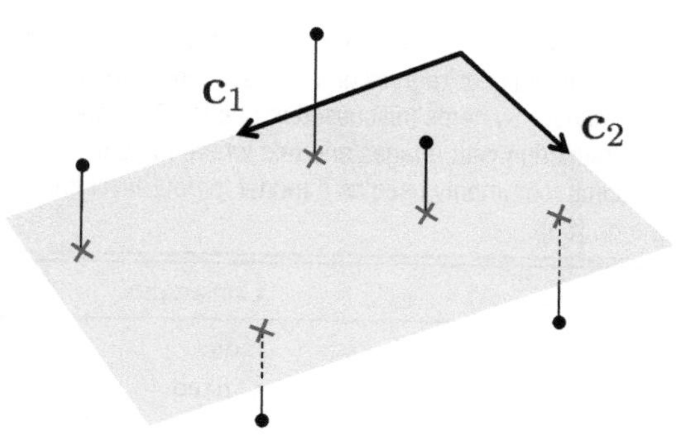

Fig. 9.5 The general PCA dimension reduction scheme visualized in 3D. We begin with three-dimensional data points (black circles) and locate a fitting set of vectors $\mathbf{c}_1$ and $\mathbf{c}_2$ that span a proper lower dimensional subspace for the data. We then project the data onto the subspace (in blue X_s).

problems. For less crucial tasks such as general data exploration where only input values are known (i.e., there is no associated output/labels), other techniques are used to reduce feature space dimension (if it is cumbersomely large). Principal component analysis (PCA), discussed in this section, is one such common technique. PCA works by simply projecting the data onto a suitable lower dimensional feature subspace, that is one which hopefully preserves the essential geometry of the original data. This subspace is found by determining one of its *spanning sets* (e.g., a basis) of vectors which spans it. The basic setup is illustrated in Fig. 9.5.

More formally, suppose that we have P data points $\mathbf{x}_1 \ldots \mathbf{x}_P$, each of dimension N. The goal with PCA is, for some user chosen dimension $K < N$, to find a set of K vectors $\mathbf{c}_1 \ldots \mathbf{c}_K$ that represent the data fairly well. Put formally, we want for each $p = 1 \ldots P$

$$\sum_{k=1}^{K} \mathbf{c}_k w_{k,p} \approx \mathbf{x}_p. \tag{9.11}$$

Note how this is analogous to the motivation for K-means discussed in Section 9.1.2, only here we wish to determine a small set of basis vectors which together explain the dataset. Stacking the desired spanning vectors column-wise into the $N \times K$ matrix $\mathbf{C}$ as $\mathbf{C} = [\mathbf{c}_1|\mathbf{c}_2|\cdots|\mathbf{c}_K]$ and denoting $\mathbf{w}_p = \begin{bmatrix} w_{1,p} & w_{2,p} & \cdots & w_{K,p} \end{bmatrix}^T$ this can be written equivalently for each p as

$$\mathbf{C}\mathbf{w}_p \approx \mathbf{x}_p. \tag{9.12}$$

Note: once $\mathbf{C}$ and $\mathbf{w}_p$ are learned the new K-dimensional feature representation of $\mathbf{x}_p$ is then the vector $\mathbf{w}_p$ (i.e., the weights over which $\mathbf{x}_p$ is represented over the spanning set). By denoting $\mathbf{W} = [\mathbf{w}_1|\mathbf{w}_2|\cdots|\mathbf{w}_P]$ the $K \times P$ matrix of weights to learn, and $\mathbf{X} = [\mathbf{x}_1|\mathbf{x}_2|\cdots|\mathbf{x}_P]$ the $N \times P$ data matrix, all P of these (and equivalently Equation (9.11)) can be written compactly as

$$\mathbf{C}\mathbf{W} \approx \mathbf{X}. \tag{9.13}$$

Table 9.1 Common matrix factorization problems (i.e., variants of PCA) subject to possible constraints on **C** and **W** including: linear regression (discussed in Chapter 3), K-means (discussed in Subsection 9.1.2), Recommender Systems (discussed in Section 9.3), nonnegative matrix factorization (often used for dimension reduction with images and text where the data is naturally nonnegative, see e.g., [45]), and sparse coding (commonly used as a model for low-level image processing in the mammalian visual cortex, see e.g., [60]).

Problem	Constraints
PCA/SVD	None
Linear regression	$\mathbf{C}$ fixed
K-means	$\mathbf{w}_i$ a standard basis vector for all i
Recommender systems	Entries in index set Ω are known i.e., $(\mathbf{CW})_\Omega = (\mathbf{X})_\Omega$
Nonnegative matrix factorization	Both $\mathbf{C}$ and $\mathbf{W}$ nonnegative
Sparse coding	k permissible nonzero entries in each $\mathbf{w}_i$ and columns of $\mathbf{C}$ have unit length

The goal of PCA, compactly stated in Equation (9.13), naturally leads to determining $\mathbf{C}$ and $\mathbf{W}$ by minimizing $\|\mathbf{CW} - \mathbf{X}\|_F^2$ i.e., by solving

$$\underset{\mathbf{C},\mathbf{W}}{\text{minimize}} \ \|\mathbf{CW} - \mathbf{X}\|_F^2 . \tag{9.14}$$

Note that this is a simpler version of the K-means problem in Equation (9.4). Note also the similarities between the PCA matrix factorization problem in Equation (9.14), and the Least Squares cost function for linear regression, K-means, recommender systems, and more: each of these problems may be thought of as variations of the basic *matrix factorization* problem in Equation (9.14) (see Table 9.1). Before discussing optimization procedures for the PCA problem we look at several examples.

Example 9.1 PCA on simulated data

In Fig. 9.6 we show the result of applying PCA on two simple 2-dimensional datasets using the solution to (9.14) shown in Equation (9.17). In the top panel dimension reduction via PCA retains much of the structure of the original data. Conversely, the more structured square dataset loses much of its original characteristic after projection onto the PCA subspace.

Example 9.2 PCA and classification data

While PCA can technically be used for preprocessing data in a predictive modeling scenario, it can cause severe problems in the case of classification. In Fig. 9.7 we illustrate

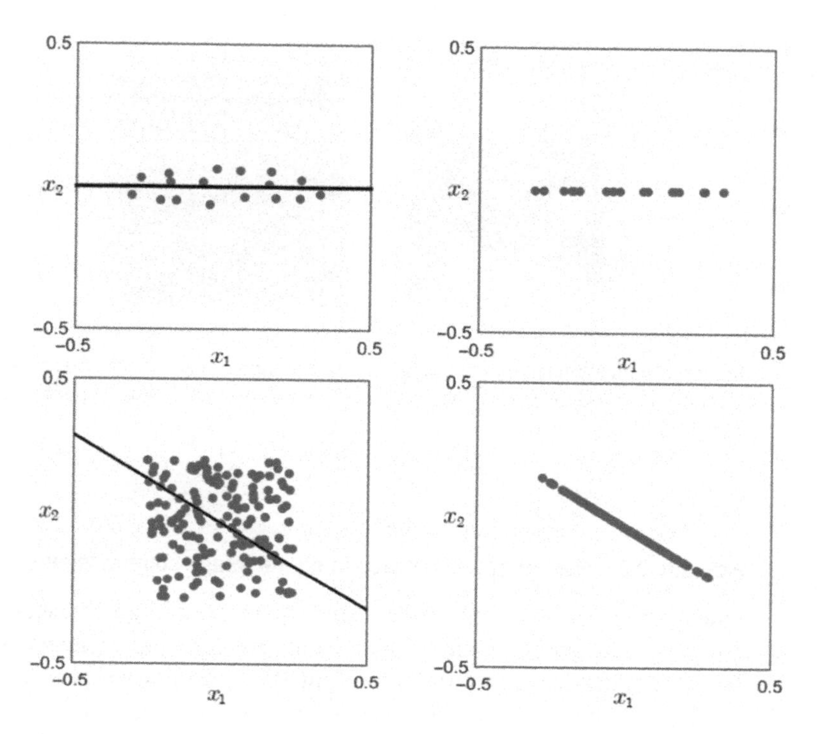

Fig. 9.6 (top panels) A simple 2-D data set (left, in blue) where dimension reduction via PCA retains much of the structure of the original data. The ideal subspace found via solving (9.14) is shown in black in the left panel, and the data projected onto this subspace is shown in blue on the right. (bottom panels) Conversely, the more structured square data-set loses much of its original structure after projection onto the PCA subspace.

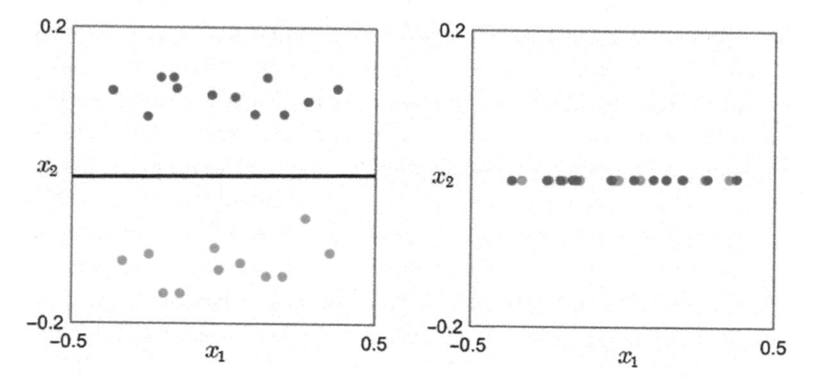

Fig. 9.7 (left) A toy classification dataset consisting of two linearly separable classes. The ideal subspace produced via PCA is shown in black. (right) Projecting the data onto this subspace (in other words reducing the feature space dimension via PCA) destroys completely the original separability of the data.

feature space dimension reduction via PCA on a simulated two-class dataset where the two classes are linearly separable. Because the ideal one-dimensional subspace in this instance runs parallel to the longer length of each class, projecting the complete dataset onto it completely destroys the separability.

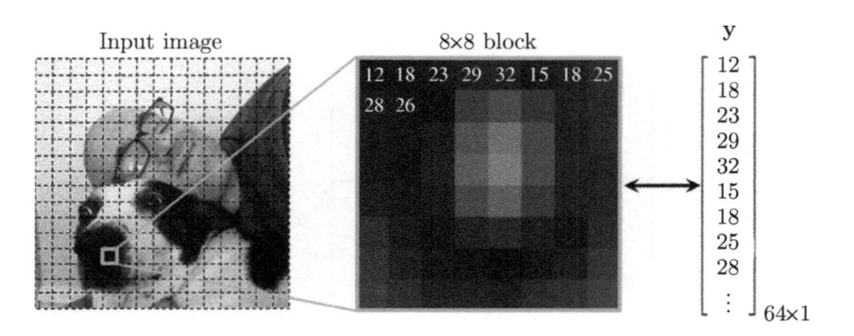

Fig. 9.8 In a prototypical image compression scheme the input image is cut into 8×8 blocks. Each block is then vectorized to make a 64×1 column vector $\mathbf{y}$ which will be input to the compression algorithm.

Example 9.3 Role of efficient bases in digital image compression

Digital image compression aims at reducing the size of digital images without adversely affecting their quality. Without compression a natural photo[2] taken by a digital camera would require one to two orders of magnitude more storage space. As shown in Fig. 9.8, in a typical image compression scheme an input image is first cut up into small square (typically 8×8 pixel) blocks. The values of pixels in each block (which are integers between 0 and 255 for an 8-bit grayscale image) are stacked into a column vector $\mathbf{y}$, and compression is then performed on these individual vectorized blocks.

The primary idea behind many digital image compression algorithms is that with the use of specific bases, we only need very few of their elements to very closely approximate any natural image. One such basis, the 8×8 discrete cosine transform (DCT) which is the backbone of the popular JPEG compression scheme, consists of two-dimensional cosine waves of varying frequencies, and is shown in Fig. 9.9 along with its analogue standard basis. Most natural image blocks can be well approximated using only a few elements of the DCT basis. The reason is, as opposed to bases with more locally defined elements (e.g., the standard basis), each DCT basis element represents a fluctuation commonly seen across the entirety of a natural image block. Therefore with just a few of these elements, properly weighted, we can approximate a wide range of image blocks. In other words, instead of seeking out a basis (as with PCA), here we have a fixed basis over which image data can be very efficiently represented (the same holds, in fact, for other natural signals as well like audio data, see e.g., Section 4.6.3).

To perform compression, DCT basis patches in Fig. 9.9 are vectorized into a sequence of $P = 64$ *fixed* basis vectors $\left\{\mathbf{c}_p\right\}_{p=1}^{P}$ in the same manner as the input image blocks. Concatenating these patches column-wise into a matrix $\mathbf{C}$ and supposing there are K

[2] Pictures of natural subjects such as cities, meadows, people, and animals, etc., as opposed to synthetic images.

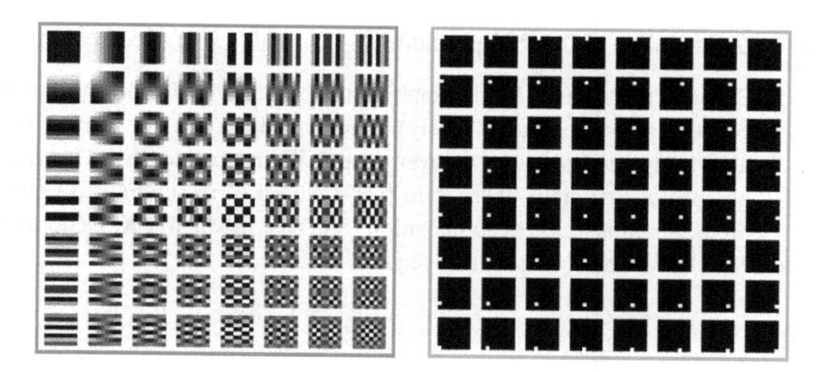

Fig. 9.9 (left) The set of 64 DCT basis elements used for compression of 8×8 image blocks. For visualization purposes pixel values in each basis patch are gray-coded so that white and black colors correspond to the minimum and maximum value in that patch, respectively. (right) The set of 64 standard basis elements, each having only one nonzero entry. Pixel values are gray-coded so that white and black colors correspond to entry values of 1 and 0, respectively. Most natural image blocks can be approximated as a linear combination of just a few DCT basis elements while the same cannot be said of the standard basis.

Fig. 9.10 From left to right, the original 256×256 input image along with its three compressed versions where we keep only the largest 20%, 5%, and 1% of the DCT coefficients to represent the image, resulting in compression by a factor of 5, 20, and 100, respectively. Although, as expected, the visual quality deteriorates as the compression factor increases, the 1% image still captures a considerable amount of information. This example is a testament to the efficiency of DCT basis in representing natural image data.

blocks in the input image, denoting by $\mathbf{x}_k$ its kth vectorized block, to represent the entire image over the basis we solve K linear systems of equations of the form

$$\mathbf{C}\mathbf{w}_k = \mathbf{x}_k. \tag{9.15}$$

Each vector $\mathbf{w}_k$ in (9.15) stores the DCT coefficients (or weights) corresponding to the image block $\mathbf{x}_k$. Most of the weights in the coefficient vectors $\{\mathbf{w}_k\}_{k=1}^{K}$ are typically quite small. Therefore, as illustrated by an example image in Fig. 9.10, setting even 80% of the smallest weights to zero gives an approximation that is essentially indistinguishable from the original image. Even setting 99% of the smallest weights to zero gives an approximation to the original data wherein we can still identify the objects in the original image. To compress the image, instead of storing each pixel value, only these few remaining nonzero coefficients are kept.

9.2.1 Optimization of the PCA problem

Because problem (9.14) is convex in each variable $\mathbf{C}$ and $\mathbf{W}$ individually (but non-convex in both simultaneously) a natural approach to solving this problem is to alternately minimize (9.14) over $\mathbf{C}$ and $\mathbf{W}$ independently.[3] However, while it is not obvious at first sight, there is in fact a closed-form solution to Equation (9.14) based on the singular value decomposition (SVD) of the matrix $\mathbf{X}$. Denoting the SVD of $\mathbf{X}$ as $\mathbf{X} = \mathbf{USV}^T$, this solution is given as

$$\begin{aligned} \mathbf{C}^{\star} &= \mathbf{U}_K \mathbf{S}_{K,K} \\ \mathbf{W}^{\star} &= \mathbf{V}_K^T, \end{aligned} \tag{9.17}$$

where $\mathbf{U}_K$ and $\mathbf{V}_K$ denote the matrices formed by the first K columns of the left and right singular matrices $\mathbf{U}$ and $\mathbf{V}$ respectively, and $\mathbf{S}_{K,K}$ denotes the upper $K \times K$ submatrix of the singular value matrix $\mathbf{S}$. Note that since $\mathbf{U}_K$ is an orthogonal matrix, the recovered basis (for the low dimensional subspace) is indeed orthogonal.

Further, it can be shown that these particular basis elements span the so-called orthogonal directions of variance (or spread) of the original dataset (see Exercise 9.3). While this characteristic is not particularly useful when using PCA as a preprocessing technique (since all we care about is reducing the dimension of the feature space, and so any basis spanning a proper subspace will suffice), it is often used in exploratory data analysis in fields like statistics and the social sciences (see e.g., factor analysis).

9.3 Recommender systems

Recommender systems are heavily used in e-commerce today, providing customers with personalized recommendations for products and services by using a consumer's previous purchasing and rating history, along with those of similar customers. For instance, a movie provider like Netflix with millions of users and tens of thousands of movies, records users' reviews and ratings (typically in the form of a number on a scale of 1–5 with 5 being the most favorable rating) in a large matrix such as the one illustrated in Fig. 9.11. These matrices are very sparsely populated, since an individual consumer has likely rated only a few of the movies available. With this data available, online movie providers can use machine learning techniques to make personalized recommendations

[3] Beginning at an initial value for both $\left(\mathbf{C}^{(0)}, \mathbf{W}^{(0)}\right)$ this produces a sequence of iterates $\left(\mathbf{C}^{(k)}, \mathbf{W}^{(k)}\right)$ where

$$\begin{aligned} \mathbf{C}^{(k)} &= \underset{\mathbf{C}}{\operatorname{argmin}} \left\| \mathbf{C}\mathbf{W}^{(k-1)} - \mathbf{X} \right\|_F^2 \\ \mathbf{W}^{(k)} &= \underset{\mathbf{W}}{\operatorname{argmin}} \left\| \mathbf{C}^{(k)}\mathbf{W} - \mathbf{X} \right\|_F^2, \end{aligned} \tag{9.16}$$

and where each may be expressed in closed form. Setting the gradient in each case to zero and solving gives $\mathbf{C}^{(k)} = \mathbf{X}\left(\mathbf{W}^{(k-1)}\right)^T \left(\mathbf{W}^{(k-1)}\left(\mathbf{W}^{(k-1)}\right)^T\right)^{\dagger}$ and $\mathbf{W}^{(k)} = \left(\left(\mathbf{C}^{(k)}\right)^T \mathbf{C}^{(k)}\right)^{\dagger} \left(\mathbf{C}^{(k)}\right)^T \mathbf{X}$ respectively, where $(\cdot)^{\dagger}$ denotes the pseudo-inverse. The procedure is stopped when the subsequent iterations do not change significantly (see exercises).

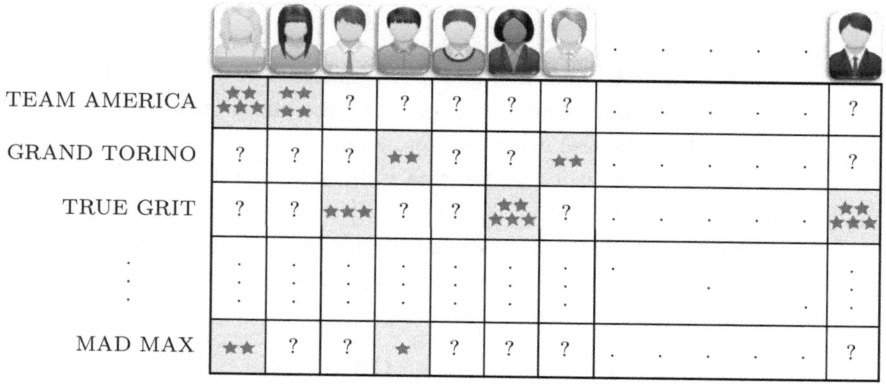

TEAM AMERICA	★★★★★	★★★★	?	?	?	?	?	. . . . . .	?
GRAND TORINO	?	?	?	★★	?	?	★★	. . . . . .	?
TRUE GRIT	?	?	★★★	?	?	★★★★★	?	. . . . . .	★★★★★
⋮	⋮	⋮	⋮	⋮	⋮	⋮	⋮	. . .	⋮
MAD MAX	★★	?	?	★	?	?	?	. . . . . .	?

Fig. 9.11 A prototypical movie rating matrix is very sparsely populated, with each user having rated only a very small number of films. In this diagram movies are listed along rows while users are listed across the columns of the rating matrix.

to customers regarding what they might like to watch next. Producing these personalized recommendations requires *completing* the movie rating matrix, or in other words filling in the many missing entries with smart guesses of how much users would like films they have not yet seen.

In completing the movie rating matrix a helpful modeling tool often used is based on the assumption that only a few factors contribute to a user's taste or interest. For instance, users typically fall into a few categories when it comes to their movie preferences. Some only like horror movies and some are only interested in action films, yet there are those who enjoy both. There are users who passionately follow documentaries, and others who completely despise romantic comedies, and so on. The relatively small number of such categories or user types compared to the total number of users or movies in a rating matrix, provides a framework to fill in the missing values. Once the matrix is completed, those movies with the highest estimated ratings are recommended to the respective users. This same concept is used broadly by digital retailers like Amazon and eBay to recommend products of all kinds to their customers.

9.3.1 Matrix completion setup

In a simple model for a recommender system we can imagine an e-vendor who sells N goods and/or services to P users. The vendor collects and stores individual user-ratings on a 1 (= bad) to R (= great) integer-valued scale for each product purchased by each customer. If a customer has not reviewed a product the entry is set to 0. Looking at the entire customer–product database as an $N \times P$ matrix $\mathbf{X}$, we assume that there are just a few, say K, fundamental factors that each customer's behavior can be explained by.

Imagine for a moment that $\mathbf{X}$ were completely filled in. Then the assumption of K fundamental factors explaining $\mathbf{X}$ translates, in linear algebra terms, to the assumption that $\mathbf{X}$ can be factorized as

$$\mathbf{CW} \approx \mathbf{X}, \tag{9.18}$$

where $\mathbf{C}$ and $\mathbf{W}$ are $N \times K$ and $K \times P$ matrices, respectively. The K columns of $\mathbf{C}$ act as a fundamental spanning set for $\mathbf{X}$ and are interpretable in this instance as the basic factors that represent customers' preferences.

Now in the realistic case where $\mathbf{X}$ is incomplete we know only a fraction of the entries of $\mathbf{X}$. Denoting by $\Omega = \{(i,j) \mid x_{ij} \neq 0\}$ the set of index pairs (i,j) of rated products, where the jth customer has rated the ith product, in the database $\mathbf{X}$ our desire is to recover a factorization for $\mathbf{X}$ given only the known (rated) entries in the set Ω. That is, for each rating with index $(i,j) \in \Omega$,

$$\mathbf{CW}|_{\Omega} \approx \mathbf{X}|_{\Omega}, \tag{9.19}$$

which for a single $(i,j) \in \Omega$ can be written as

$$\mathbf{c}^i \mathbf{w}_j \approx x_{ij}, \tag{9.20}$$

where $\mathbf{c}^i$ is the ith row of $\mathbf{C}$. We then aim to learn $\mathbf{C}$ and $\mathbf{W}$ which minimize $\left(\mathbf{c}^i \mathbf{w}_j - x_{ij}\right)^2$ over all index pairs in Ω, i.e., by solving the optimization problem

$$\underset{\mathbf{C},\mathbf{W}}{\text{minimize}} \sum_{(i,j)\in\Omega} \left(\mathbf{c}^i \mathbf{w}_j - x_{ij}\right)^2. \tag{9.21}$$

9.3.2 Optimization of the matrix completion model

The matrix completion problem in (9.21) can be solved following an alternating minimization approach, in a similar sense as with K-means. The gradient of the objective function $g(\mathbf{C},\mathbf{W}) = \sum_{(i,j)\in\Omega} \left(\mathbf{c}^i \mathbf{w}_j - x_{ij}\right)^2$ with respect to the pth column of $\mathbf{W}$ is given by

$$\nabla_{\mathbf{w}_p} g = 2 \sum_{(i,p)\in\Omega} \left(\mathbf{c}^i \mathbf{w}_p - x_{ip}\right) \left(\mathbf{c}^i\right)^T. \tag{9.22}$$

Setting the gradient to zero, we can recover the optimal $\mathbf{w}_p$ by solving the following linear system of equations:

$$\left(\sum_{(i,p)\in\Omega} \mathbf{c}^i \left(\mathbf{c}^i\right)^T \right) \mathbf{w}_p = \sum_{(i,p)\in\Omega} x_{ip} \left(\mathbf{c}^i\right)^T. \tag{9.23}$$

Similarly, the optimal $\mathbf{c}^n$ can be found as a solution to

$$\mathbf{c}^n \left(\sum_{(n,j)\in\Omega} \mathbf{w}_j \mathbf{w}_j^T \right) = \sum_{(n,j)\in\Omega} x_{nj} \mathbf{w}_j^T. \tag{9.24}$$

By alternately solving these linear system until the values of $\mathbf{C}$ and $\mathbf{W}$ do not change very much we can solve (9.21).

Fig. 9.12 (left panel) Simulated matrix $\mathbf{X}$ with integer-valued entries between 1 and 20. (middle panel) Matrix $\mathbf{X}$ after 95% of entries are removed. (right panel) Recovered matrix $\mathbf{X}^{\star}$ resembles the original matrix $\mathbf{X}$ and matches the original very well.

Example 9.4 Matrix completion of simulated dataset

Here we show an experiment with a simulated data matrix $\mathbf{X}$ with $N = 100$ rows (or products) and $P = 200$ columns (or users). Each entry in $\mathbf{X}$ is an integer-valued rating between 1 and 20, and the complete matrix $\mathbf{X}$ has a rank of $K = 5$, or in other words a matrix rightly approximated using $N \times K$ matrix $\mathbf{C}$ and $K \times P$ matrix $\mathbf{W}$. In the left panel of Fig. 9.12 we show the original matrix $\mathbf{X}$ with each entry color-coded depending on its value. The middle panel shows the matrix $\mathbf{X}$ after 95% of its entries have been randomly removed. Applying matrix completion to this corrupted version of $\mathbf{X}$, by using alternating minimization, we can recover the original matrix (with a small root mean square error of 0.05), as illustrated in the right panel of Fig. 9.12.

9.4 Summary

In this chapter we have described several methods for dimension reduction, beginning in Section 9.1 describing commonly used data dimension reduction techniques (or ways of properly reducing the size of a dataset). Random subsampling, described first, is the most commonly used method for slimming down a regression/classification dataset that simply involves keeping a random selection of points from an original dataset. Here one typically keeps as much of the original data as computational resources permit. A popular alternative to random subsampling, also used for various "data analysis" tasks beyond predictive modeling, is the K-means clustering method. This approach involves the computation of cluster "centroids" of a dataset that best describe its overall structure.

In Section 9.2 we described a classic technique for feature dimension reduction (that is shrinking the dimension of each data point) referred to as principal component analysis (PCA). While not particularly useful for predictive modeling (see Example 9.2 for how it can in fact be destructive when applied to classification data), PCA is the fundamental matrix factorization problem which can help frame our understanding of a wide array of models/problems including: Least Squares for linear regression, K-means, and more. For example in the final section of the chapter we detailed a common matrix factorization approach to recommender systems, or algorithms that recommend products/services to a common base of users.

9.5 Exercises

Section 9.1 exercises

Exercises 9.1 Code up K-means

In this exercise you will reproduce the results shown in Fig. 9.4 by coding up the K-means algorithm (shown in Algorithm 9.1).

a) Place your K-means code in the function

$$[\mathbf{C}, \mathbf{W}] = \text{your\_}K\text{\_means}\left(\mathbf{X}, K\right), \tag{9.25}$$

located inside the wrapper *kmeans_demo* (this wrapper together with the assocated dataset *kmeans_demo_data.csv* may be downloaded from the book website). All of the additional code necessary to generate the associated plots is already provided in the wrapper. Here $\mathbf{C}$ and $\mathbf{W}$ are the centroid and assignment matrices output by the algorithm, while $\mathbf{X}$ and K are the data matrix and number of desired centroids, respectively.

b) Run the wrapper with $K = 2$ centroids using the initialization $\mathbf{C} = \begin{bmatrix} 0 & 0 \\ -0.5 & 0.5 \end{bmatrix}$. This should reproduce the successful run of K-means shown in the bottom panels of the figure.

c) Run the wrapper with $K = 2$ centroids using the initialization $\mathbf{C} = \begin{bmatrix} 0 & 0 \\ 0 & 0.5 \end{bmatrix}$. This should reproduce the unsuccessful run of K-means shown in the top panels of the figure.

Section 9.2 exercises

Exercises 9.2 Code up PCA

In this exercise you will reproduce the results shown in Fig. 9.6 by coding up PCA.

a) Implement a singular value decomposition approach to PCA described in Section 9.2.1, placing the resulting code in the function

$$[\mathbf{C}, \mathbf{W}] = \text{your\_PCA}\left(\mathbf{X}, K\right), \tag{9.26}$$

located inside the wrapper *PCA_demo* (this wrapper together with the assocated dataset *PCA_demo_data.csv* may be downloaded from the book website). Here $\mathbf{C}$ and $\mathbf{W}$ are the spanning set and weight matrices output by the algorithm, while $\mathbf{X}$ and K are the data matrix and number of desired basis elements, respectively.

All of the additional code necessary to generate the associated plots is already provided in the wrapper. Run the wrapper to ensure that you have coded up PCA correctly.

b) Using the wrapper and data from part a) implement the alternating directions solution to PCA described in footnote 3, once again placing this code inside the function *your_PCA* described previously.

Exercises 9.3 Deriving principal components as orthogonal directions of variance

In this exercise you will show how to derive principal component analysis as the orthogonal directions of largest variance of a dataset. Given P points $\left\{\mathbf{x}_p\right\}_{p=1}^{P}$ of dimension N we may calculate the variance in a unit direction $\mathbf{d}$ (i.e., how much the dataset spreads out in the direction $\mathbf{d}$) with respect to the data as the average squared inner product of the data against $\mathbf{d}$,

$$\frac{1}{P}\sum_{p=1}^{P}\left\langle \mathbf{x}_p, \mathbf{d}\right\rangle^2. \tag{9.27}$$

This can be written more compactly as

$$\frac{1}{P}\|\mathbf{X}^T\mathbf{d}\|_2^2 = \frac{1}{P}\mathbf{d}^T\mathbf{X}\mathbf{X}^T\mathbf{d}. \tag{9.28}$$

Note that the outer product $\mathbf{X}\mathbf{X}^T$ is a symmetric positive semi-definite matrix.

a) Compute the largest direction of variance of the data, i.e., the unit vector $\mathbf{d}$ that maximizes the value $\mathbf{d}^T\mathbf{X}\mathbf{X}^T\mathbf{d}$. *Hint: use the eigen-decomposition of XX^T.*

b) Compute the second largest direction of variance of the matrix $\mathbf{X}\mathbf{X}^T$, i.e., the unit vector $\mathbf{d}$ that maximizes the value of $\mathbf{d}^T\mathbf{X}\mathbf{X}^T\mathbf{d}$ but where $\mathbf{d}$ is also orthogonal to the first largest direction of variance. *Hint: use the eigen-decomposition of XX^T.*

c) Conclude from part a) and b) that the orthogonal directions of variance of the data are precisely the singular value solution given in Equation (9.17).

Section 9.3 exercises

Exercises 9.4 Code up the matrix completion recommender system

In this exercise you will reproduce the matrix completion recommender system results shown in Fig. 9.12.

Code up the alternating minimization algorithm described in Section 9.3.2, placing the resulting code in the function

$$[\mathbf{C},\ \mathbf{W}] = \text{matrix\_complete}\left(\mathbf{X},\ K\right), \tag{9.29}$$

located inside the wrapper *recommender_demo* (this wrapper and the assocated dataset *recommender_demo_data.csv* may be downloaded from the book website). Here $\mathbf{C}$ and

$\mathbf{W}$ are the spanning set and weight matrices output by the algorithm, while $\mathbf{X}$ and K are the data matrix and number of desired basis elements, respectively.

All of the additional code necessary to generate the associated plots is already provided in the wrapper. Run the wrapper to ensure that you have coded up the matrix completion algorithm correctly.

Part IV

Appendices

A: Basic vector and matrix operations

A.1 Vector operations

Vector addition: The addition of two N-dimensional vectors

$$\mathbf{a}_{N\times 1} = \begin{bmatrix} a_1 \\ a_2 \\ \vdots \\ a_N \end{bmatrix} \quad \text{and} \quad \mathbf{b}_{N\times 1} = \begin{bmatrix} b_1 \\ b_2 \\ \vdots \\ b_N \end{bmatrix}, \tag{A.1}$$

is defined as the entry-wise sum of $\mathbf{a}$ and $\mathbf{b}$, resulting in a vector of the same dimension denoted by

$$\mathbf{a} + \mathbf{b} = \begin{bmatrix} a_1 + b_1 \\ a_2 + b_2 \\ \vdots \\ a_N + b_N \end{bmatrix}. \tag{A.2}$$

Subtraction of two vectors is defined in a similar fashion.

Vector multiplication by a scalar: Multiplying a vector $\mathbf{a}$ by a scalar γ returns a vector of the same dimension whose every element is multiplied by γ

$$\gamma\mathbf{a} = \begin{bmatrix} \gamma a_1 \\ \gamma a_2 \\ \vdots \\ \gamma a_N \end{bmatrix}. \tag{A.3}$$

Vector transpose: The transpose of a *column vector* $\mathbf{a}$ (with vertically stored elements) is a *row vector* with the same elements which are now stored horizontally, denoted by

$$\mathbf{a}^T = \begin{bmatrix} a_1 & a_2 & \cdots & a_N \end{bmatrix}. \tag{A.4}$$

Similarly the transpose of a row vector is a column vector, and we have in general that $\left(\mathbf{a}^T\right)^T = \mathbf{a}$.

Inner product of two vectors: The inner product (or dot product) of two vectors **a** and **b** (of the same dimensions) is simply the sum of their component-wise product, written as

$$\mathbf{a}^T\mathbf{b} = \sum_{n=1}^{N} a_n b_n. \tag{A.5}$$

The inner product of **a** and **b** is also often written as $\langle \mathbf{a}, \mathbf{b} \rangle$.

Inner product rule and correlation: The inner product rule between two vectors provides a measurement of

$$\mathbf{a}^T\mathbf{b} = \|\mathbf{a}\|_2 \, \|\mathbf{b}\|_2 \cos(\theta), \tag{A.6}$$

where θ is the angle between the vectors **a** and **b**. Dividing both vectors by their length gives the *correlation* between the two vectors

$$\frac{\mathbf{a}^T\mathbf{b}}{\|\mathbf{a}\|_2 \, \|\mathbf{b}\|_2} = \cos(\theta), \tag{A.7}$$

which ranges between -1 and 1 (when the vectors point in completely opposite or parallel directions respectively).

Outer product of two vectors: The outer product of two vectors $\mathbf{a}_{N\times 1}$ and $\mathbf{b}_{M\times 1}$ is an $N \times M$ matrix **C** defined as

$$\mathbf{C} = \begin{bmatrix} a_1b_1 & a_1b_2 & \cdots & a_1b_M \\ a_2b_1 & a_2b_2 & \cdots & a_2b_M \\ \vdots & \vdots & \ddots & \vdots \\ a_Nb_1 & a_Nb_2 & \cdots & a_Nb_M \end{bmatrix}. \tag{A.8}$$

The outer product of **a** and **b** is also written as $\mathbf{ab}^T$. Unlike the inner product, the outer product does not hold the commutative property meaning that the outer product of **a** and **b** does *not* necessarily equal the outer product of **b** and **a**.

A.2 Matrix operations

Matrix addition: The addition of two $N \times M$ matrices

$$\mathbf{A}_{N\times M} = \begin{bmatrix} a_{1,1} & a_{1,2} & \cdots & a_{1,M} \\ a_{2,1} & a_{2,2} & \cdots & a_{2.M} \\ \vdots & \vdots & \ddots & \vdots \\ a_{N,1} & a_{N,2} & \cdots & a_{N,M} \end{bmatrix} \text{ and } \mathbf{B}_{N\times M} = \begin{bmatrix} b_{1,1} & b_{1,2} & \cdots & b_{1,M} \\ b_{2,1} & b_{2,2} & \cdots & b_{2.M} \\ \vdots & \vdots & \ddots & \vdots \\ b_{N,1} & b_{N,2} & \cdots & b_{N,M} \end{bmatrix}, \tag{A.9}$$

is defined again as the entry-wise sum of **A** and **B**, given as

$$\mathbf{A} + \mathbf{B} = \begin{bmatrix} a_{1,1} + b_{1,1} & a_{1,2} + b_{1,2} & \cdots & a_{1,M} + b_{1,M} \\ a_{2,1} + b_{2,1} & a_{2,2} + b_{2,2} & \cdots & a_{2.M} + b_{2,M} \\ \vdots & \vdots & \ddots & \vdots \\ a_{N,1} + b_{N,1} & a_{N,2} + b_{N,2} & \cdots & a_{N,M} + b_{N,M} \end{bmatrix}. \tag{A.10}$$

Subtraction of two matrices is defined in a similar fashion.

Matrix multiplication by a scalar: Multiplying a matrix $\mathbf{A}$ by a scalar γ returns a matrix of the same dimension whose every element is multiplied by γ

$$\gamma\mathbf{A} = \begin{bmatrix} \gamma a_{1,1} & \gamma a_{1,2} & \cdots & \gamma a_{1,M} \\ \gamma a_{2,1} & \gamma a_{2,2} & \cdots & \gamma a_{2.M} \\ \vdots & \vdots & \ddots & \vdots \\ \gamma a_{N,1} & \gamma a_{N,2} & \cdots & \gamma a_{N,M} \end{bmatrix}. \tag{A.11}$$

Matrix transpose: The transpose of an $N \times M$ matrix $\mathbf{A}$ is formed by putting the transpose of each column of $\mathbf{A}$ into the corresponding row of $\mathbf{A}^T$, giving the $M \times N$ transpose matrix as

$$\mathbf{A}^T = \begin{bmatrix} a_{1,1} & a_{2,1} & \cdots & a_{N,1} \\ a_{1,2} & a_{2,2} & \cdots & a_{N,2} \\ \vdots & \vdots & \ddots & \vdots \\ a_{1,M} & a_{2,M} & \cdots & a_{M,N} \end{bmatrix}. \tag{A.12}$$

Again, we have $\left(\mathbf{A}^T\right)^T = \mathbf{A}$.

Matrix multiplication: The product of two matrices $\mathbf{A}_{N\times M}$ and $\mathbf{B}_{M\times P}$ is an $N \times P$ matrix defined via the sum of M outer product matrices, as

$$\mathbf{C} = \mathbf{AB} = \sum_{m=1}^{M} \mathbf{a}_m \mathbf{b}^m, \tag{A.13}$$

where $\mathbf{a}_m$ and $\mathbf{b}^m$ respectively denote the mth column of $\mathbf{A}$ and the mth row of $\mathbf{B}$.

The pth column of $\mathbf{C}$ can be found via multiplying $\mathbf{A}$ by the pth column of $\mathbf{B}$,

$$\mathbf{c}_p = \mathbf{Ab}_p = \sum_{m=1}^{M} \mathbf{a}_m b_{m,p}. \tag{A.14}$$

The nth row of $\mathbf{C}$ can be found via multiplying the nth row of $\mathbf{A}$ by $\mathbf{B}$,

$$\mathbf{c}^n = \mathbf{a}^n\mathbf{B} = \sum_{m=1}^{M} a_{n,m} \mathbf{b}^m. \tag{A.15}$$

The (n, p) th entry of $\mathbf{C}$ is found by multiplying the nth row of $\mathbf{A}$ by the pth column of $\mathbf{B}$,

$$c_{n,p} = \mathbf{a}^n \mathbf{b}_p. \tag{A.16}$$

Note that vector inner and outer products are special cases of matrix multiplication.

Entry-wise product: The entry-wise product (or Hadamard product) of two matrices $\mathbf{A}_{N\times M}$ and $\mathbf{B}_{N\times M}$ is defined as

$$\mathbf{A} \circ \mathbf{B} = \begin{bmatrix} a_{1,1}b_{1,1} & a_{1,2}b_{1,2} & \cdots & a_{1,M}b_{1,M} \\ a_{2,1}b_{2,1} & a_{2,2}b_{2,2} & \cdots & a_{2.M}b_{2,M} \\ \vdots & \vdots & \ddots & \vdots \\ a_{N,1}b_{N,1} & a_{N,2}b_{N,2} & \cdots & a_{N,M}b_{N,M} \end{bmatrix}. \tag{A.17}$$

In other words, the (n, m) th entry of $\mathbf{A} \circ \mathbf{B}$ is simply the product of the (n, m) th entry of $\mathbf{A}$ and the (n, m) th entry of $\mathbf{B}$.

B: Basics of vector calculus

B.1 Basic definitions

Throughout this section suppose that $g(\mathbf{w})$ is a scalar valued function of the $N \times 1$ vector $\mathbf{w} = \begin{bmatrix} w_1 & w_2 & \cdots & w_N \end{bmatrix}^T$.

A *partial derivative* is the derivative of a multivariable function with respect to one of its variables. For instance, the partial derivative of g with respect to w_i is written as

$$\frac{\partial}{\partial w_i} g(\mathbf{w}). \tag{B.1}$$

The *gradient* of g is then the vector of all partial derivatives denoted as

$$\nabla g(\mathbf{w}) = \begin{bmatrix} \frac{\partial}{\partial w_1} g(\mathbf{w}) \\ \frac{\partial}{\partial w_2} g(\mathbf{w}) \\ \vdots \\ \frac{\partial}{\partial w_N} g(\mathbf{w}) \end{bmatrix}. \tag{B.2}$$

For example, the gradient for the linear function $g_1(\mathbf{w}) = \mathbf{w}^T \mathbf{b}$ and quadratic function $g_2(\mathbf{w}) = \mathbf{w}^T \mathbf{A} \mathbf{w}$ can be computed as $\nabla g_1(\mathbf{w}) = \mathbf{b}$ and $\nabla g_2(\mathbf{w}) = \left(\mathbf{A} + \mathbf{A}^T\right) \mathbf{w}$, respectively.

The *second order partial derivative* of g with respect to variables w_i and w_j is written as

$$\frac{\partial^2}{\partial w_i \partial w_j} g(\mathbf{w}), \tag{B.3}$$

or equivalently as

$$\frac{\partial^2}{\partial w_j \partial w_i} g(\mathbf{w}). \tag{B.4}$$

The *Hessian* of g is then the square symmetric matrix of all second order partial derivatives of g, denoted as

$$\nabla^2 g(\mathbf{w}) = \begin{bmatrix} \frac{\partial^2}{\partial w_1 \partial w_1} g(\mathbf{w}) & \frac{\partial^2}{\partial w_1 \partial w_2} g(\mathbf{w}) & \cdots & \frac{\partial^2}{\partial w_1 \partial w_N} g(\mathbf{w}) \\ \frac{\partial^2}{\partial w_2 \partial w_1} g(\mathbf{w}) & \frac{\partial^2}{\partial w_2 \partial w_2} g(\mathbf{w}) & \cdots & \frac{\partial^2}{\partial w_2 \partial w_N} g(\mathbf{w}) \\ \vdots & \vdots & \ddots & \vdots \\ \frac{\partial^2}{\partial w_N \partial w_1} g(\mathbf{w}) & \frac{\partial^2}{\partial w_N \partial w_2} g(\mathbf{w}) & \cdots & \frac{\partial^2}{\partial w_N \partial w_N} g(\mathbf{w}) \end{bmatrix}. \tag{B.5}$$

B.2 Commonly used rules for computing derivatives

Here we give five rules commonly used when making gradient and Hessian calculations.

1. The derivative of a sum is the sum of derivatives
If $g(w)$ is a sum of P functions $g(w) = \sum_{p=1}^{P} h_p(w)$, then $\frac{d}{dw}g(w) = \sum_{p=1}^{P} \frac{d}{dw}h_p(w)$.

2. The chain rule
If g is a composition of functions of the form $g(w) = h(r(w))$, then the derivative $\frac{d}{dw}g(w) = \frac{d}{dt}h(t)\,\frac{d}{dw}r(w)$ where t is evaluated at $t = r(w)$.

3. The product rule
If g is a product of functions of the form $g(w) = h(w)\,r(w)$, then the derivative $\frac{d}{dw}g(w) = \left(\frac{d}{dw}h(w)\right)r(w) + h(w)\left(\frac{d}{dw}r(w)\right)$.

4. Various derivative formulae
For example, if $g(w) = w^c$ then $\frac{d}{dw}g(w) = cw^{c-1}$, or if $g(w) = \sin(w)$ then $\frac{d}{dw}g(w) = \cos(w)$, or if $g(w) = c$ where c is some constant then $\frac{d}{dw}g(w) = 0$, etc.

5. The sizes/shapes of gradients and Hessians
Remember: if $\mathbf{w}$ is an $N \times 1$ column vector then $\nabla g(\mathbf{w})$ is also an $N \times 1$ column vector, and $\nabla^2 g(\mathbf{w})$ is an $N \times N$ symmetric matrix.

B.3 Examples of gradient and Hessian calculations

Here we show detailed calculations for the gradient and Hessian of various functions employing the definitions and common rules previously stated. We begin by showing several first and second derivative calculations for scalar input functions, and then show the gradient/Hessian calculations for the analogous vector input versions of these functions.

Example B.1 Practice derivative calculations: scalar input functions

Below we compute the first and second derivatives of three scalar input functions $g(w)$ where w is a scalar input. Note that throughout we will use two notations for a scalar derivative, $\frac{d}{dw}g(w)$ and $g'(w)$ interchangeably.

a) $g(w) = \frac{1}{2}qw^2 + rw + d$ where q, r, and d are constants
Using derivative formulae and the fact that the derivative of a sum is the sum of derivatives, we have

$$g'(w) = qw + r \tag{B.6}$$

and

$$g''(w) = q. \tag{B.7}$$

b) $g(w) = -\cos\left(2\pi w^2\right) + w^2$
Using the chain rule on the $\cos(\cdot)$ part, the fact that the derivative of a sum is the sum of derivatives, and derivative formulae, we have

$$g'(w) = \sin\left(2\pi w^2\right) 4\pi w + 2w. \tag{B.8}$$

Likewise taking the second derivative we differentiate the above (additionally using the product rules) as

$$g''(w) = \cos\left(2\pi w^2\right)(4\pi w)^2 + \sin\left(2\pi w^2\right) 4\pi + 2. \tag{B.9}$$

c) $g(w) = \sum_{p=1}^{P} \log\left(1 + e^{-a_p w}\right)$ where $a_1 \ldots a_P$ are constants
Call the pth summand $h_p(w) = \log\left(1 + e^{-a_p w}\right)$. Then using the chain rule since $\frac{d}{dt}\log(t) = \frac{1}{t}$ and $\frac{d}{dw}\left(1 + e^{-a_p w}\right) = -a_p e^{-a_p w}$ together, we have $\frac{d}{dw}h_p(w) = \frac{1}{1+e^{-a_p w}}\left(-a_p e^{-a_p w}\right) = -\frac{a_p e^{-a_p w}}{1+e^{-a_p w}} = -\frac{a_p}{e^{a_p w}+e^{a_p w}e^{-a_p w}} = -\frac{a_p}{1+e^{a_p w}}$. Now using this result, and since the derivative of a sum is the sum of the derivatives, and $g(w) = \sum_{p=1}^{P} h_p(w)$, we have $\frac{d}{dw}g(w) = \sum_{p=1}^{P} \frac{d}{dw}h_p(w)$ and so

$$g'(w) = -\sum_{p=1}^{P} \frac{a_p}{1 + e^{a_p w}}. \tag{B.10}$$

To compute the second derivative let us again do so by first differentiating the above summand-by-summand. Denote the pth summand above as $h_p(w) = \frac{a_p}{1+e^{a_p w}}$. To compute its derivative we must apply the product and chain rules once again, we have $h_p'(w) = -\frac{a_p}{\left(1+e^{a_p w}\right)^2} a_p e^{a_p w} = -\frac{e^{a_p w}}{\left(1+e^{a_p w}\right)^2} a_p^2$. We can then compute the full second derivative as $g''(w) = -\sum_{p=1}^{P} h_p'(w)$, or likewise

$$g''(w) = \sum_{p=1}^{P} \frac{e^{a_p w}}{\left(1 + e^{a_p w}\right)^2} a_p^2. \tag{B.11}$$

Example B.2 Practice derivative calculations: vector input functions

Below we compute the gradients and Hessians of three vector input functions $g(\mathbf{w})$ where $\mathbf{w}$ is an $N \times 1$ dimensional input vector. The functions discussed here are analogous to the scalar functions discussed in the first example.

a) $g(\mathbf{w}) = \frac{1}{2}\mathbf{w}^T\mathbf{Q}\mathbf{w} + \mathbf{r}^T\mathbf{w} + d$, here $\mathbf{Q}$ is an $N \times N$ symmetric matrix, $\mathbf{r}$ is an $N \times 1$ vector, and d is a scalar.

Note that $g(\mathbf{w})$ here is the vector version of the function shown in a) of the first example. We should therefore expect the final shape of the gradient and Hessian to generally match the first and second derivatives we found there.

Writing out g in terms of the individual entries of $\mathbf{w}$ we have $g(\mathbf{w}) = \frac{1}{2}\sum_{n=1}^{N}\sum_{m=1}^{N} w_n Q_{nm} w_m + \sum_{n=1}^{N} r_n w_n + d$, then taking the jth partial derivative we have, since the derivative of a sum is the sum of derivatives, $\frac{\partial}{\partial w_j} g(\mathbf{w}) = \frac{1}{2}\sum_{n=1}^{N}\sum_{m=1}^{N} \frac{\partial}{\partial w_j}(w_n Q_{nm} w_m) + \sum_{n=1}^{N} \frac{\partial}{\partial w_j}(r_n w_n)$, where d vanishes since it is a constant and $\frac{\partial}{\partial w_j} d = 0$. Now evaluating each derivative we apply the product rule to each $w_n Q_{nm} w_m$ (and remembering that all other terms in w_k where $k \neq j$ are constant and thus vanish when taking the w_j partial derivative), and we have

$$\frac{\partial}{\partial w_j} g(\mathbf{w}) = \frac{1}{2}\left(\sum_{n=1}^{N} w_n Q_{nj} + \sum_{m=1}^{N} Q_{jm} w_m\right) + r_j. \tag{B.12}$$

All together the gradient can then be written compactly as

$$\nabla g(\mathbf{w}) = \frac{1}{2}\left(\mathbf{Q} + \mathbf{Q}^T\right)\mathbf{w} + \mathbf{r}, \tag{B.13}$$

and because $\mathbf{Q}$ is symmetric this is equivalently

$$\nabla g(\mathbf{w}) = \mathbf{Q}\mathbf{w} + \mathbf{r}. \tag{B.14}$$

Note how the gradient here takes precisely the same shape as the corresponding scalar derivative shown in (B.6).

To compute the Hessian we compute mixed partial derivatives of the form $\frac{\partial^2}{\partial w_i \partial w_j} g(\mathbf{w})$. To do this efficiently we can take the partial $\frac{\partial}{\partial w_i}$ of Equation (B.12), since $\frac{\partial^2}{\partial w_i \partial w_j} g(\mathbf{w}) = \frac{\partial}{\partial w_i}\left(\frac{\partial}{\partial w_j} g(\mathbf{w})\right)$, which gives

$$\frac{\partial^2}{\partial w_i \partial w_j} g(\mathbf{w}) = \frac{1}{2}\left(Q_{ij} + Q_{ji}\right). \tag{B.15}$$

All together we then have that the full Hessian matrix is

$$\nabla^2 g(\mathbf{w}) = \frac{1}{2}\left(\mathbf{Q} + \mathbf{Q}^T\right), \tag{B.16}$$

and because $\mathbf{Q}$ is symmetric this is equivalently

$$\nabla^2 g(\mathbf{w}) = \mathbf{Q}. \tag{B.17}$$

Note how this is exactly the vector form of the second derivative given in (B.7).

b) $g(\mathbf{w}) = -\cos\left(2\pi\mathbf{w}^T\mathbf{w}\right) + \mathbf{w}^T\mathbf{w}$

First, note that this is the vector input version of b) from the first example, therefore we should expect the final shape of the gradient and Hessian to generally match the first and second derivatives we found there.

Writing out g in terms of individual entries of $\mathbf{w}$ we have $g(\mathbf{w}) = -\cos\left(2\pi\sum_{n=1}^{N} w_n^2\right) + \sum_{n=1}^{N} w_n^2$, now taking the jth partial we have

$$\frac{\partial}{\partial w_j} g(\mathbf{w}) = \sin\left(2\pi\sum_{n=1}^{N} w_n^2\right) 4\pi w_j + 2w_j. \tag{B.18}$$

From this we can see that the full gradient then takes the form

$$\nabla g(\mathbf{w}) = \sin\left(2\pi\mathbf{w}^T\mathbf{w}\right) 4\pi\mathbf{w} + 2\mathbf{w}. \tag{B.19}$$

This is precisely the analog of the first derivative of the scalar version of this function shown in Equation (B.8) of the previous example.

To compute the second derivatives we can take the partial $\frac{\partial}{\partial w_i}$ of Equation (B.18), which gives

$$\frac{\partial^2}{\partial w_i \partial w_j} g(\mathbf{w}) = \begin{cases} \cos\left(2\pi\sum_{n=1}^{N} w_n^2\right)(4\pi)^2 w_i w_j + \sin\left(2\pi\sum_{n=1}^{N} w_n^2\right) 4\pi + 2 & \text{if } i = j \\ \cos\left(2\pi\sum_{n=1}^{N} w_n^2\right)(4\pi)^2 w_i w_j & \text{else.} \end{cases} \tag{B.20}$$

All together then, denoting $\mathbf{I}_{N\times N}$ the $N \times N$ identity matrix, we may write the Hessian as

$$\nabla^2 g(\mathbf{w}) = \cos\left(2\pi\mathbf{w}^T\mathbf{w}\right)(4\pi)^2\mathbf{w}\mathbf{w}^T + \left(2 + \sin\left(2\pi\mathbf{w}^T\mathbf{w}\right)4\pi\right)\mathbf{I}_{N\times N}. \tag{B.21}$$

Note that this is analogous to the second derivative, shown in Equation (B.9), of the scalar version of the function.

c) $g(\mathbf{w}) = \sum_{p=1}^{P} \log\left(1 + e^{-\mathbf{a}_p^T\mathbf{w}}\right)$ where $\mathbf{a}_1 \ldots \mathbf{a}_P$ are $N \times 1$ vectors

This is the vector-input version of c) from the first example, so we should expect similar patterns to emerge when computing derivatives here.

Denote by $h_p(\mathbf{w}) = \log\left(1 + e^{-\mathbf{a}_p^T\mathbf{w}}\right) = \log\left(1 + e^{-\sum_{n=1}^{N} a_{pn}w_n}\right)$ one of the summands of g. Then using the chain rule, twice the jth partial can be written as

$$\frac{\partial}{\partial w_j} h_p(\mathbf{w}) = \frac{1}{1 + e^{-\mathbf{a}_p^T\mathbf{w}}} e^{-\mathbf{a}_p^T\mathbf{w}}\left(-a_{pj}\right). \tag{B.22}$$

Since $\frac{1}{1+e^{-\mathbf{a}_p^T\mathbf{w}}} e^{-\mathbf{a}_p^T\mathbf{w}} = \frac{1}{e^{\mathbf{a}_p^T\mathbf{w}} + e^{\mathbf{a}_p^T\mathbf{w}} e^{-\mathbf{a}_p^T\mathbf{w}}} = \frac{1}{1+e^{\mathbf{a}_p^T\mathbf{w}}}$, we can rewrite the above more compactly as

$$\frac{\partial}{\partial w_j} h_p(\mathbf{w}) = -\frac{a_{pj}}{1 + e^{\mathbf{a}_p^T\mathbf{w}}} \tag{B.23}$$

and summing over p gives

$$\frac{\partial}{\partial w_j} g(\mathbf{w}) = -\sum_{p=1}^{P} \frac{a_{pj}}{1 + e^{\mathbf{a}_p^T \mathbf{w}}}. \tag{B.24}$$

The full gradient of g is then given by

$$\nabla g(\mathbf{w}) = -\sum_{p=1}^{P} \frac{\mathbf{a}_p}{1 + e^{\mathbf{a}_p^T \mathbf{w}}}. \tag{B.25}$$

Note the similar shape of this gradient compared to the derivative of the scalar form of the function, as shown in Equation (B.10).

Computing the second partial derivatives from equation (B.24), we have

$$\frac{\partial^2}{\partial w_i \partial w_j} g(\mathbf{w}) = \sum_{p=1}^{P} \frac{e^{\mathbf{a}_p^T \mathbf{w}}}{\left(1 + e^{\mathbf{a}_p^T \mathbf{w}}\right)^2} a_{pi} a_{pj}, \tag{B.26}$$

and so we may write the full Hessian compactly as

$$\nabla^2 g(\mathbf{w}) = \sum_{p=1}^{P} \frac{e^{\mathbf{a}_p^T \mathbf{w}}}{\left(1 + e^{\mathbf{a}_p^T \mathbf{w}}\right)^2} \mathbf{a}_p \mathbf{a}_p^T. \tag{B.27}$$

Note how this is the analog of the second derivative of the scalar version of the function shown in Equation (B.11).

C: Fundamental matrix factorizations and the pseudo-inverse

C.1 Fundamental matrix factorizations

In this section we discuss the fundamental matrix factorizations known as the singular value decomposition and the eigenvalue decomposition of square symmetric matrices, and end by describing the so-called pseudo-inverse of a matrix. Mathematical proofs showing the existence of these factorizations can be found in any linear algebra textbook.

C.1.1 The singular value decomposition

The singular value decomposition (SVD) is a fundamental factorization of matrices that arises in a variety of contexts: from calculating the inverse of a matrix and the solution to the Least Squares problem, to a natural encoding of matrix rank. In this section we review the SVD, focusing especially on the motivation for its existence. This motivation for the SVD is to understand, in the simplest possible terms, how a given $M \times N$ matrix $\mathbf{A}$ acts on N-dimensional vectors $\mathbf{w}$ via the multiplication $\mathbf{Aw}$. We refer to this as *parsimonious representation* or, in other words, the drive to represent $\mathbf{Aw}$ in the simplest way possible. For ease of exposition we will assume that the matrix $\mathbf{A}$ has at least as many rows as it has columns, i.e., $N \leq M$, but what follows generalizes easily to the case when $N > M$.

Through the product $\mathbf{Aw} = \mathbf{y}$ the matrix $\mathbf{A}$ sends the vector $\mathbf{w} \in \mathbb{R}^N$ to $\mathbf{y} \in \mathbb{R}^M$. Using any two sets of linearly independent vectors which span $\mathbb{R}^N$ and $\mathbb{R}^M$, denoted as $\mathbb{V} = \{\mathbf{v}_1, \mathbf{v}_2, \ldots, \mathbf{v}_N\}$ and $\mathbb{U} = \{\mathbf{u}_1, \mathbf{u}_2, \ldots, \mathbf{u}_M\}$ respectively, we can decompose an arbitrary $\mathbf{w}$ over $\mathbb{V}$ as

$$\mathbf{w} = \sum_{n=1}^{N} \alpha_n \mathbf{v}_n, \tag{C.1}$$

for some coefficients α_n for $n = 1 \ldots N$. Further, since for each n the product $\mathbf{Av}_n$ is some vector in $\mathbb{R}^M$, each product itself can be decomposed over $\mathbb{U}$ as

$$\mathbf{Av}_n = \sum_{m=1}^{M} \beta_{n,m} \mathbf{u}_m, \tag{C.2}$$

for some coefficients $\beta_{n,m}$ for $m = 1 \ldots M$. Together these two facts allow us to decompose the action of $\mathbf{A}$ on an arbitrary vector $\mathbf{w}$ in terms of how $\mathbf{A}$ acts on the individual $\mathbf{v}_n$s as

$$\mathbf{A}\mathbf{w} = \mathbf{A}\left(\sum_{n=1}^{N}\alpha_n\mathbf{v}_n\right) = \sum_{n=1}^{N}\alpha_n\mathbf{A}\mathbf{v}_n = \sum_{n=1}^{N}\sum_{m=1}^{M}\alpha_n\beta_{n,m}\mathbf{u}_m. \tag{C.3}$$

This representation would be much simpler if $\mathbb{U}$ and $\mathbb{V}$ were such that $\mathbf{A}$ acted on each $\mathbf{v}_n$ via direct proportion, sending it to a weighted version of one of the $\mathbf{u}_m$s. In other words, if $\mathbb{U}$ and $\mathbb{V}$ existed such that

$$\mathbf{A}\mathbf{v}_n = s_n\mathbf{u}_n \quad \text{for all } n, \tag{C.4}$$

this would considerably simplify the expression for $\mathbf{A}\mathbf{w}$ in (C.3), giving instead

$$\mathbf{A}\mathbf{w} = \sum_{n=1}^{N}\alpha_n s_n\mathbf{u}_n. \tag{C.5}$$

If such a pair of bases for $\mathbf{A}$ indeed exists, (C.4) can be written equivalently in matrix form as

$$\mathbf{A}\mathbf{V} = \mathbf{U}\mathbf{S}, \tag{C.6}$$

where $\mathbf{V}$ and $\mathbf{U}$ are $N \times N$ and $M \times M$ matrices formed by concatenating the respective basis vectors column-wise, and $\mathbf{S}$ is an $M \times N$ matrix with the s_i values on its diagonal (and zero elsewhere). That is,

$$\mathbf{A}\begin{bmatrix} | & | & & | \\ \mathbf{v}_1 & \mathbf{v}_2 & \cdots & \mathbf{v}_N \\ | & | & & | \end{bmatrix} = \begin{bmatrix} | & | & & | \\ \mathbf{u}_1 & \mathbf{u}_2 & \cdots & \mathbf{u}_M \\ | & | & & | \end{bmatrix}\begin{bmatrix} s_1 & & & \\ & s_2 & & \\ & & \ddots & \\ & & & s_N \\ & & & \\ & & & \end{bmatrix}. \tag{C.7}$$

If, in addition, the basis matrices were *orthogonal*,[1] we can rearrange (C.6) for $\mathbf{A}$ alone giving the factorization

$$\mathbf{A} = \mathbf{U}\mathbf{S}\mathbf{V}^T. \tag{C.8}$$

This ideal factorization can in fact be shown to hold rigorously (see e.g., [81]) and is referred to as the singular value decomposition of $\mathbf{A}$. The matrices $\mathbf{U}$ and $\mathbf{V}$ each have orthonormal columns (meaning the columns of $\mathbf{U}$ all have unit length and are orthogonal to each other, and likewise for $\mathbf{V}$) and are typically referred to as *left* and *right singular matrices* of $\mathbf{A}$, with the *real nonnegative values* along the diagonal of $\mathbf{S}$ referred to as *singular values*.

Any matrix $\mathbf{A}$ may be factorized as $\mathbf{A} = \mathbf{U}\mathbf{S}\mathbf{V}^T$ where $\mathbf{U}$ and $\mathbf{V}$ have orthonormal columns and $\mathbf{S}$ is a diagonal matrix containing the (real and nonnegative) singular values of $\mathbf{A}$ along its diagonal.

[1] A square matrix $\mathbf{Q}$ is called orthogonal if $\mathbf{Q}^T\mathbf{Q} = \mathbf{Q}\mathbf{Q}^T = \mathbf{I}$.

The SVD can also be written equivalently as a weighted sum of outer product matrices

$$\mathbf{A} = \sum_{i=1}^{\min(N,M)} \mathbf{u}_i s_i \mathbf{v}_i^T. \tag{C.9}$$

Note that we can (and do) assume that the singular values are placed in descending order along the diagonal of $\mathbf{S}$.

C.1.2 Eigenvalue decomposition

When $\mathbf{A}$ is square and symmetric, i.e., when $N = M$ and $\mathbf{A} = \mathbf{A}^T$, there is an additional factorization given by

$$\mathbf{A} = \mathbf{E}\mathbf{D}\mathbf{E}^T, \tag{C.10}$$

where $\mathbf{E}$ is an $N \times N$ matrix with orthonormal columns referred to as *eigenvectors*, and $\mathbf{D}$ is a diagonal matrix whose diagonal elements *are always real numbers* and are referred to as *eigenvalues*.

A square symmetric matrix $\mathbf{A}$ may be factorized as $\mathbf{A} = \mathbf{E}\mathbf{D}\mathbf{E}^T$ where $\mathbf{E}$ is an orthogonal matrix of eigenvectors and $\mathbf{D}$ a diagonal matrix of all real eigenvalues.

We may also write this spectral decomposition equivalently as a sum of N weighted outer product matrices:

$$\mathbf{A} = \sum_{i=1}^{N} d_i \mathbf{e}_i \mathbf{e}_i^T. \tag{C.11}$$

This factorization can be motivated analogously to the SVD in the case of square symmetric $\mathbf{A}$, and is therefore highly related to $\mathbf{A}$'s SVD (for a proof of this fact, commonly referred to as the spectral theorem of symmetric matrices, see e.g., [81]). Specifically, when $\mathbf{A}$ is additionally positive (semi) definite one can show that this factorization is precisely the SVD of $\mathbf{A}$.

Note also that a symmetric matrix is invertible if and only if it has all nonzero eigenvalues. In this case the inverse of $\mathbf{A}$, denoted as $\mathbf{A}^{-1}$, can be written as $\mathbf{A}^{-1} = \mathbf{E}^T\mathbf{D}^{-1}\mathbf{E}$ where $\mathbf{D}^{-1}$ is a diagonal matrix containing the reciprocal of the eigenvalues in $\mathbf{D}$ along its diagonal.

A square symmetric matrix $\mathbf{A}$ is invertible if and only if it has all nonzero eigenvalues.

C.1.3 The pseudo-inverse

Here we describe the so-called pseudo-inverse solution to the linear system of equations

$$\mathbf{A}\mathbf{w} = \mathbf{b}, \tag{C.12}$$

where $\mathbf{w}$ is an $N \times 1$ vector, $\mathbf{A}$ is an $M \times N$, and $\mathbf{b}$ an $M \times 1$ vector, and where we assume the system has at least one solution. By taking the SVD of $\mathbf{A}$ as $\mathbf{A} = \mathbf{U}\mathbf{S}\mathbf{V}^T$, and removing all columns of $\mathbf{U}$ and $\mathbf{V}$ associated to any zero singular values, we may then write a solution to this system using the fact that the columns of $\mathbf{U}$ and $\mathbf{V}$ are orthonormal,

$$\mathbf{w} = \mathbf{V}\mathbf{S}^{-1}\mathbf{U}^T\mathbf{b}. \tag{C.13}$$

Note that since $\mathbf{S}$ is a diagonal matrix the matrix $\mathbf{S}^{-1}$ is also diagonal, containing the reciprocal of the nonzero singular values along its diagonal. The matrix $\mathbf{A}^{\dagger} = \mathbf{V}\mathbf{S}^{-1}\mathbf{U}^T$ is referred to as the pseudo-inverse of the matrix $\mathbf{A}$, and we generally write the solution above as

$$\mathbf{w} = \mathbf{A}^{\dagger}\mathbf{b}. \tag{C.14}$$

When $\mathbf{A}$ is square and invertible the pseudo-inverse equals the matrix inverse itself, i.e., $\mathbf{A}^{\dagger} = \mathbf{A}^{-1}$. Otherwise, if there are infinitely many solutions to the system $\mathbf{A}\mathbf{w} = \mathbf{b}$ then the pseudo-inverse solution provides the *smallest* solution to this system.

The smallest solution to the system $\mathbf{A}\mathbf{w} = \mathbf{b}$ is given by $\mathbf{w} = \mathbf{A}^{\dagger}\mathbf{b}$ where $\mathbf{A}^{\dagger}$ is the pseudo-inverse of $\mathbf{A}$.

D: Convex geometry

D.1 Definitions of convexity

In this section we describe two further definitions of convexity (beyond the second order definition described in Section 2.1.3) that can be used for verifying convexity of a given scalar valued function $g : \mathbb{R}^N \rightarrow \mathbb{R}$. Additional care must be taken when the domain of g is not the entire $\mathbb{R}^N$ but a subset $\mathcal{D}$ of it. Specifically, the domain of a convex function g must be a convex set itself. A set $\mathcal{D}$ is convex if for any $\mathbf{w}_1$ and $\mathbf{w}_2$ in it, $\mathcal{D}$ also contains the line segment connecting $\mathbf{w}_1$ and $\mathbf{w}_2$. This line segment can be expressed via

$$\lambda \mathbf{w}_1 + (1 - \lambda)\, \mathbf{w}_2, \tag{D.1}$$

where each value for λ in the unit interval $[0,\ 1]$ uniquely corresponds to one point on the line segment. Examples of a convex and a non-convex set are illustrated in Fig. D.1.

D.1.1 Zeroth order definition of a convex function

A function g is convex if and only if any line segment connecting two points on the graph of g lies *above* its graph. Figure D.2 illustrates this definition of a convex function.

Stating this geometric fact algebraically, g is convex if and only if for all $\mathbf{w}_1$ and $\mathbf{w}_2$ in the domain of g and all $\lambda \in [0,\ 1]$, we have

$$g\left(\lambda \mathbf{w}_1 + (1 - \lambda)\, \mathbf{w}_2\right) \leq \lambda g\left(\mathbf{w}_1\right) + (1 - \lambda)\, g\left(\mathbf{w}_2\right). \tag{D.2}$$

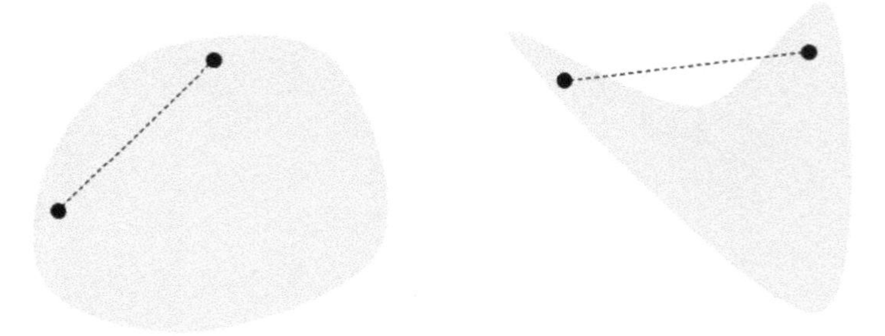

Fig. D.1 (left) A convex set contains the line segment connecting any two points inside it. (right) A non-convex set does not satisfy this property.

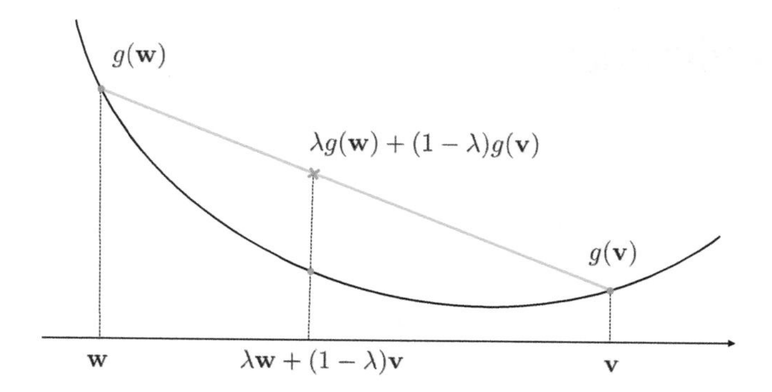

Fig. D.2 The line segment (shown in orange) connecting two points on the graph of a convex function.

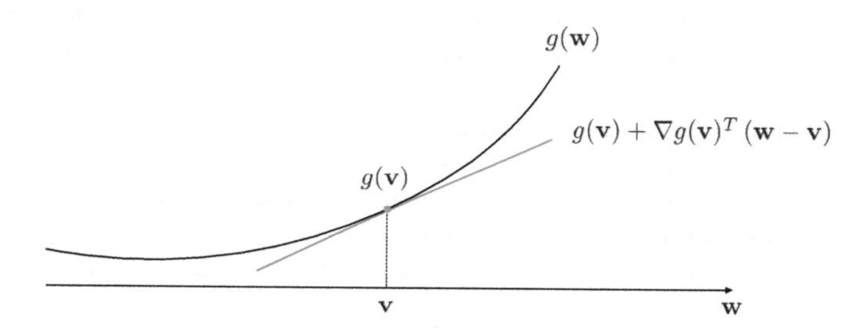

Fig. D.3 A differentiable convex function is one whose tangent plane at any point $\mathbf{v}$ lies below its graph.

D.1.2 First order definition of a convex function

A *differentiable* function g is convex if and only if at each point $\mathbf{v}$ in its domain the tangent plane (generated by its first order Taylor approximation) lies *below* the graph of g. This definition is shown in Fig. D.3.

Algebraically, it says that a differentiable g is convex if and only if for all $\mathbf{w}$ and $\mathbf{v}$ in its domain, we have

$$g\left(\mathbf{w}\right) \geq g\left(\mathbf{v}\right) + \nabla g\left(\mathbf{v}\right)^T \left(\mathbf{w} - \mathbf{v}\right). \tag{D.3}$$

Note that this definition of convexity only applies for differentiable functions.

References

[1] Gabriella Csurka *et al.* Visual categorization with bags of keypoints. *Workshop on Statistical Learning in Computer Vision*, ECCV, volume 1, no. 1–22, 2004.

[2] Jianguo Zhang *et al.* Local features and kernels for classification of texture and object categories: A comprehensive study. *International Journal of Computer Vision* **73**(*2*) 213–238, 2007.

[3] David G. Lowe. Distinctive image features from scale-invariant keypoints. *International Journal of Computer Vision*, **60**(*2*) 91–110, 2004.

[4] Svetlana Lazebnik, Cordelia Schmid, and Jean Ponce. Beyond bags of features: Spatial pyramid matching for recognizing natural scene categories. *Computer Vision and Pattern Recognition, 2006 IEEE Computer Society Conference on*, volume 2. IEEE, 2006.

[5] Jianchao Yang *et al.* Linear spatial pyramid matching using sparse coding for image classification. *Computer Vision and Pattern Recognition, 2009. CVPR 2009. IEEE Conference on*. IEEE, 2009.

[6] Geoffrey Hinton *et al.* Deep neural networks for acoustic modeling in speech recognition: The shared views of four research groups. *Signal Processing Magazine, IEEE* **29**(*6*) 82–97, 2012.

[7] Yoshua Bengio, Ian Goodfellow, and Aaron Courville. *Deep learning*. An MIT Press book in preparation. Draft chapters available at *http://www.iro.umontreal.ca/~bengioy/dlbook* (2014).

[8] Karen Simonyan and Andrew Zisserman. Very deep convolutional networks for large-scale image recognition. *arXiv preprint arXiv:1409.1556*(2014).

[9] Yann LeCun, Yoshua Bengio, and Geoffrey Hinton. Deep learning. *Nature* **521**(*7553*) 436–444, 2015.

[10] Alex Krizhevsky, Ilya Sutskever, and Geoffrey E. Hinton. Imagenet classification with deep convolutional neural networks. *Advances in Neural Information Processing Systems*. 2012.

[11] Bernhard Schölkopf and Alexander J. Smola. *Learning with Kernels: Support Vector Machines, Regularization, Optimization, and Beyond.* MIT Press, 2002.

[12] World economic outlook database, https://www.imf.org/external/pubs/ft/weo/2013/02/weodata/index.aspx.

[13] Anelia Angelova, Yaser Abu-Mostafa, and Pietro Perona. Pruning training sets for learning of object categories. In *Computer Vision and Pattern Recoanition, 2005. CVPR 2005. IEEE Computer Society Conference on*, volume 1, pp. 494–501. IEEE, 2005.

[14] Sitaram Asur and Bernardo A Huberman. Predicting the future with social media. In *IEEE/WIC/ACM International Conference on Web Intelligence and Intelligent Agent Technology (WI-IAT), 2010*, volume 1, pp. 492–499. IEEE, 2010.

[15] Horace Barlow. Redundancy reduction revisited. *Network: Computation in Neural Systems*, **12**(*3*) 241–253, 2001.

[16] Horace B Barlow. The coding of sensory messages. In *Current Problems in Animal Behaviour*, pp. 331–360, 1961.

[17] Yoshua Bengio, Yann LeCun, *et al.* Scaling learning algorithms towards AI. *Large-scale Kernel Machines*, **34**(*5*), 2007.

[18] Dimitri P Bertsekas. Incremental gradient, subgradient, and proximal methods for convex optimization: A survey. In *Optimization for Machine Learning*, 2010, 1–38, MIT Press, 2011.

[19] Christopher M Bishop. *Neural Networks for Pattern Recognition.* Oxford University Press, 1995.

[20] Christopher M Bishop *et al. Pattern Recognition and Machine Learning*, volume 4. Springer, 2006.

[21] Léon Bottou. Large-scale machine learning with stochastic grant descent. In *Proceedings of COMPSTAT'2010*, pp. 177–186. Springer, 2010.

[22] Léon Bottou and Chih-Jen Lin. Support vector machine solvers. *Large Scale Kernel Machines*, pp. 301–320, MIT Press, 2007.

[23] Stephen Boyd, Neal Parikh, Eric Chu, Borja Peleato, and Jonathan Eckstein. Distributed optimization and statistical learning via the alternating direction method of multipliers. *Foundations and Trends® in Machine Learning*, **3**(*1*) 1–122, 2011.

[24] Stephen Poythress Boyd and Lieven Vandenberghe. *Convex Optimization.* Cambridge University Press, 2004.

[25] Hilton Bristow and Simon Lucey. Why do linear svms trained on hog features perform so well? *arXiv preprint arXiv:1406.2419*, 2014.

[26] Paul R Burton, David G Clayton, Lon R Cardon, *et al.* Genome-wide association study of 14,000 cases of seven common diseases and 3,000 shared controls. *Nature*, **447**(*7145*) 661–678, 2007.

[27] Olivier Chapelle. Training a support vector machine in the primal. *Neural Computation*, **19**(*5*) 1155–1178, 2007.

[28] George Cybenko. Approximation by superpositions of a sigmoidal function. *Mathematics of Control, Signals and Systems*, **2**(*4*) 303–314, 1989.

[29] Navneet Dalal and Bill Triggs. Histograms of oriented gradients for human detection. In *Computer Vision and Pattern Recognition, 2005. CVPR 2005. IEEE Computer Society Conference on*, volume 1, pp. 886–893. IEEE, 2005.

[30] Richard O Duda, Peter E Hart, and David G Stork. *Pattern Classification.* John Wiley & Sons, 2012.

[31] Jeremy Elson, John R Douceur, Jon Howell, and Jared Saul. Asirra: a captcha that exploits interest-aligned manual image categorization. In *ACM Conference on Computer and Communications Security*, pp. 366–374. Citeseer, 2007.

[32] Markus Enzweiler and Dariu M Gavrila. Monocular pedestrian detection: Survey and experiments. *Pattern Analysis and Machine Intelligence, IEEE Transactions on*, **31**(*12*) 2179–2195, 2009.

[33] Carmen Fernandez, Eduardo Ley, and Mark FJ Steel. Model uncertainty in cross-country growth regressions. *Journal of Applied Econometrics*, **16**(*5*) 563–576, 2001.

[34] Jerome Friedman, Trevor Hastie, Robert Tibshirani, *et al.* Additive logistic regression: a statistical view of boosting (with discussion and a rejoinder by the authors). *The Annals of Statistics*, **28**(2) 337–407, 2000.

[35] Galileo Galilei. *Dialogues Concerning Two New Sciences*. Dover, 1914.

[36] Xavier Glorot, Antoine Bordes, and Yoshua Bengio. Deep sparse rectifier networks. In *Proceedings of the 14th International Conference on Artificial Intelligence and Statistics. JMLR W&CP Volume*, volume 15, pp. 315–323, 2011.

[37] James Douglas Hamilton. *Time Series Analysis*, volume 2. Princeton University Press, 1994.

[38] Kurt Hornik, Maxwell Stinchcombe, and Halbert White. Multilayer feedforward networks are universal approximators. *Neural Networks*, **2**(*5*) 359–366, 1989.

[39] Dilawar (http://math.stackexchange.com/users/1674/dilawar). Largest eigenvalue of a positive semi-definite matrix is less than or equal to sum of eigenvalues of its diagonal blocks. Mathematics Stack Exchange. URL:http://math.stackexchange.com/q/144890 (version: 2012-05-14).

[40] Xuedong Huang, Alex Acero, Hsiao-Wuen Hon, *et al. Spoken Language Processing*, volume 18. Prentice Hall, 2001.

[41] Judson P Jones and Larry A Palmer. An evaluation of the two-dimensional gabor filter model of simple receptive fields in cat striate cortex. *Journal of Neurophysiology*, **58**(*6*) 1233–1258, 1987.

[42] Alex Krizhevsky, Ilya Sutskever, and Geoffrey E Hinton. Imagenet classification with deep convolutional neural networks. In *Advances in Neural Information Processing Systems*, pp. 1097–1105, NIPS, 2012.

[43] Yann LeCun and Yoshua Bengio. Convolutional networks for images, speech, and time series. *The Handbook of Brain Theory and Neural Networks*, 3361(10), MIT Press, 1995.

[44] Yann LeCun, Koray Kavukcuoglu, and Clément Farabet. Convolutional networks and applications in vision. In *Circuits and Systems (ISCAS), Proceedings of 2010 IEEE International Symposium on*, pp. 253–256. IEEE, 2010.

[45] Daniel D Lee and H Sebastian Seung. Algorithms for non-negative matrix factorization. In *Advances in Neural Information Processing Systems*, pp. 556–562, MIT Press, 2001.

[46] Donghoon Lee, Wilbert Van der Klaauw, Andrew Haughwout, Meta Brown, and Joelle Scally. Measuring student debt and its performance. *FRB of New York Staff Report*, (668), 2014.

[47] Moshe Lichman. UCI Machine Learning Repository, [http://archive.ics.uci.edu/ml]. Irvine, CA: University of California, School of Information and Computer Science, 2013.

[48] Jianqiang Lin, Sang-Mok Lee, Ho-Joon Lee, and Yoon-Mo Koo. Modeling of typical microbial cell growth in batch culture. *Biotechnology and Bioprocess Engineering*, **5**(*5*) 382–385, 2000.

[49] Zhiyun Lu, Avner May, Kuan Liu, *et al.* How to scale up kernel methods to be as good as deep neural nets. *arXiv preprint arXiv:1411.4000*, 2014.

[50] David G Luenberger. *Linear and Nonlinear Programming*. Springer, 2003.

[51] David J C MacKay. Introduction to gaussian processes. *NATO ASI Series F Computer and Systems Sciences*, **168** 133–166, 1998.

[52] David J C MacKay. *Information Theory, Inference and Learning Algorithms*. Cambridge University Press, 2003.

[53] Saturnino Maldonado-Bascon, Sergio Lafuente-Arroyo, Pedro Gil-Jimenez, Hilario Gomez-Moreno, and Francisco López-Ferreras. Road-sign detection and recognition based on support vector machines. *Intelligent Transportation Systems, IEEE Transactions on*, **8**(2):264–278, 2007.

[54] Christopher D Manning and Hinrich Schütze. *Foundations of Statistical Natural Language Processing*. MIT Press, 1999.

[55] Stjepan Marčelja. Mathematical description of the responses of simple cortical cells. *JOSA*, **70**(*11*) 1297–1300, 1980.

[56] Valerii Mayer and Ekaterina Varaksina. Modern analogue of ohm's historical experiment. *Physics Education*, **49**(*6*) 689, 2014.

[57] Gordon E Moore. Cramming more components onto integrated circuits. *Proceedings of the IEEE*, **86**(*1*): 82–85, 1998.

[58] Isaac Newton. *The Principia: Mathematical Principles of Natural Philosophy*. University of California Press, 1999.

[59] Jorge Nocedal and S Wright. *Numerical Optimization, Series in Operations Research and Financial Engineering*. Springer-Verlag, 2006.

[60] Bruno A Olshausen and David J Field. Sparse coding with an overcomplete basis set: A strategy employed by v1? *Vision Research*, **37**(*23*) 3311–3325, 1997.

[61] Brad Osgood. The Fourier transform and its applications. Electrical Engineering Department, Stanford University, 2009.

[62] Reggie Panaligan and Andrea Chen. Quantifying movie magic with google search. *Google Whitepaper–Industry Perspectives+ User Insights*, 2013.

[63] Jooyoung Park and Irwin W Sandberg. Universal approximation using radial-basis-function networks. *Neural Computation*, **3**(*2*) 246–257, 1991.

[64] Jeffrey Pennington, Felix Yu, and Sanjiv Kumar. Spherical random features for polynomial kernels. In *Advances in Neural Inforamtion Processing Systems*, pages 1837–1845, NIPS, 2015.

[65] Simon J D Prince. *Computer Vision: Models, Learning, and Inference*. Cambridge University Press, 2012.

[66] Ning Qian. On the momentum term in gradient descent learning algorithms. *Neural Networks*, **12**(*1*) 145–151, 1999.

[67] Lawrence R Rabiner and Biing-Hwang Juang. *Fundamentals of Speech Recognition*, volume 14, Prentice-Hall, 1993.

[68] Ali Rahimi and Benjamin Recht. Random features for large-scale kernel machines. In *Advances in Neural Inforamtion Processing Systems*, pp. 1177–1184, NIPS, 2007.

[69] Ali Rahimi and Benjamin Recht. Uniform approximation of functions with random bases. In *Communication, Control, and Computing, 2008 46th Annual Allerton Conference on*, pp. 555–561. IEEE, 2008.

[70] Ryan Rifkin and Aldebaro Klautau. In defense of one-vs-all classification. *The Journal of Machine Learning Research*, **5** 101–141, 2004.

[71] Walter Rudin. *Principles of Mathematical Analysis*, volume 3. McGraw-Hill, 1964.

[72] Xavier X Sala-i Martin. I just ran two million regressions. *The American Economic Review*, pp. 178–183, 1997.

[73] Jonathan Shewchuk. An introduction to the conjugate gradient method without the agonizing pain, http://www-2.cs.cmu.edu/jrs/jrspapers, 1994.

[74] Elias M Stein and Rami Shakarchi. *Fourier Analysis: An Introduction*, volume 1. Princeton University Press, 2011.

[75] Samuele Straulino. Reconstruction of Galileo Galilei's experiment: the inclined plane. *Physics Education*, **43**(*3*) 316, 2008.

[76] Silke Szymczak, Joanna M Biernacka, Heather J Cordell, *et al.* Machine learning in genome-wide association studies. *Genetic Epidemiology*, **33**(*S1*) S51–S57, 2009.

[77] Yichuan Tang. Deep learning using linear support vector machines. arXiv preprint arXiv:1306.0239, 2013.

[78] Andrea Vedaldi and Brian Fulkerson. Vlfeat: An open and portable library of computer vision algorithms. In *Proceedings of the International Conference on Multimedia*, pp. 1469–1472. ACM, 2010.

[79] Pierre Verhulst. Notice sur la loi que la population poursuit dans son accroissement. *Correspondance Mathématique et Physique* **10**: 113–121. Technical report, Retrieved 09/08, 2009.

[80] Patrik Waldmann, Gábor Mészáros, Birgit Gredler, Christian Fürst, and Johann Sölkner. Evaluation of the lasso and the elastic net in genome-wide association studies. *Frontiers in Genetics*, **4**, 2013.

[81] Horn, Roger A and Johnson, Charles R. *Matrix analysis*. Cambridge University Press, 2012.

Index